Advances in 21st Century Human Settlements

Indexed by SCOPUS

This Series focuses on the entire spectrum of human settlements – from rural to urban, in different regions of the world, with questions such as: What factors cause and guide the process of change in human settlements from rural to urban in character, from hamlets and villages to towns, cities and megacities? Is this process different across time and space, how and why? Is there a future for rural life? Is it possible or not to have industrial development in rural settlements, and how? Why does 'urban shrinkage' occur? Are the rural areas urbanizing or is that urban areas are undergoing 'ruralisation' (in form of underserviced slums)? What are the challenges faced by 'mega urban regions', and how they can be/are being addressed? What drives economic dynamism in human settlements? Is the urban-based economic growth paradigm the only answer to the quest for sustainable development, or is there an urgent need to balance between economic growth on one hand and ecosystem restoration and conservation on the other – for the future sustainability of human habitats? How and what new technology is helping to achieve sustainable development in human settlements? What sort of changes in the current planning, management and governance of human settlements are needed to face the changing environment including the climate and increasing disaster risks? What is the uniqueness of the new 'socio-cultural spaces' that emerge in human settlements, and how they change over time? As rural settlements become urban, are the new 'urban spaces' resulting in the loss of rural life and 'socio-cultural spaces'? What is leading the preservation of rural 'socio-cultural spaces' within the urbanizing world, and how? What is the emerging nature of the rural-urban interface, and what factors influence it? What are the emerging perspectives that help understand the human-environment-culture complex through the study of human settlements and the related ecosystems, and how do they transform our understanding of cultural landscapes and 'waterscapes' in the 21st Century? What else is and/or likely to be new vis-à-vis human settlements – now and in the future? The Series, therefore, welcomes contributions with fresh cognitive perspectives to understand the new and emerging realities of the 21st Century human settlements. Such perspectives will include a multidisciplinary analysis, constituting of the demographic, spatio-economic, environmental, technological, and planning, management and governance lenses.

If you are interested in submitting a proposal for this series, please contact the Series Editor, or the Publishing Editor:

Bharat Dahiya (bharatdahiya@gmail.com) or
Loyola D'Silva (loyola.dsilva@springer.com)

More information about this series at http://www.springer.com/series/13196

Timothy Gbenga Nubi · Isobel Anderson ·
Taibat Lawanson · Basirat Oyalowo
Editors

Housing and SDGs in Urban Africa

 Springer

Editors
Timothy Gbenga Nubi
University of Lagos
Lagos, Nigeria

Isobel Anderson
Faculty of Social Sciences
University of Stirling
Stirling, Stirlingshire, UK

Taibat Lawanson
University of Lagos
Lagos, Nigeria

Basirat Oyalowo
University of Lagos
Lagos, Nigeria

ISSN 2198-2546   ISSN 2198-2554 (electronic)
Advances in 21st Century Human Settlements
ISBN 978-981-33-4426-6   ISBN 978-981-33-4424-2 (eBook)
https://doi.org/10.1007/978-981-33-4424-2

This Springer imprint is published by the registered company Springer Nature Singapore Pte Ltd.
The registered company address is: 152 Beach Road, #21-01/04 Gateway East, Singapore 189721,
Singapore

For all those
who have strived through the years
for a greater Africa

Foreword

For too long, African countries have been the poster child of spatial dysfunctionality, as poverty and slum proliferation rates, infrastructure deficiencies and a plethora of structural inefficiencies continue to define the region. Yet, African nations are in fact experiencing unprecedented rates of urbanisation and citizens are responding to existential challenges in a variety of ways, essentially recreating cities to serve their purposes.

In September 2000, leaders of over 150 countries—many of them African heads of state—signed the Millennium Declaration, thus affirming their commitment to attainment of sustainable development. By 2015, the eight Millennium Development Goals (MDGs) were expanded to 17 Sustainable Development Goals (SDGs), reflecting more global needs and, more specifically, concerns for environmental protection, social justice and economic development. The SDGs are the current development agenda designed to address the multiplicity of issues threatening global well-being.

The SDGs align directly with Africa's development priorities as articulated in Agenda 2063, dubbed 'The Africa We Want'. Therefore, it is imperative that African nations localise the goals, mainstream them into substantive policies and ensure they are implemented at subnational and local levels. This publication, coming at the outset of the 'UN Decade of Action' for the attainment of all 17 SDGs, is indeed timely.

The Centre for Housing and Sustainable Development at the University of Lagos, as the African Research Universities Alliance (ARUA) Centre of Excellence for Urbanisation and Habitable Cities, has consistently explored the concept of sustainability on various dimensions. Through research, advocacy and community engagement, the Centre actively works towards achieving the various targets and indicators of the SDGs.

In line with the Centre's objectives, therefore, this book focuses on the African built environment and housing sector, especially given the nexus between housing and all SDGs. Comprehensive in its outlook, this edited volume is an attempt to establish the nexus between a functional, equitable and accessible housing sector and national development. Using case studies from various African cities, it interrogates extant assumptions, challenges the status quo and provides evidence for strategic repositioning of city systems and processes. In addition, the authors have stressed the fact that the development challenges of African cities are interconnected and so require systematic and holistic responses.

By interrogating old ideas and testing new ones empirically, this volume presents nuanced perspectives on the African housing sector as it relates to local, regional and international development. It is my fervent desire that the book will precipitate wide-ranging re-evaluation of the SDGs' progress in Africa, while also serving as a valued knowledge resource in the urgent quest for a socially just and environmentally friendly economic development model that citizens and governments can truly own.

Professor Oluwatoyin Temitayo Ogundipe
Vice Chancellor
University of Lagos
Lagos, Nigeria

Professor Oluwatoyin Temitayo Ogundipe is Professor of Botany in the Department of Botany, Faculty of Science, University of Lagos, Nigeria. He holds a Ph.D. in Botany (Ife) and MBA (UNILAG). He has attended trainings at the University of Johannesburg, South Africa; University of Reading and University of Cambridge, both in the UK; Harvard University, USA; and Kunming Institute of Botany, China. His areas of research include molecular plant taxonomy/biosystematics, forensic botany, cytogenetics, ethnobotany, paleobotany and ecological conservation.

Since joining the University of Lagos in 1990, he has held different administrative positions. He was Head of Department of Botany, Sub-Dean of the Faculty of Science and Dean, School of Postgraduate Studies, where he attracted partnerships with industry resulting in vast improvements in infrastructure at the university. He was also Director of the Academic Planning Unit. In 2016, he was appointed the Deputy Vice Chancellor (Academics and Research).

Since his appointment as Vice Chancellor on 12 November 2017, the university has recorded a significant rise in demand-driven research and the university–industry relationship has been blooming. The university in fact patented 11 inventions within 18 months of his emergence as Vice Chancellor, with three of the patents being prototyped. His administration has been able to attract over 7 billion naira in research grants, including the grant for the African Research Network for Urbanisation and Habitable Cities, which is to provide a strategic platform for developing research capacities in African institutions in the drive towards achieving the Sustainable Development Goals. He has attracted funding for research through the Lagos State Science Research and Innovation Council (LASRIC). The LASRIC fund is aimed at making Lagos the hub of innovation and technology in the country, with emphasis on the promotion of Science, Technology, Engineering and Mathematics (STEM).

He instituted the Prof. Ogundipe Innovation Challenge (POIC), which, with cash rewards, aims to motivate young academics to undertake research and innovation projects. He has also been working assiduously to promote entrepreneurship among students of the university. Moreover, he has identified staff homeownership as one of the priorities of his administration. In this regard, the university is collaborating with Family Homes Fund, an initiative of the Federal Government of Nigeria, to provide homes for staff from grade level 9 and below. Under his leadership, the University of Lagos TV station has debuted with news and entertainment coverage across West Africa.

He has over 90 publications in accredited academic journals in addition to co-authoring eight books. Two of his research collaborations are undergoing patenting.

Preface

Several agreements and policies have been signed at the international and regional levels to address the housing and urban development challenges facing Africa. Not surprisingly, even though these agendas are unique in their own rights with set goals and targets, they converge at the urbanisation nexus. This is because it is recognised that cities are supposed to advance economic, social and environmental development and present the laboratory for tackling most challenges confronting humanity today. Accordingly, in most of these agreements, the need to guide the growth and development of African cities has been paramount. From the Sustainable Development Goals (SDGs) and its precursor Millennium Development Goals (MDGs) to Agenda 2063 of the African Union and the Addis Ababa Action Agenda on Financing for Development, there has been a common attempt to set up policy frameworks for resolving the challenges facing Africa today. As we enter the last decade towards Agenda 2030 and given the challenges still dogging African cities, it has become imperative to empirically determine how far the continent's priorities align with global goals, as well as to undertake grounded interrogation of the linkages between these goals and policymaking in Africa.

The motivation for embarking on this book project came during a special strategic session of the management team of the University of Lagos Centre for Housing and Sustainable Development. As the African Research Universities Alliances (ARUA) Centre of Excellence for Urbanisation and Habitable Cities, with a strategic focus on African development, the Centre deemed it necessary to articulate, from a mainly African perspective, the linkages between the African housing sector and the global goals in housing. This thinking was due, first, to the centrality of housing as a dynamic and complex sector of any economy and, second, to the need to recognise the inherent linkages between various facets of housing and the socio-economic and environmental well-being of households, communities and even nations. We therefore decided to midwife a pan-African edited volume that would offer a platform for documenting often under-represented African scholarship on this all-important global discourse.

We invited chapter proposals exploring the interconnections, interactions and linkages between the SDGs and housing through original research, practice experience, case studies, desk-based research and other knowledge media. Expectedly, we received contributions from academics, practitioners, policy actors and activists from within and outside Africa. The chapters address housing and SDGs linkages in African cities/countries and offer best practices, policy transfer and knowledge sharing for stakeholders in Africa.

While integrated and multidisciplinary approaches were strongly encouraged, authors were urged to present ideas in a systematic manner that is accessible to a general audience. Authors were also encouraged to support the capacity development drive of the Centre for Housing and Sustainable Development by collaborating with co-authors across institutional, gender and generational backgrounds. All submitted proposals underwent a double-blind peer review process. Selected proposals were then developed into full papers and subjected to double-blind peer review, resulting in the chapters that are included in this volume.

We therefore thank all the 24 contributors for sharing their research and perspectives with us. Authors come from a wide range of backgrounds: architecture, African studies, building, construction economics, development studies, ecotoxicology and conservation, ecosystem analysis (ESA), environmental management, housing studies, human resource management, real estate, as well as urban and regional planning, among others.

We also want to appreciate Prof. Bharat Dahiya, the 'Advances in 21st Century Human Settlements Series' editor, who has been quite enthusiastic and supportive of this book. We equally thank the team in Springer Nature, especially series publishing editor Loyola D'Silva and production editor Sanjievkumar Mathiyazhagan, who both worked closely with us during the production process.

Our intention has been to show, via multidisciplinary research, the importance of housing to national development, urban management and attainment of the SDGs. We are convinced that the African housing challenge is best approached by engendering a reorientation of urbanisation in African cities along the lines of sustainability. In pursuit of this, we have embarked on the book project to present new ideas and to subject deep assumptions to reality checks in the context of African cities in a way that would be relevant for users in the academic, policy, development and civil society spaces.

Indeed, we hope that the book will promote broader understanding of the African urban reality and provoke deeper discourse on the diverse approaches to achieving sustainable development in Africa.

Lagos, Nigeria   Timothy Gbenga Nubi
Stirling, UK   Isobel Anderson
Lagos, Nigeria   Taibat Lawanson
Lagos, Nigeria   Basirat Oyalowo

Contents

Editors and Contributors

About the Editors

Professor Timothy Gbenga Nubi leads the Centre for Housing and Sustainable Development at the University of Lagos. He has been Member of the Technical Board of Nigeria's Federal Housing Authority (FHA). He also played a major role in the formation and development of Real Estate Development Association of Nigeria (REDAN), where he served as Executive Member (South-west Coordinator) and is currently, Member of the Board of Trustees. He has also served on the Advisory Council of Habitat for Humanity, Nigeria, and as well as that of First World Communities Ltd.

Between 2015 and 2017, he was Dean of the Faculty of Environmental Science, University of Lagos, as well as Chairman of the University's Housing Unit. He was Founding Director, University of Lagos Centre for Housing and Sustainable Development. The Centre, which was established with a grant from the African Development Bank, conducts research and delivers capacity-building courses and academic programmes in the field of housing and real estate development and management. Under his leadership, the Centre won the rights to host a Centre of Excellence in Urbanisation and Habitable Cities in Africa under the auspices of the African Research Universities Alliance (ARUA) and the UK Research and Innovation's (UKRI) Global Challenge Research Fund. He has a wealth of experience in engaging with government, communities, NGOs and private-sector organisations both locally and internationally and has over 60 publications. He is currently leading funded research under the ESRC and AHRC of the UKRI GCRF.

Professor Isobel Anderson is Chair in Housing Studies in the Faculty of Social Sciences at the University of Stirling, UK, where she has worked since 1994. Having held a range of leadership roles, she currently leads the Home, Housing and Community Research Programme. She is Chief Examiner for the M.Sc./Diploma in Housing Studies and chairs the University's Academic Panel for Postgraduate Research Students.

Her main research and Ph.D. supervision interests are in homelessness and access to housing, sustainable housing and communities, inequality and social exclusion, housing and health/well-being, participation and empowerment, international comparative housing studies and the use of evidence for policy and practice. She has held more than 40 research awards from research councils, charities and government bodies and has published widely for scholarly as well as practice audiences. She was previously UK researcher for the European Observatory on Homelessness (convened by FEANTSA, the EU association of national homelessness agencies) and remains on the international advisory committee of *The European Journal of Homelessness.*

She has been active in the European Network for Housing Research (ENHR) throughout her career. She founded the Working Group on Welfare Policy, Homelessness and Social Exclusion (WELPHASE), jointly coordinating it from 2004 to 2013. Since 2015, she has been joint coordinator of the working group on Housing in Developing Countries. She has collaborated with colleagues in Cuba, South Africa and Turkey, in addition to being privileged to be a UK partner of the Centre for Housing and Sustainable Development at the University of Lagos, Nigeria.

Dr. Taibat Lawanson is Associate Professor of Urban Planning at the University of Lagos, Nigeria, where she leads the Pro-Poor Development Research Cluster and serves as Co-Director at the Centre for Housing and Sustainable Development. She holds a Ph.D. in Urban and Regional Planning from the Federal University of Technology, Akure, Nigeria.

Her research work is in the broad areas of urban informality, pro-poor development, governance and environmental justice. She is particularly interested in how formal and informal urban systems synthesise in emerging African contexts, especially Lagos. She has authored over 60 scholarly articles and received research funding from DFID, British Academy, UKRI, Africa Multiple of the University of Bayreuth and Cambridge-Africa ALBORADA Research Fund among others. She is published in leading urban study journals including Habitat International and Area Development and Policy. She is on the editorial board of Urban Forum and serves as International Corresponding Editor of Urban Studies. She is also Member of the international advisory board of UNHABITAT's flagship 'State of the World Cities'.

She is Member of the International Society of City and Regional Planners, a proud alumnus of the Rocke-feller Foundation Bellagio Centre and a pioneer World Social Science Fellow of the International (Social) Science Council.

Basirat Oyalowo researches into contemporary issues in housing studies, informality and urban sustainability, with an interest in decolonisation, comparative African studies and mixed methods research. She lectures in the Department of Estate Management at the University of Lagos, where she obtained her Ph.D. with a thesis on the co-operative societies and housing supply in Lagos. Earlier, she had earned her master's degree in Housing Policy and Management from the University of Northumbria at Newcastle, where she graduated in 2006 as the Best Full-Time Student in Housing Programme in the North-East, awarded by the Chartered Institute of Housing, North-East Branch, UK. She is also on the management team of the University of Lagos Centre for Housing and Sustainable Development, where she drives the Centre's grants response, research, capacity-building and postgraduate programmes, as well as housing advocacy activities. In the last decade, she contributed chapters on housing and human capital development to the 25-year Regional Development Plan of the Ogun State Government of Nigeria, aspects of which she was actively engaged in implementing, as an academic where she has linked research with teaching and practice. She was Lead Facilitator, Informality Discovery Working Group, and later Member of the Strategy Writing Group for the preparation of the Lagos Resilience Strategy. Individually and as part of a team, she has won capacity-building grants under the ESRC and AHRC of the UK Research and Innovation Global Challenge Research Fund (UKRI GCRF). She is a registered Estate Surveyor and Valuer.

Contributors

Irene Appeaning Addo Institute of African Studies, University of Ghana, Accra, Ghana

Olumuyiwa Bayode Adegun Department of Architecture, Federal University of Technology, Akure, Nigeria

John Ogbonnaya Agwu Department of Urban and Regional Planning, Imo State University, Owerri, Nigeria

Akinmayowa Akin-Otiko Institute of Africa and Diaspora Studies, University of Lagos, Lagos, Nigeria

Akeem Ayofe Akinwale Department of Employment Relations and Human Resource Management, University of Lagos, Lagos, Nigeria

Isobel Anderson Faculty of Social Sciences, University of Stirling, Stirling, UK

Andrew Ebekozien School of Housing, Building and Planning, Universiti Sains Malaysia, George Town, Malaysia;
Bekos Energy Service Nigeria Limited, Ikorodu, Nigeria and Bowen Partnership, Quantity Surveying Consultant Firm, Benin City, Nigeria

Johnson Bade Falade Gotosearch.com Ltd, Ajah, Lagos, Nigeria

Eziyi O. Ibem Department of Architecture, Covenant University, Ota, Nigeria

Foluke O. Jegede Department of Architecture, Covenant University, Ota, Nigeria

Hikmot A. Koleoso Department of Estate Management/Centre for Housing and Sustainable Development, University of Lagos, Lagos, Nigeria

Adedoyin Kehinde Lasisi Environmental Management Department, Lagos State Ministry of the Environment and Water Resources, Lagos, Nigeria

Taibat Lawanson Department of Urban and Regional Planning/Centre for Housing and Sustainable Development, University of Lagos, Lagos, Nigeria

Lochner Marais Centre for Development Support, University of the Free State, Bloemfontein, South Africa

Emmanuel Musoke Mutyaba Uganda Martyrs University, Kampala, Uganda

John Ntema Department of Human Settlements, University of Fort Hare, Alice, South Africa

Timothy Nubi Department of Estate Management/Centre for Housing and Sustainable Development, University of Lagos, Lagos, Nigeria

Opeyemi Anne Ogunkoya Ecotoxicology and Conservation Unit, Department of Zoology, Faculty of Science, University of Lagos, Lagos, Nigeria

Oluwaseun James Oguntuase Lagos State University, Lagos, Nigeria

Oluwafemi Ayodeji Olajide Department of Urban and Regional Planning, University of Lagos, Lagos, Nigeria

Esther Iyanuoluwa Olaniran Ecotoxicology and Conservation Unit, Department of Zoology, Faculty of Science, University of Lagos, Lagos, Nigeria

Saidat Damola Olanrewaju Department of Architecture, Federal University of Technology, Akure, Nigeria

Abiodun Anthony Olowoyeye ADIGO Limited, Whitburn Bathgate, UK

Adedapo A. Oluwatayo Department of Architecture, Covenant University, Ota, Nigeria

Basirat Oyalowo Department of Estate Management/Centre for Housing and Sustainable Development, University of Lagos, Lagos, Nigeria

Thomas-Benjamin Seiler Aachen Biology and Biotechnology, Department of Ecosystem Analysis (ESA), Institute for Environmental Research (Biology V), RWTH Aachen University, Aachen, Germany

Temitope Olawunmi Sogbanmu Ecotoxicology and Conservation Unit, Department of Zoology, Faculty of Science, University of Lagos, Lagos, Nigeria

Abimbola Windapo University of Cape Town, Cape Town, South Africa

Africa's Housing Sector as a Pathway to Achieving the SDGs

Timothy Nubi and Isobel Anderson

Keywords African cities · Housing policy · SDGs · Urban development · Urbanization

1 Introduction: The Urban Development Challenge for Africa

Adequate, affordable housing has long been recognised as central to global development strategies [5]. However, the inclusion of a specific 'urban goal' for sustainable cities and communities in the 2030 Agenda for Sustainable Development [17] has been recognised as a new milestone in the effort to integrate the residential environment into the sustainable development debate. A further required step, however, is fuller recognition of the myriad of ways in which a healthy housing sector can also contribute to meeting the other 16 integrated Sustainable Development Goals (SDGs) and their associated targets by 2030. As the first five-year milestone of implementation of Agenda 2030 is reached, this book presents new scholarship in the African context that demonstrates the centrality of quality homes to achieving the SDGs and the New Urban Agenda, which was ratified in 2016 at the HABITAT III conference in Quito [14–16].

No doubt, there is an urgent need for scholarship in this direction. Cities are supposed to advance economic, social and environmental development and present

T. Nubi
Department of Estate Management/Centre for Housing and Sustainable Development, University of Lagos, Lagos, Nigeria
e-mail: tnubi@unilag.edu.ng

I. Anderson (✉)
Faculty of Social Sciences, University of Stirling, Stirling FK9 4LA, UK
e-mail: isobel.anderson@stir.ac.uk

the laboratory for resolving most of the challenges confronting humanity today. However, with high and rapidly rising urbanisation rates, African cities are not seen to be fulfilling this role. Rather, they are characterised by increasing slum proliferation and housing shortages as well as inadequate urban infrastructure that reduces business formation, competitiveness and productivity. All of these inhibit efficient inter- and intra-city human, material and resource flows. Similarly, African city management has for too long been based on outdated, irrelevant colonial-era urban planning dogmas. Consequently, there have been discordant city layouts and poor spatial connectivity, congestion and gridlocks, sprawls, inefficiency of infrastructure systems and decreased productivity. All of these have created negative externalities for the social, environmental and economic sustainability of urban areas. Therefore, it is now imperative to question the old ways of viewing and managing the city, if we are to chart a fresh path towards more competitive and yet more sustainable cities in Africa.

This collection draws on evidence reviews and new empirical studies to scrutinise the ways in which housing policy and provision can provide a pathway to achieving the SDGs in the African context. A key tenet of UN Agenda 2030 was that no one would be left behind in the effort to end poverty and hunger as well as to share wealth and address inequality. To this end, sustainable urbanisation was viewed as crucial to people's quality of life. Although implementation of Agenda 2030 is envisaged as being deliverable through a 'global partnership', paragraph 41 indicates that individual countries have the primary responsibility for their own economic and social development. As such, the chapters in this collection explore the shared and contrasting experiences on housing provision in African cities, alongside examining the deficiencies in government policy from various perspectives, considering alternative housing finance models and, very importantly, questioning—through empirical analysis—the various assumptions that had been believed to limit Africa's progress in achieving adequate housing for its teeming urban populace. In all these, linkages between housing and the SDGs are explored and specific areas of connection identified.

2 Housing and the Sustainable Development Goals

Housing issues are of course most directly relevant to SDG11, which seeks to "make cities and human settlements inclusive, safe, resilient and sustainable." Under SDG11, Target 11.1 is to ensure access for all to adequate, safe and affordable housing and basic services, and to upgrade 'slums' by 2030 [17]. Reducing the proportion of the urban population living in slums, informal settlements or inadequate housing is a key indicator of progress.

Subsequently, the New Urban Agenda emerged as a detailed declaration and implementation plan, with a similar goal of 'leaving no one behind'. Paragraph 32 of the Implementation Plan committed signatories to promoting the development of age- and gender-responsive housing policies, which were integrated with the employment, education, healthcare and social integration sectors, as well as with all levels of government. Policies were to incorporate the provision of "adequate, affordable, accessible, resource efficient, safe, resilient, well-connected and well-located housing" (Paragraph 32). Paragraph 33 committed to stimulating the supply of affordable and accessible housing for different income groups, including those in marginalised communities and vulnerable situations, e.g., homeless persons. The UN HABITAT's New Urban Agenda (NUA) offers a shared vision in which all people have equal rights to housing and other benefits of the world's cities (2017a). Calling for a 'paradigm shift', the NUA proposed standards and principles for the planning, construction, development, management and improvement of urban areas, including housing and residential neighbourhoods, from national urban policies to local implementation. The approach applied equally to all civil-society organisations, emphasising links between urbanism, economic opportunities and an improved quality of life. It levers for the transformative change of city planning, land readjustment programmes and basic services as well as housing and public space, while recognising the importance of monitoring progress [16].

The New Urban Agenda's integrated approach recognises the important connections among housing, well-being and people's capacity to flourish and contribute to economic and social development. It is in this sense that the quality of homes and neighbourhoods underpin the capacity of societies to achieve the 17 integrated goals (Fig. 1). Chapters in this book make explicit connections with Goals 3 (Health), 6 (Water and Sanitation), 7 (Energy), 9 (Infrastructure), 11 (Sustainable communities), 13 (Climate Action), 14 (Life below water), 15 (Life on Land) and 16 (Institutions). There is undoubtedly equal scope to make connections between access to housing

Fig. 1 The sustainable development goals. *Source* United Nations [18]

and eradicating poverty, hunger and inequality (1, 2, 10), as well as supporting education (4), gender equality (5), work (8), responsible consumption (12), while strong
partnerships (17) are required to coordinate housing and urban development. Key to
optimising these intersections is the need for integrated research often in the form of
"interdisciplinary collaborations between natural and social scientists and transdisciplinary team building that brings together academic and non-academic (practice-
based) researchers to investigate sustainability challenges of mutual interest" [13,
p. 787].

The measurement of progress towards the SDGs is central to implementation
of Agenda 2030 but it is also important to consider how this implementation can
be conceptualised for analysis. Arguably, implementation depends largely on mobil-
ising structures of governance (at global, national and local levels) through a classical
policy-making approach [8]. In practice, however, implementation may be much
more incremental, while the attachment of measureable targets to the broad aspi-
rational goals reflects evidence-based policy models that emerged in the 2000s. In
their review of the conceptualisation of informality in housing, d'Alençon et al. [1]
acknowledged the need for a global/theoretical conception embracing national, city
and neighbourhood levels of implementation. Therefore, they suggested better understanding of governance frameworks involving actors associated with 'informality',
as well as of the political economy underpinning specific urban realities (p. 64).

In a policy paper directed towards developing the NUA [12], it was argued that
inadequate housing and unequal access remain central characteristics of rapid urbanisation, at least as partly explained by the failure to adopt effective measures due to
the fact that key stakeholders benefit from the status quo. In order to disrupt that
status quo, the paper suggested, it is necessary to make the case for more radical
interventions including market regulation, use of taxation to influence affordability
and widen access, taking account of formal and informal housing and stressing social
justice perspectives, as well as for adopting post-colonial theory and de-colonialising
methods [12, p. 7–8]. In the global north, housing policy outcomes have also been at
least partly explained by the embeddedness of institutional structures (often underpinned by neoliberal agendas), thus creating a path dependency that resists change
[2, 3].

Kaika [9] noted that the NUA's call for "safe, resilient, sustainable and inclusive cities" essentially remains path-dependent. This includes the use of techno-
managerial solutions (such as smart cities), the indicators for monitoring and
the overall institutional frameworks, which were viewed as belonging to a failed
paradigm. Kaika characterised the approach as merely 'vaccinating' citizens in order
to help them absorb further inequality. This interpretation sees the NUA as mediating,
rather than alleviating, the effects of global inequality. Where communities rejected
such strategies, there was more possibility to disrupt path dependency and establish
alternative approaches to accessing housing, healthcare, sanitation, etc. Real social
innovation, then, was to be found in dissent, rather than in consensus [9], while
policy development from within countries in the south was seen as a requirement for
disrupting policy pathways. The notion of path dependency and how prior trajectories can be disrupted or shifted to new directions lends itself to a wider consideration

of how housing can be a pathway to meeting the wider set of integrated development goals.

3 Housing as a Pathway to Achieving the SDGs

Scholarship to date has connected housing to the fields of some SDGs more than others. For example, housing has long been understood as a social determinant of health [6, 10]. Globally, home and housing are at the heart of communities and represent a key sphere in which public policy can affect people of all socio-economic cadres. The World Health Organisation's (WHO) Global Strategy and Action Plan on Ageing and Health shows the intergenerational nature of the SDGs [4]. The World Health Organisation housing and health guidelines on the health benefits of improved housing emphasise avoidance of overcrowding, appropriate indoor temperature and indoor safety and accessibility [21] . The guidelines also recognise the importance of housing affordability, security and the surrounding environment in achieving overall well-being.

The residential setting also underpins community activities, including community resilience and recognition of 'asset-based' approaches to residents' capacity to enjoy life and cope with its challenges [20]. In a resilience model, individuals, households and communities are seen as having resources to cope with difficulties, in contrast to a 'deficit' model of illness or housing disadvantage. Resilience, as capacity to deal with difficulties and even thrive in overcoming adversity, has also been recognised by the World Health Organisation as an important factor in lifelong health and well-being [11]. However, the concept of resilience, has also been criticised for failing to sufficiently challenge the prevailing structures that cause and sustain disadvantage [9].

Nevertheless, there is still scope to better document housing as a determinant of other factors such as economic, social and cultural opportunities and outcomes. Communities may build resilience to disadvantage, while institutionalism and path dependency may help explain why change is sometimes only incremental, even where progressive strategies are in place. It is therefore important to consider how far historical policies (colonial and post-colonial) determine contemporary national and local implementation of Agenda 2030 in African countries. While ultimately there may be no single conceptual approach that best explains the intersection of housing and the SDGs at the community level and national development, the triangulation of new research findings with the existing scientific literature and policy/practice can shed new light on evolving debates. Within the wider context of global sustainable development and city governance, then, our contributions analyse the ways in which a wide range of factors comes into play in shaping not only the development of settlements but also changes in sustainable economic and social development.

The evidence in the following chapters suggests that the benefits of housing improvements are not yet fully recognised in the wider spheres of the SDGs and Agenda 2030. There are also potential research areas that have not been addressed in this collection. Nevertheless, the range of issues examined demonstrates the continuing need to better integrate interventions to achieve the global goals.

To an extent, agenda 2030 and the New Urban Agenda can be interpreted as rational planning tools for sustainable development but, where implementation is highly constrained, arguably path-dependent and incremental. National programmes and local projects can bring about real improvements in quality of life. The importance of community settings and practices to good health and education, for example, remains core to better understanding of community interpretations of sustainability and development and thereby to developing culturally effective responses and interventions.

The 2030 Agenda is ambitious, if also somewhat idealistic. All participating nations have signed up to 'no one being left behind'. In a largely neoliberal world focused on economic competitiveness, however, this has rarely, if ever, been a realistic goal in most national agendas. The importance of housing to the other SDGs may be widely recognised but national and local agendas need to demonstrate integrated commitment to improved housing along with other mechanisms for achieving the SDGs. While there may be a danger that incrementalism ultimately implies almost no or limited progress, the international priority to the UN Sustainable Development Goals offers continuing opportunities to better address the global challenge of meeting the housing and community needs of low-income groups in an environmentally sustainable way, which in turn will support progress regarding the other goals.

4 Contemporary Housing and Urbanization in Africa

Contemporary African cities are defined by multiple issues that have challenged optimisation, competitiveness and efficiency in urban areas. Globally acceptable tools such as the Sustainable Development Goals provide the critical directions for change that can bring about resolution of these challenges for the benefit of all Africans. However, there is urgent need to incorporate these goals into sectoral concerns to catalyse change. Housing represents a substantial proportion of urban land use. As both an economic good and a social service, housing transcends the brick and mortar with which it is built. Indeed, homes underpin the fundamental quality of life and all other economic and social activities for work, play and living, thereby incorporating the immediate environment and neighbourhood, as well as the individual's dwelling. The science and practice of the planning, supply, production and management of housing in its micro and macro environment are avenues for incorporating and operationalising all of the SDGs, with a particular focus on the residential environment.

This book's focus on Africa and African perspectives is deliberate. There is a dearth of collections of scholarly works dedicated wholly to African issues and that comes out of the work done by African scholars and practitioners with collaborators from within Africa and elsewhere. The challenges facing African cities are monumental, as they play host to some of the most daunting statistics on the impact of infrastructure and economic development on livelihoods, municipal efficiency and national growth and development. As the SDGs seek to promote environmentally conscious and socially just development, the quest for development in Africa often overrides consideration of what has been termed the tripod of sustainability (economic, social and environmental sustainability). This volume brings together scholarly research and argument that cuts across and intertwines this tripod into options that can deliver on the promise of the SDGs. The project is an initiative of the Centre for Housing and Sustainable Development at the University of Lagos, which identified the gap in grounded research that links the housing sector with the SDGs in African cities.

It is worth reiterating that the chapters in this volume critically examine the various interconnections between housing and the SDGs, exploring how the SDGs can be used as a platform for addressing issues surrounding housing, such as affordability and accessibility, and how housing and its neighbourhood impact the environment. Thus, externalities of housing neighbourhoods on pollution, waste management and slum proliferation are brought forward. The volume also addresses contemporary issues in housing production, housing investment and finance, housing governance, housing supply and urban design processes as avenues for achieving the goals and targets of the SDGs. The critical analysis of taken-for-granted dogmas, such as the right to housing and the relevance of the SDGs, also finds expression in the chapters.

In achieving the book's core objective of interrogating the connections between housing and the SDGs, the chapters represent the outcomes of engaging stakeholders in the epistemological, practical, theoretical and methodological issues associated with housing as a platform for achieving the SDGs in Africa. The book therefore explores the interconnections, interactions and linkages between the SDGs and housing through original research, practice, experience, case studies, desk-based research and other knowledge media.

The remainder of this introductory chapter outlines the range of contributions in the book in the context of progress towards the SDGs and the ways in which housing within African countries contributes to sustainability and development, beyond the SDG11 accommodation target. The contributions have a strong focus on evidence from Nigeria, as well as contributions from South Africa, Ghana and Uganda. Many of the materials will have relevance/transferability more widely in Africa, even as there are also wider comparative insights from outside Africa.

5   Housing, Urbanisation and SDGs in Africa

The chapters in Sect. 2 explore housing policy and governance in the African context. Chapter "Global Goal, Local Context: Pathways to Sustainable Urban Development in Lagos, Nigeria" (Taibat Lawanson, Basirat Oyalowo and Timothy Nubi) examines the global goals in local contexts through a focus on pathways to sustainable urban development in Lagos. Anthony Olowoyeye addresses the theme of housing management in Chapter "The Road not Taken: Policy and Politics of Housing Management in Africa". Housing's centrality to meeting the Sustainable Development Goals is reflected in unequal patterns of infrastructure development, health outcomes and social injustice. The chapter identifies strategic and historical causes of the problems of post-independence housing systems in Africa having 'not taken' the road recommended by the Economic Commission for Africa in 1963, with the consequences of this missed opportunity still being felt. A key conclusion is that until robust policies enable housing to become a key driver of economic, social and infrastructural development, in a system that works for all, meeting the SDG goals will remain difficult across Africa. Chapter "Learning from Experience: An Exposition of Singapore's Home Ownership Scheme and Imperatives for Nigeria" presents our international comparative analysis, in a consideration of potential learning from the experience of homeownership in Singapore (Hikmot Koleoso & Basirat Oyalowo). Starting with Sustainable Development Goal 11.1 as a declaration of the need to ensure access to adequate, safe and affordable housing for all citizens, the authors address poor access to housing as a cause of poverty in developed and developing countries. They examine national homeownership structures as means of economic and social development and as a potential platform for eradicating poverty, as in the transformative experience of Singapore from its underdeveloped status into an international economic hub. The study analyses the key policy tools and features of Singapore's housing reforms and interventions to reflect on what Nigeria (or other African nations) might adapt for a sustainable housing policy to contribute to achieving SDG11.1. The importance of government's political and financial commitment to homeownership is supported by evidence of socioeconomic and cultural needs as well as financial ability. Successful features of the Singapore scheme that might be considered in African countries include design, affordability, legislation, regulation, allocation, physical management and regeneration of properties within the homeownership sector.

The third section of our collection focuses on housing quality and health, with findings from Nigeria and South Africa. John Ntema, Isobel Anderson and Lochner Marais present their findings on health outcomes in Mangaung Upgraded Informal Settlement in South Africa (Chapter "Housing and Possible Health Implications in Upgraded Informal Settlements: Evidence from Mangaung Township, South Africa"). The close association between housing and health underpins their consideration of SDG3 (health and well-being) and SDG6 (standards in water and sanitation provision, as well as the housing-oriented SDG11. Their evidence supports international and South African findings that the health of residents in upgraded informal settlements is connected more to infrastructural development and neighbourhood access to social amenities than to physical housing conditions. New evidence from the analysis of Mangaung households' perceptions of health found that lack of basic sanitation and water, as well as of primary health clinics, were factors constraining health improvements. Thus, policies to ensure improved health services and sanitation in upgrading programmes are needed to contribute more effectively to achieving the SDGs on health and housing as well as water and sanitation. Similarly, in Chapter "Housing, Health and Well-Being of Slum Dwellers in Nigeria: Case Studies of Six

Cities", SGDs 3, 6 and 11, as well as 5 (gender equality) and 17 (partnerships), are considered in relation to the housing and health of slum dwellers in six Nigerian cities (Johnson Bade Falade). This study analysed the roles of slum housing in meeting the physiological and psychological needs of residents and in protecting them from disease. The study assessed the contributions of 26 indicators to meeting the health and well-being needs of residents. The results varied across the six slums and the analysis highlights the implications for promoting healthy housing and urban renewal, as well as for realising the targets of Agenda 2030 and the New Urban Agenda.

The distinctive theme of housing and the environment is explored in Section Four of the volume. Chapter "Adverse Impact of Human Activities on Aquatic Ecosystems: Investigating the Environmental Sustainability Perception of Stakeholders in Lagos and Ogun States, Nigeria", by Temitope Sogbanmu, Opeyemi Ogunkoya, Esther Olaniran, Adedoyin Lasisi and Thomas Seiler, examines environmental sustainability and the impact of human activities on aquatic ecosystems in Nigeria's Lagos and Ogun states. SDG 14, which addresses the need to support life below water, is thus the focus, given the realisation that in urban Africa rising population rates pose environmental challenges for the management of aquatic resources. The authors' research exposes stakeholders' perceptions of environmental risk to air quality, water quality and impact on aquatic animals. The need for improved education, communication and implementation of evidence-based policies for the management of these ecosystems is highlighted as crucial to the planning of human settlements near aquatic ecosystems. Air quality is the focus of Chapter "Meeting the Sustainable Development Goals: Considerations for Household and Indoor Air Pollution in Nigeria and Ghana" (Irene Appeaning Addo and Oluwafemi Olajide)—a feature of good health (SDG3), climate action (SDG13), and life on land (SDG15), as well as being central to sustainable settlements and homes. Indoor air pollution in Africa is recognised as one of the leading causes of pulmonary diseases and death, given the high incidence of use of biomass fuel for cooking. This study on Nigeria and Ghana revealed a high incidence of indoor air pollution in sub-Saharan Africa (related to widespread home-based enterprises, use of solid fuels and poor ventilation), yet indoor air pollution is hardly recognised at policy and institutional levels, with little emphasis on monitoring and abatement. The study therefore recommends further research and action on indoor air pollution to drive the achievement of improved air quality by 2030.

The critical question of how human rights are embedded in the Sustainable Development Goals is explored in Section Five on the right to housing. Chapter "A Study of Housing, Good Health and Well-Being in Kampala, Uganda", by Mutyaba Emmanuel Musoke, is a narrative of the linkages between housing and well-being in Uganda. Starting from the premise that adequate housing is a necessity for health and well-being, the chapter develops the theoretical argument that housing is a human right that needs to be respected to achieve the promotion of health and well-being (SDG3). The review confirms that the home should be a place where people feel safe and have a sense of belonging, in addition to being a place that supports physical and emotional well-being as well as productivity. John Ntema presents a South African case study that examines aspects of relocation and informal settlement upgrading in Mangaung Township, Free State Province (Chapter "Relocation and Informal Settlements Upgrading in South Africa: The Case Study of Mangaung Township, Free State Province"). In post-apartheid South Africa, only 28% of all settlement upgrading projects improved existing communities, while the remainder were green-field projects to which residents in informal settlements were relocated. The mixed-methods study contributes new evidence on households' perceptions of basic service infrastructure, amenities and governance in a relocation site in Mangaung Town-ship (Bloemfontein). Challenges included promotion of housing development at the expense of inclusive, sustainable communities and undermining of participa-tory project planning and design and limiting contributions to SDGs 3 (health), 4 (education), 6 (water and sanitation) and 11 (cities and communities).

Resources to achieve the sustainable development goals are a critical factor in delivering progress. Thus, the four chapters in Section 6 analyse the role of housing investment and finance. In Chapter "Green Bonds and Green Buildings: New Options for Achieving Sustainable Development in Nigeria", Oluwaseun Oguntuase and Abimbola Windapo review the potential of green bonds and buildings for achieving sustainable and affordable housing and reaching a number of SDGs, including 3 (Health), 6 (Water and Sanitation), 7 (Affordable and Clean Energy), 12 (Responsible Consumption and Production) and 13 (Climate Action). The analysis reflects how the development of the Nigerian green bond market presents an investment opportunity for green buildings to achieve sustainable development. Chapter "Homeownership in a Sub-Saharan Africa City: Exploring Self-Help Via Qualitative Insight to Achieve Sustainable Housing" considers self-help financing for homeownership. Andrew Ebekozien examines the lived experiences of Nigerian middle-income groups in Lagos. The analysis highlights barriers to, and strategies for, becoming homeowners, identifying residential mobility among middle-income earners as contributing to the city's expansion. Land purchase and building approval processes were common obstacles, while organised self-help housing provision merited further government policy support. Pro-poor policy and regulatory frameworks, it is suggested, will improve housing outcomes. Micro-finance support tools such as "soft housing-loans" with less prohibitive conditions and supporting local building schemes can contribute to achieving more sustainable homeownership for all by 2030. Chapter "Exchange Rate and Housing Deficit Trends in Nigeria: Descriptive and Inferential Analyses" presents an analysis of macroeconomic strategy in Nigeria. Using archival data from 1960–2019, John Ogbonnaya Agwu explores the dynamics between the exchange rate, rental values and the housing deficit. The results support previous findings that aggregated macroeconomic indices influence the real estate sector and that exchange rates may be a catalyst for stimulating housing development and reducing the housing deficit. In Chapter "Analysing Hernando de Soto's The Mystery of Capital in the Nigerian Poverty Equation", Akeem Ayofe Akinwale challenges the application of de Soto's [7] work on "why capitalism triumphs in the West and fails everywhere else" to the poverty equation in Nigeria. The review sets the literature around Soto's ideas on poverty alleviation against the backdrop of Nigeria's capitalist ideologies and anti-poverty programmes and also draws on Sensemaking Theory [19] to consider alternative strategies for property rights and access to capital among the poor in developing countries.

Three chapters on Urban Design make up Section 7. In Chapter "Beyond a Mere Living Space: Meaning and Morality in Traditional Yoruba Architecture before Colonialism", Akin-Otiko reflects on the meanings and morality of pre-colonial traditional Yoruba architecture, showing how post-colonial housing responses departed from traditional designs and meanings of home, thereby impacting negatively on attempts to solve the housing problem in Nigeria. It is argued that traditional meanings should be reincorporated into strategies to solve Nigeria's housing problems. The themes of urban sprawl versus densification and housing in Nigerian cities are explored in Chapter "Urban Sprawl and Housing: A Case for Densification in Nigerian Cities" (Saidat Olanrewaju and Olumuyiwa Adegun). It is argued that urban sustainability and resilient communities cannot be achieved without significantly changing the way housing and other urban spaces are planned, designed and developed. Reviewing research to date, the chapter concludes that the urban sprawl in Nigeria has undesirable economic, environmental and social impacts, including for housing. Compact urban housing development, including densification policies, is proposed as an urban form better suited to achieving sustainability in Nigerian cities. Neighbourhood design and security in mass housing schemes in Lagos is the focus of Chapter "Environmental Planning in Mass Housing Schemes: Strategies for Achieving Inclusive and Safe Urban Communities", by Foluke O. Jegede, Eziyi O. Ibem and Adedapo A. Oluwatayo. This chapter presents findings from a study on the influence of residential neighbourhood planning and design of housing units on the security of lives and property in 12 public housing estates in Lagos Metropolis. The results reveal how the layout of estates, housing design and construction features all influenced residents' perception of security. The study also identifies areas that need to be strengthened.

We close our volume with conclusions from Basirat Oyalowo and Taibat Lawanson on how housing and urbanisation in Africa can contribute to a sustainable future and achievement of the SDGs (Section 8, Chapter "Housing and the SDGs in African Cities: Towards a Sustainable Future"). Given that the [18] SDG platform monitors progress across all 17 sustainable development goals, this collection adds valuable new scholarship on the role of housing in the African context. We hope that readers will become further galvanised in their efforts towards future housing research and policy development across African nations, as well as engaging in more integrated actions to strengthen the housing sector as a pathway to achieving truly sustainable and socially just development for all Africans.

References

1. d'Alençon PA, Smith H, Álvarez de Andrés E, Cabrera C, Fokdal J, Lombard M, Mazzolini A, Michelutti E, Moretto L, Spire A (2018) Interrogating informality: conceptualizations, practices and policies in the light of the new urban agenda. Habitat Int 75:59–66

2. Anderson I, Dyb E, Finnerty J (2016) The 'arc of prosperity' revisited: homelessness policy change in North Western Europe. Soc. Incl 4(4):108–124
3. Bengtsson B, Ruonavaara H (2010) Introduction to a special issue on path dependence in housing. Housing Theor Soc 27(3):193–203
4. Bluestone (2016) World Health Organization (WHO) Global strategy and action plan on ageing and health (2016–2020)
5. Cociña C, Frediani A, Acuto M, Levy C (2019) Knowledge translation in global urban agendas: a history of research-practice encounters in the Habitat conferences. World Dev 122:130–141
6. Dahlgren G, Whitehead M (1991) Policies and strategies to promote social equity in health. Institute for Futures Studies, Stockholm, Sweden
7. de Soto H (2001) The mystery of capital: why capitalism triumphs in the west and fails everywhere else. Bantam Press, Black Swan Books, London Edition
8. Hill M, Irving Z (2020) Exploring the world of social policy: an international approach. Policy Press, Bristol
9. Kaika M (2017) 'Don't call me resilient again!': the New Urban Agenda as immunology … or … what happens when communities refuse to be vaccinated with 'smart cities' and indicators. Environ Urbanization 29(1):89–102. https://doi.org/10.1177/0956247816684763
10. Marmot M (2010) Fairer society, healthy lives. The marmot review. UCL. Institute of Health Equity
11. Medical Research Council, UK (2010) Lifelong health and wellbeing. https://mrc.ukri.org/research/initiatives/lifelong-health-wellbeing/
12. N-AERUS (2016) Recommendations for the New Urban Agenda. Cities Alliance
13. Patel Z, Greyling S, Simon D, Arfvidsson H, Moodley N, Primo N, Wright C (2017) Local responses to global sustainability agendas: learning from experimenting with the urban sustainable development goal in Cape Town. Sustain Sci 12:785–797. https://doi.org/10.1007/s11625-017-0500-y
14. Habitat UN (2015) Housing at the Centre of the new urban agenda. UN HABITAT, Nairobi
15. UN Habitat (2017a) The new urban agenda HABITAT III Quito
16. UN HABITAT (2017b) Action framework for implementation of the new urban agenda
17. United Nations (2015) Transforming our world: the 2030 agenda for sustainable development. United Nations, Geneva
18. United Nations (2020) UN sustainable development. https://sdgs.un.org/
19. Weick KE (1995) Sensemaking in organizations. Sage Publications, Thousand Oaks, CA
20. Windle G, Bennett K, Noyes JA (2011) A methodological review of resilience measurement scales. BioMed Central. Health Qual Life Outcomes 9:8
21. World Health Organization (WHO) (2018) WHO housing and health guidelines. WHO, Geneva

Global Goal, Local Context: Pathways to Sustainable Urban Development in Lagos, Nigeria

Taibat Lawanson, Basirat Oyalowo, and Timothy Nubi

Abstract This chapter argues that at the core of the opportunity to deliver sustainable urban development in Nigeria is the lack of capacity to mainstream global goals at the municipal level. The chapter thus address the following questions: Who is responsible for local implementation of the global development agenda? What mechanisms should be in place for local implementation? How can local capacities be strengthened for effective delivery of sustainable urban development? In doing this, we critically assess the challenges of localising the global agenda by focusing on SDG11 and its application in Lagos, Nigeria's largest city. A content analysis of SDG11 and the Lagos State Development Plan reveals major gaps in the development approach of the state, while other findings reveal that paucity of data, weak institutional capacity as well as poor governance strategies are major impediments to mainstreaming SDG11 in Lagos. The chapter concludes by recommending some approaches to conciliate the global agenda with local exigencies, such as local capacity building and inclusive development.

Keywords SDG · Urban · Nigeria · Municipal · Capacity

T. Lawanson (✉)
Department of Urban and Regional Planning/Centre for Housing and Sustainable Development,
University of Lagos, Lagos, Nigeria
e-mail: tlawanson@unilag.edu.ng

B. Oyalowo · T. Nubi
Department of Estate Management/Centre for Housing and Sustainable Development, University
of Lagos, Lagos, Nigeria
e-mail: boyalowo@unilag.edu.ng

T. Nubi
e-mail: tnubi@unilag.edu.ng

1   Introduction

Global development agreements have become the tool for reaching consensus and providing directions for policies, programmes and projects to be implemented at regional, national and sub-national levels for holistic economic and social development. In September 2000, the heads of state of 150 United Nations (UN) member countries signed the Millennium Declaration in the American state of New York, thus affirming their commitment to the attainment of eight Development Goals with 18 specific targets to be achieved by 2015 and to be measured with 48 indicators. These goals are believed to be the most comprehensive set of goals for global development and reflect the need to significantly improve the quality of human life in developing countries.

Like most pacts on international development, the MDGs are meant to be broken down to national and sub-national levels for implementation. In the review of the Millennium Development Goals, the need for incorporating environmental concern, economic development and social justice into global development goals produced the Sustainable Development Goals (SDGs) in 2015. Oleribe and Taylor-Robinson [32] stated that Nigeria failed to meet any of the MDG targets due to a multiplicity of health system issues as well as political and systemic challenges, including a top-down approach to implementation. As reflected in its precursor, sustainable urban development became a global imperative in the Sustainable Development goals (SDG). Goal 11 is dedicated to making cities inclusive, safe, resilient and sustainable. This eleventh goal is thus the focus of this chapter, given that Lagos is not only a megacity but also one of the fastest-growing cities in Africa.

Furthermore, in 2016, the New Urban Agenda (NUA) was signed by member countries of the United Nations [15–17], including Nigeria. The NUA affirmed "a global commitment to sustainable urban development as a critical step for realising sustainable development in an integrated and coordinated manner at the global, regional, national, sub-national and local levels, with the participation of all relevant actors [45]. The NUA was thus developed as a tool for implementing the SDGs. Even though signed at the national level, many of the targets of both the SDGs and NUA are implementable at the sub-national and municipal levels [40–43]. It is, therefore, necessary to examine these agreements in the light of development priorities at the sub-national and municipal levels.

Current urbanisation trends in many African cities are accompanied by significant social and environmental challenges such as poverty, informality, deplorable state of infrastructure, poor basic services, conflicts in resource control, as well as rising inequality and social exclusion. Lagos, Nigeria's commercial centre, is home to a population of over 23 million residents [23] who daily grapple with challenges of infrastructure, overpopulation, poverty and inequality, all of which are key issues that the SDGs are meant to address. Being a signatory to several development agreements, including the SDGs protocol, Nigeria can only successfully meet the set targets by aligning local and international development priorities. Ensuring such an alignment will facilitate the seamless localisation and achievement of goals. However, when there are contradictions, the policy 'import' could have unintended negative consequences.

How these issues are addressed will determine the effectiveness of any global urban development agenda. This chapter, therefore, investigates the challenges of localising global agendas while also exploring the contents of the urban focus of (SDG11) and its applicability to the Nigerian local urban context, specifically Lagos. It outlines the functions and capacities of local governance institutions by addressing the following questions:

i.   Who is responsible for local implementation and how successful have they been?
ii.   What mechanisms are available for local implementation?
iii. How successful are local practices that align with the SDGs?
iv.   How can local capacities be strengthened for effective delivery of sustainable urban development in Nigeria?

The chapter is based on a qualitative research design, with content analysis of two major policy documents, the UN's Sustainable Development Goals (SDGs) and the Lagos State Development Plan 2012–2025 [44]. The SDGs document is available on the UN's dedicated website. The LSDP 2012–2025 is a sub-national strategic document that is a roadmap for the infrastructural, economic and social development of Lagos for the 13-year period. Following general procedures associated with the content analysis methodology, the areas of interest in each document are identified. The goals, targets and indicators of SDG 11 are identified as the specific content to be analysed from the SDGs, while the indicators of the LSDP are the major focus of the analysis. An alignment of the monitoring indicators of SDG 11 and the LSDP indicators is then done. The reporting is done in a tabular form, with additional data from secondary sources. After this, a deductive process is adopted as the basis for drawing out the findings.

2 Assessing the Effectiveness of Local Implementation of SDGs

The sustainable city concept stresses the need to promote environmental safety, social inclusiveness and economic productivity at once. This concept becomes operational in cities through implementation of SDG 11, the Urban SDG, which is themed "making cities and human settlements inclusive, safe, resilient and sustainable." According to Satterthwaite [40–43], for global development agreements such as the SDGs to work, they must be relevant on the urban scale—to urban governments, urban dwellers, politicians, civil servants and civil society groups. They must also be within the capacity of urban governments to implement.

While there are many SDGs-related programmes by civil-society actors, until June 2019,[1] there was no targeted state or local government programme attempting to

---

[1] The Lagos State Government instituted an Office for SDGs and Investment in June 2019.

address any of the SDG-related targets or indicators. However, a perusal of the LSDP 2012–2025, which is the most comprehensive plan to address the urban development challenges of Lagos, reveals a clear alignment with many SDG goals. Therefore, the following section will attempt to establish the linkage between SDG11 and the operative urban development priorities of Lagos as articulated in the LSDP 2012–2025 (Table 1). In this way, we can determine if Satterthwaite's position holds true for Lagos.

There are clear alignments between the targets of SDG11 and the LSDP 2012–2025. Even though the Lagos State Government did not categorically set out to implement the SDGs, many activities of the various ministries and agencies remain relevant. The Lagos Global Office, hitherto focused on attracting international investments to the state, has been expanded to cover implementation of the SDGs. A Special Adviser to the Governor on SDGs and Investment was appointed with broad responsibilities to promote the SDGs and investment in Lagos and in collaboration with relevant agencies, to monitor and measure outcomes and impacts of all SDGs-related projects. The impact of this office is yet to be evaluated, as the office is still in the infancy stage.

Nevertheless, the analysis above shows that the capacity of the Lagos State institutional framework to achieve her SDGs-related targets remains suboptimal. Local/municipal governments were marginally involved in the activities highlighted. This is largely because the Nigerian local government system in structurally incapacitated due to constitutional limitations as well as technical know-how [4]. This position corroborates the view by Hardoy [18] that while SDG11 is relevant to the urban scale, encapsulating many provisions of municipal development, the capacity of sub-national governments to implement is still precarious.

3   Mechanisms for Local Implementation

Extant literature highlights three important elements that must be in place for effective implementation of any urban development agenda [11], viz:

i.   Governance structures
ii. Means of implementation
iii. Evidence base and practical guidance

Aspects of these three elements are now discussed in the context of Lagos and the local implementation of SDGs.

4   Governance Structures

The New Urban Agenda recognises the need to "strengthen urban governance with sound institutions and mechanisms that empower and include urban stakeholders

Table 1 SDG-LSDP linkages

SDG 11 monitoring framework target	Monitoring indicator	Key relevant LSDP indicator (2012–2025)	Main actions taken (2015–2019)	Comments
11.1 By 2030, ensure access for all to adequate, safe and affordable housing and basic services and upgrade slums	Proportion of urban population living in slums, informal settlements, or inadequate housing	• 100% of citizens will have access to decent housing • At least 25% of citizenry will own their home by improving access to affordable mortgages • Annual house building rates to be progressively accelerated over the plan period to achieve 20,000 units p.a. by 2015, 50,000 units by 2020 and 80,000 units by 2025	Investment of $40.9 million in Slum upgrading projects in 9 communities in the Lagos Metropolitan Development and Governance Programme (LMDGP) • Introduction of the Lagos home ownership and mortgage scheme (Lag HOMS)	Total number of housing units in Lagos HOMS was 5008 housing (Lagos State Ministry of Housing 2017) LMDGP was adjudged moderately unsatisfactory [50]. At the closure of the project, most facilities constructed in different slums had fallen apart. The project also caused forced eviction of residents in the areas of intervention

(continued)

Table 1 (continued)

SDG 11 monitoring framework target	Monitoring indicator	Key relevant LSDP indicator (2012–2025)	Main actions taken (2015–2019)	Comments
11.2 By 2030, provide access to safe, affordable, accessible and sustainable transport systems for all, improving road safety, notably by expanding public transport, with special attention to the needs of those in vulnerable situations, women, children, persons with disabilities and older persons	Proportion of the population that has convenient access to public transport disaggregated by age group, sex and persons with disabilities	• By 2025, 90% of population lives within 15 mins walking distance to public transportation • By 2025, the capacity of the public transport sector increases from handling 7 million passenger trips per day to 12 million passenger trips per day • By 2025, there is a 50% fall in the numbers of unroadworthy vehicles on the Lagos State registered vehicles on the roads • By 2025, there is a 70% increase in traffic system management awareness by the public • By 2015, there is a well-articulated and adopted transport policy and strategic management framework	LAMATA strategic transport master plan Bus reform initiative: transport reform bill 2017 Projects including traffic improvement scheme, provision of traffic infrastructure, public road safety programme, road channelisation, clearing of waterways and construction of jetties, rail road construction as well as construction of bridges and the Oshodi transport interchange project Lagos rail mass transit project: LAMATA Light rail (blue line) project, a 27-km rail project running from Okokomaiko to Marina The Lagos State Ferry Service capacity was expanded with the purchase of fourteen ferries	820 high-capacity buses were injected into the BRT system in 2019 The blue line light rail is still uncompleted • Traffic congestion is increasing due to increase in motor vehicle movement without a corresponding improvement in transportation facilities • There is no adequate transportation modal interaction • The operation of the BRT has been affected by lack of vehicular maintenance and structural decadence. Also Time has not been incorporated into its functionalities • The road mode of transportation is still been over utilized while the rail has not fully functioned and the water transport system is under-utilised • The development plan recognises the need for a transportation policy but up until now, there has not been one

(continued)

Table 1 (continued)

SDG 11 monitoring framework target	Monitoring indicator	Key relevant LSDP indicator (2012–2025)	Main actions taken (2015–2019)	Comments
11.3 By 2030, enhance inclusive and sustainable urbanisation and capacity for participatory, integrated and sustainable human settlement planning and management in all countries	• Percentage of cities with a direct participation structure of civil society in urban planning and management which operate regularly and democratically	• By 2025, increase in participation of women in politics and decision making • By 2020, 60% of the youth and women population are economically empowered and effectively participate in state development programmes	• Office of civic engagement was formed • Lagos State Youth Policy (2016)	• The Lagos State Youth Policy was instituted in 2016 (Lagos State Ministry of Youth and Social Development 2016) • Office of Civic Engagement was established to strengthen citizen participation in governance • Women are poorly represented in the governance framework with only 4 female legislators among 40 in the state house of assembly

(continued)

Table 1 (continued)

SDG 11 monitoring framework target	Monitoring indicator	Key relevant LSDP indicator (2012–2025)	Main actions taken (2015–2019)	Comments
11.4 Strengthen efforts to protect and safeguard the world's cultural and natural heritage	Share of national (or municipal) budget which is dedicated to preservation, protection and conservation of national cultural heritage including World Heritage sites	• Increase the influx of tourists to the state by 15% p.a. • Increased revenue generation for the state from tourism to 10% of GDP • Revive all existing and functional tourism sites • 50% increase in citizenry participation of cultural events • 10 new rural community tourist infrastructures developed • All museums and historic sites refurbished	• Upgraded Tinubu Square Fountain (2017) • Reconstruction and construction of new monuments • Redesigning of under-bridges with art • Redesign of Onikan area as a heritage district	Poor conservation of heritage buildings that resulted in demolition of Ilojo Bar and Onikan swimming pool Over-commercialisation of beaches

(continued)

Table 1 (continued)

SDG 11 monitoring framework target	Monitoring indicator	Key relevant LSDP indicator (2012–2025)	Main actions taken (2015–2019)	Comments
11.5 By 2030, significantly reduce the number of deaths and the number of people affected and substantially decrease the direct economic losses relative to global gross domestic product caused by disasters, including water-related disasters, with a focus on protecting the poor and people in vulnerable situations	Number of deaths, missing people, injured, relocated or evacuated due to disasters per 100,000 people	• Reduce incidence of flooding in Lagos State from 40 to 20% of urbanised and semi-urbanised areas of Lagos by 2015 and eliminate all by 2025 • Develop a storm water drainage master plan for the entire Lagos and implement by 2025 • Develop a regulatory framework for wetlands management • Sustain the continuous monitoring of flooding problems through the establishment of flood disaster early warning and advocacy mechanisms through the drainage offices within the 20 LGAs and 37 LDCs	• Establishment of the Lagos State emergency management agency • Demolition of shanties in Ilubirin and along all creeks and waterways that constituted health hazards • De-silting a total of 179 drainages across the state • Protected shores lines at Alaguntan/Okunkobo/Olomometa in Ojo • Infrastructure storm drainage with 49 km in place and 65% overall water system 14 km in place	• Drainage master plan is yet to be passed into law • Increased incidents of flooding, especially in areas close to dredging/land reclamation sites

(continued)

Table 1 (continued)

SDG 11 monitoring framework target	Monitoring indicator	Key relevant LSDP indicator (2012–2025)	Main actions taken (2015–2019)	Comments
11.6 By 2030, reduce the adverse per capita environmental impact of cities, including by paying special attention to air quality and municipal and other waste management	Percentage of urban solid waste regularly collected and with adequate final discharge with regard to the total waste generated by the city • Annual mean levels of fine particulate matter (i.e. PM2.5 and PM10) in cities (population weighted)	·To reduce industrial, vehicular and commercial emissions to comply with WHO air quality control guidelines by 2025 • To ensure that 50% of industries in the State install functional and efficient effluent treatment plants • Improve on efficiency of waste collection by 80% in the state • Upgrading of dumpsites to landfill sites by 2013 • Improve refuse collection within the State's shoreline by 2013 • Improve on solid waste sorting by 50% by 2013 • To put in place effective institutional framework for sustainable sewage infrastructure by 2013	Lagos State environmental management protection law, 2017 ("EMPL 2017") • Reposition and rebranded the KAI brigade into Lagos environmental sanitation corps • Sustained the management of 13,000 metric tons of waste	The law resulted in the disruption of the waste management value chain Air quality is not effectively monitored and is exacerbated by emissions from poorly regulated public buses Waste sorting is not prioritised Open defecation is still a challenge, especially close to the lagoon and beaches

(continued)

Table 1 (continued)

SDG 11 monitoring framework target	Monitoring indicator	Key relevant LSDP indicator (2012–2025)	Main actions taken (2015–2019)	Comments
11.7 By 2030, provide universal access to safe, inclusive and accessible, green and public spaces, in particular for women and children, older persons and persons with disabilities	The average share of the built-up area of cities that is open space in public use for all disaggregated by age group, sex and persons with disabilities	• To establish recreational parks in 25% of the 57 LG/LDCs in the State by 2015 and thereafter complete all additional parks by 2025 • To ensure proper maintenance of the 6 existing recreational parks state-wide and thereafter of all newly completed parks • Access to basic infrastructure, utilities and services to be 15 min walking distance	Lagos State Parks and Gardens Law No. 13 of 2011 was enacted and took off July 2012 (Lagos State Government, Ministry of Environment, 2012) LASPARK has established, captured and maintained 327 parks and gardens state-wide Also, over 6,203,553 trees have been planted till date Total number of people that have visited and used the parks are 288,681	The government has allocated spaces for greens and public interactions/integrations. However, majority of these spaces are not accessible to the public especially the poor, people living with disabilities and the elderly

(continued)

Table 1 (continued)

SDG 11 monitoring framework target	Monitoring indicator	Key relevant LSDP indicator (2012–2025)	Main actions taken (2015–2019)	Comments
11 a Support positive economic, social and environmental links between urban, peri-urban and rural areas by strengthening national and regional development planning	Number of countries that are developing and implementing a National Urban Policy or Regional Development Plans that (a) respond to population dynamics, (b) ensure balanced territorial development, and (c) increase local fiscal space	To commission and complete a new Regional Master Plan for Lagos State by 2015 To have District Master Plans in place by 2015 To have a complete coverage of Neighbourhood Plans in place by 2020 Promote 15 new Growth Poles/corridors to accommodate the future expansion of the city	Oshodi Isolo model city plan and Ikorodu Regional Master Plan (2016–2036) were commissioned Master plan and model city plans commissioned and/or completed, though those due for review are not yet done	Epe and Ikorodu regional masterplans are yet to be passed into law Model city plans are not being implemented and reviewed effectively, while lower order plans are yet to be developed

(continued)

Table 1 (continued)

SDG 11 monitoring framework target	Monitoring indicator	Key relevant LSDP indicator (2012–2025)	Main actions taken (2015–2019)	Comments
11 b By 2020, substantially increase the number of cities and human settlements adopting and implementing integrated policies and plans towards inclusion, resource efficiency, mitigation and adaptation to climate change, resilience to disasters, and develop and implement, in line with the Sendai Framework for Disaster Risk Reduction	Percentage of cities that are implementing risk reduction and resilience strategies aligned with accepted international frameworks (such as the successor to the Hyogo framework for action on disaster risk reduction) that include vulnerable and marginalized groups in their design, implementation and monitoring	• To produce a State Policy Framework on Climate Change • To continue to organise State summits on climate change • To conduct a study on the vulnerability level of Lagos State to Sea level rise and develop appropriate mitigation measures and emergency procedures for dealing with inundations • 95% compliance with environmental sector laws and regulations	Lagos State Climate Change Policy, 2012 Setting up of the Lagos state Resilience Office	Climate change policy is not being implemented Resilience office is largely administrative in nature Climate change summit has not been convened since (2014)

(continued)

Table 1 (continued)

SDG 11 monitoring framework target	Monitoring indicator	Key relevant LSDP indicator (2012–2025)	Main actions taken (2015–2019)	Comments
11 c Support least developed countries, including through financial and technical assistance, in building sustainable and resilient buildings utilizing local materials	Number of jobs in the construction industry of LDCs involved in the manufacture of local building materials, out of the total number of jobs in the construction industry	Reduction of building costs through championing the use of alternative materials and technology Enhance capacity to attract investment and create jobs, improve and replace all substandard housing	Setting up of Lagos State employment trust fund	There is no specific intervention to adopt local building materials or create construction sector specific jobs

*Analysis between 2015 (when the SDGs started) and May 2019 (tenure of the immediate past state government) because the new government that instituted the SDG office had not fully taken off as at the time of writing this article

Source

[a]Lagos State Development Pan (2012–2025)

[b]https://sustainabledevelopment.un.org/topics/sustainabledevelopmentgoals

[c]Lagos Bureau of Statistics: Compendium of Lagos State Statistics (2015–2018)

[d]Ministerial Press Briefing of various Lagos State Government Ministries and Agencies (2016–2019)

[e]Akinwunmiambode.com

[f]Comments are based on authors conclusions

in urban development plans to enable social inclusion, inclusive, and sustainable economic growth and environmental protection".

Since Nigeria's independence in 1960, the official structure of government has been based on three distinct administrative levels—national, state and local government—each with defined spheres of jurisdiction and constitutional functions. Of particular importance to the implementation of both the NUA and the SDGs is the role of local governments, as all the 774 constitutionally recognised local government area headquarters in Nigeria are regarded as urban centres in the 1999 Constitution of the Federal Republic of Nigeria [13] and in the National Urban Development Policy of 2006 [27].

Local governments are generally modelled to serve three purposes. First, they are a mechanism for democratic participation and inclusive governance. Second, they are an efficient service delivery tool for providing social services and basic infrastructure. Three, they are a tool for national development and a medium through which the grassroots can share in the national wealth [34]. Efficient functioning of the urban (local) governance system is central to achieving inclusive and sustainable urban service delivery and, by extension, the new urban agenda and SDG 11 [4].

As observed by Pieterse [37], good urban governance involves complex interactions between various actors—government (local, state and federal), civil organisations, individuals, households—all working together to reduce urban challenges and enhance opportunities. However, the contestation between various levels of government and the marginalisation of the local government system in Nigeria makes this an onerous task. According to Pieterse [37], the municipal governance system is often rendered incapacitated to perform its statutory responsibilities of urban service provision [37].

No doubt, local governments are typically weak in terms of logistics and human resources, even in primate cities like Lagos [25]. With state governments exerting fiscal control, most local governments function as mere administrative extensions of the state, a situation that has become common in the country [20]. In Lagos, for instance, the 2003 Lagos Local Government (Administrative) Law empowers the state government to oversee the affairs of the entire local government system through the State Ministry of Local Government Affairs. It also restricts the sources from which local governments can raise funds to much less than what the Nigerian Constitution allows.

The urban planning hierarchy is also tripartite, given the provisions of the Nigerian Urban and Regional Planning Law of 1992 [12] Decree 88 of 1992, which allows for a Federal Planning Commission, state planning board and local planning authorities. However, 28 years later, there is still no national planning commission, with only 13 states having domesticated the law [30] and fewer still implementing it. The local planning authorities are mostly redundant, lacking capacity where they exist at all [2, 24]. The situation in Lagos State, where there is a functional Urban and Regional Planning Law 2010, as well as a vibrant planning administration framework, is such that the hierarchy of operation is highly centralised [5].

The 2010 law allows for the creation of "Local Planning Permit Offices in cooperation with the Local Governments and Local Development Areas for the discharge of its functions at the Local Government level with the approval of the Governor on recommendation of the Commissioner." This limits the role of local planning offices to development control alone [3]. Furthermore, the failure of the law to recognise that some of the constitutional functions of the local government (e.g., the establishment and maintenance of cemeteries, slaughter houses, markets, motor parks, public conveniences, as well as the control and regulation of outdoor advertising, etc.) presupposes that local planning authorities would be established to coordinate physical planning and development within their areas of jurisdiction. Thus, this takes away the opportunities for participatory community-led development and municipal-level planning and illustrates how the Nigerian governance framework fails in translating national policies to local realities, further underscoring the importance of providing structures for localising international agreements and policies.

The urban planning regulatory framework is also largely autocratic and based on a 'top-down' perception of the planner-citizen relationship. A typical example is the Lagos State Urban and Regional Planning Law of 2010, which is replete with words like *demolish, demolition, pull down, seal up, ejected, zero tolerance* and *forfeiture*, reminiscent of Nigeria's colonial and military past and perpetuating the long-held belief that development control is primarily for enforcement rather than a set of activities for ensuring building quality and environmental safety [5]. Also indicative of this mindset is the failure to set up an Appeal Committee as provided for in the 2010 law [22] to provide an avenue for residents to seek redress on the actions of the planning agencies.[2]

Both the SDGs and the NUA emphasise that urban areas and urban planning will be expected to deal with all the key global issues. However, this urban paradigm shift is very different from current planning practices in Nigeria [4], as it refers directly to the acceptance and inclusion of informality, rural–urban continuum, density, mixed use, public transport, environmental priorities and plans shaped by a connected and high-quality public space. It also includes the promotion and preservation of cultural heritage, as well as upholding of the principles of an inclusive, rights-based approach to urban development.

5   Means of Implementation

One of the most pressing challenges for addressing the SDGs in urban areas is having urban governments with the technical and financial capacity required to act. Available data on municipal finance in Nigerian cities shows limited investment capacity, with budgets being as low as US$20 per capita in some cases (see Table 2).

In many cases, cities are burdened with large populations from rural–urban migration, unemployment and dearth of robust economic bases. In Nigeria, both the state

---

[2]The Physical Planning and Building Controls Appeals committee was inaugurated in December 2019.

Table 2 Budgetary analysis for selected Nigerian cities

City (2013)	Own source revenue ($)	Own source revenue/capita ($)	Budget ($)	Budget/capita ($)
Lagos	2,000,000,000	91	3,000,000,000	145
Ibadan	420,600,000	1	7,900,000	20
Oyo	705	3	11,500,000	0
Gombe	393,500	1	22,600,000	66

Source UN-Habitat Global Municipal Database [46]

and local governments typically rely on federal allocations, which have been dwindling and insufficient to meet these needs [6]. It is, therefore, necessary to strengthen the financial base of urban governments to improve implementation capacities by encouraging the development of alternative but equitable revenue bases. Lagos, by virtue of her status as Nigeria's economic and financial hub, has a broad revenue base and is able to generate significant funds across sectors, including the informal economy.

Drawing on the work of Rose [39], Benson [7] observes that the growing participation of national governments in international and supranational forms of governance ensures that policy transfer occurs through different mechanisms such as copying, adaptation, hybridisation, synthesis and inspiration. Invariably, these mechanisms lead to various depths of localisation. In copying, for instance, the level of localisation is low, as governments enact programmes that are already in effect in other countries, a process described as 'copying-and-pasting'. Bezmez [8] and Nubi et al. [31] observe the inefficiency of this approach in reference to regeneration projects, noting that in developing countries such as Turkey and Tallinn, the "copying-and-pasting" of foreign approaches to waterfront regeneration has led to lack of community acceptance, stiff resistance by civil-society organisations and, eventually, failure to realise project objectives.

What is required, therefore, is understanding the successes and failures of different aspects of policies and then using these to develop new programmes that are suitable for the local context. Policymakers are, therefore, required to first expand their horizon, and to be aware of policymaking processes and outcomes in comparable contexts. They are required to understand deeply their own local contexts and be imbued with the skills to determine, at any one time and for any given problem, the appropriate approach to use.

Regarding the international development agenda, the policy transfer continuum demands greater understanding of what may be described as the baseline stage of the recipient country, that is, previous policy commitments at both the local and international levels. They will also work within constraints relating to the internal political environment, such as the depth of the need and demand for change in the policy area, the complexity of the policy to be transferred, the clarity of the processes, indicators and benchmarks of the policy, as well as the relevance of the policy to the circumstances of local stakeholders.

In their examination of the transfer of neoliberal policy to the housing sector of developing countries, Olunubi and Oyalowo [33] recommend the following for successful localisation: an understanding of the operation and effectiveness of the policy in its home country, local institutional factors, structures and processes in both originating and recipient countries, the role of agents (politicians, policy staff) in championing change transfer and redesigning to fit the local context and still meet desired objectives. Apart from financial capacities and policy fit, there is also a need to strengthen the capacity of the urban governance system. In doing this, municipal governments are encouraged and supported to act through membership of city networks. This aligns directly with SDG 17, which stresses the importance of global partnerships for sustainable development.

City networks provide opportunities for peer learning, capacity building and facilitating the participation of this often poorly represented constituency in global processes under the aegis of the Global Taskforce of Local and Regional Governments, which is the umbrella network for facilitating the participation of local governments in UN processes and undertake joint advocacy relating to international policy processes, particularly the climate change agenda, sustainable development goals and Habitat III. Nigerian participation is practically nonexistent, although the Lagos State government is the most prominent Nigerian actor on these networks. Sadly, there is not much evidence around the state of any benefits from belonging to such networks (see Table 3).

6   Evidence Base and Practical Guidance

When the system allows for participatory local processes, it is easier to meet the long list of needs by a diverse range of groups, for only by investing in local processes will the SDGs and the New Urban Agenda be achieved. The SDGs stress the need for monitoring progress and outlining a long list of indicators to do so [48]. However, disaggregated data at the municipal level needed to design and implement required interventions are quite limited. Nigeria mostly relies on data collected by national and international agencies that do not produce local-level datasets [28]. State and local government surveys are not usually consistent and often do not have sample sizes large enough to provide relevant disaggregated data to urban governments [21]. While censuses should provide this, figures from the 2006 Nigerian Population Census data are yet to be fully released [29], even though the next one is several years overdue.

Providing the evidence base for urban decision-making and ensuring innovative solutions to challenges require interdisciplinary collaborations using a co-production framework. It is essential that urban knowledge and urban solutions are co-created by representatives of the political, research, business and civil-society communities. Co-production in the context of urban development is an opportunity for relevant stakeholders to work collaboratively to enhance understanding of urban issues, inform appropriate actions, produce defensible policy and increase the legibility of

Table 3 Nigerian membership of urban governance networks

Organisation	Objective	Nigerian membership
(Local Governments for Sustainability)—ICLEI	Global network of 1500 cities and local governments committed to building a sustainable future	• Ido LGA, Oyo state • Isoko South LGA • Lagos State Government • Ajeromi Ifelodun LGA • Mushin LGA
Cities Climate Leadership Group—C40	Network of 90 large cities committed to addressing climate change and building a sustainable urban future	• Lagos State Government
Rockefeller Foundation Resilient Cities Network—100RC	Network of 300 cities dedicated to helping cities around the world become more resilient to the physical social and economic challenges that are a growing part of the 21st Century	• Enugu State Government • Lagos State Government
Commonwealth Local Government Forum—CLGF	Network of 200 cities across the Commonwealth working to promote and strengthen democratic local government and to encourage the exchange of best practices through conferences, technical assistance projects and research	• ALGON
The Network Of Regional Governments For Sustainable Development—NRG4SD	Global network of 50 states, regions and provinces in the field of climate change, biodiversity and sustainable development, particularly following the mandates of UN Conventions and agendas	• Cross River State Government
United Cities and Local Governments	Network of 240,000 sub-national bodies supporting international cooperation between cities and their associations, and facilitating programmes, networks and partnerships to build the capacities of local governments	• Office of the Secretary to the Federal Republic of Nigeria
Global Covenant of Mayors	International alliance of 9149 cities and local governments with a shared long-term vision of promoting and supporting voluntary action to combat climate change and move to a low emission, resilient society	• Lagos State Government • Isoko South LGA

Source ICLEI—Local Governments for Sustainability [19]; C40 Cities [9], 100 Resilient Cities [1], Commonwealth Local Government Forum [10]; Regions 4 Sustainable Development [38]; United Cities and Local Governments [47]; Global Covenant of Mayors for Climate and Energy [14]

policy [35]. Furthermore, co-production of knowledge by academics, practitioners and civil society actors is critical to developing workable city-level solutions to pressing African urban development problems [26].

7 Strengthening Local Capacity for Effective Delivery

In strengthening local capacity for effective delivery of the SDGs and NUA, it is important to frame the narrative within the context of the three necessary elements discussed earlier: governance structures, means of implementation and evidence base. Therefore, it is recommended that local government reforms be instituted as a first step, as this will necessitate promoting constitutional amendments that establish the autonomy of the local government system and recognise it as a key lever of governance. This will be followed by capacity building for local actors–politicians, local government officials and civil society groups—through well-resourced programmes aimed at improving the capacity of community leaders and public institutions to engage in dialogue and support a collaborative approach to development. Peer learning through membership of city networks and inter-municipal cooperation is also important.

In providing avenues to localise policies further, there is a need to "practice urbanisation at scale". This requires acknowledging multi-level settlement structures—from the hamlet to the megacity, and importantly, the rural–urban continuum. Thus, the establishment of appropriate platforms for service delivery, citizen engagement and regional development at various levels of human agglomeration is key. Local planning authorities should be restored, the local government system should be reinvigorated and opportunities for community-led projects should be explored. Disaggregated and localised data collection and needs assessment should be collected in the bid to ensure that contextualised solutions are co-produced at the required geographic scale.

The strength of legal institutional frameworks needs to be deepened. It is necessary that the appropriate institutional and regulatory frameworks for delivering the SDGs and NUA are outlined and enabled. Bureaucratic challenges should be addressed and an integrated governance system that leverages on technology can be put in place. Furthermore, there is need to revise the extant planning laws and policies and to institutionalise them for proper implementation. It is also important that the national legal and institutional frameworks are adapted to specific local contexts while being flexible enough to respond to changing urban dynamics.

In all of these, it is imperative to establish collaborative governance, a process based on the principles of subsidiarity and decentralisation and that is a necessary prerequisite for inclusive service delivery and attainment of the New Urban Agenda and Sustainable Development Goals. Collaborations are encouraged across vertical (between different levels of government) and horizontal (within the same level, e.g. between ministries or between local governments) dimensions. Furthermore, partnerships with actors from civil society, academia and the private sector are also encouraged.

Accordingly, human capital development in the urban space is required in order to strengthen professionalism and capacity building for urban actors. Localisation of the SDGs and the New Urban Agenda will require a broad process of capacity building. Indeed, there is need for well-resourced capacity-building programmes to support the transformative process of training public employees, as well as local leaders from civil-society organisations. A system-wide capacity-building alliance between national and local governments, like-minded partners (e.g., academia and NGOs), as well as civil-society networks and international organisations, will be crucial for fostering capacity building [15–17]. Revitalising urban planning education and practice to reflect local realities is also necessary for changing the professional landscape, as well as improving the capacity and efficiency of urban management and administrative systems [36, 49].

8 Conclusion

This chapter has highlighted the need to embrace global agreements as the starting point on the journey to sustainable development. It has also revealed that for these global agreements to be effective, they have to be viewed from a contextualised local lens. It is necessary to ensure that the global agreements align with local governance priorities and targets. Furthermore, they must be interpreted in a manner that is actionable at the basic neighbourhood level while being potentially scalable within a multilevel governance framework. The SDGs and NUA are a veritable platform for achieving sustainable urban development, since they provide clear measurable targets for catalysing action towards a preferred urban future. However, these targets can only be achieved when local governance capacities are strengthened, local urban priorities are addressed and all actors at all levels, and from all sectors, work together in conceptualising and implementing urban solutions that leave no one behind.

References

1. 100 Resilient Cities (2020) www.100resilientcities.org. Accessed 16 May 2020
2. Abubakar I, Doan P (2017) Building new capital cities in Africa: lessons for new satellite towns in developing countries. African Stud 76(4):546–565
3. Adediran A (2017) Dialectics of urban planning in Nigeria. Paper delivered at the 5th Urban Dialogue of the Department of Urban and Regional Planning, University of Lagos
4. Agunbiade E, Olajide O (2016) Urban governance and turning African cities around: Lagos case study. Partnership for African Social and Governance Research Working Paper No. 019, Nairobi, Kenya
5. Agunbiade M, Ewedairo K (2014) A review of the Lagos State Urban and Regional planning and development law 2010: an international perspective. In: Smith II (ed) Essays on Lagos State urban and regional planning and development law 2010. Faculty of Law, University of Lagos, Lagos
6. Babatunde A (2010) People's attitude toward property tax payment in Minna. In: Laryea S, Leiringer R, Hughes W (eds) Proceedings West Africa built environment research (WABER) conference, 27–28 July 2010, Accra, Ghana, pp 435-441
7. Benson D (2009) Constraints on policy transfer. CSERGE Working Paper EDM, 9–13. University of East Anglia, The Centre for Social and Economic Research on the Global Environment (CSERGE), Norwich

8. Bezmez D (2009) The politics of urban waterfront regeneration: the case of Haliç (the Golden Horn), Istanbul. Int J Urban Reg Res 31(4):759–81. https://hdl.handle.net/10419/48824 Accessed 16 May 2020
9. C40 Cities (2020) www.c40.org. Accessed 16 May 2020
10. Commonwealth Local Government Forum (2020) www.clgf.org.uk. Accessed 16 May 2020
11. Fabre EA (2017) Local implementation of the SDGs and the New Urban Agenda: Towards a Swedish national urban policy. Stockholm, Global Utmaning
12. Federal Government of Nigeria (1992) The Nigerian urban and regional planning law. Decree 88 of 1992. The Federal Government of Nigeria, Abuja
13. Federal Government of Nigeria (1999) 1999 constitution of the federal Republic of Nigeria. Federal Government of Nigeria, Abuja
14. Global Covenant of Mayors for Climate and Energy (2020) www.globalcovenantofmayors.org. Accessed 16 May 2020
15. Habitat III (2016a) HABITAT III, 17–20 Oct 2016. In: The United Nations conference on housing and sustainable urban development. https://habitat3.org/the-new-urban-agenda/prepar atory-process/national-participation/nigeria/#national-report-executive-summar. Accessed 16 May 2020
16. Habitat III (2016b) HABITAT III policy paper 3: National Urban Policy. https://habitat3.org/ wp-content/uploads/Habitat%20III%20Policy%20Paper%203.pdf. Accessed 16 May 2020
17. Habitat III (2016c) HABITAT III policy paper 4: urban governance, capacity and institutional development. https://habitat3.org/wp-content/uploads/PU4-HABITAT-III-POLICY-PAPER.pdf. Accessed 16 May 2020
18. Hardoy J (2017) An urban development agenda for Latin American cities: integrating global and local challenges.https://ugecviewpoints.wordpress.com. Accessed 16 May 2020
19. ICLEI—Local Governments for Sustainability (2020) www.iclei.org. Accessed 16 May 2020
20. Khemani S (2001) Fiscal federalism and service delivery in Nigeria: the role of states and local governments. Prepared for the Nigerian PER Steering Committee, 24
21. Lagos Bureau of Statistics (2017) Abstract of local government statistics. https://mepb.lagoss tate.gov.ng/wp-content/uploads/sites/29/2018/06/Abstract-of-LG-Statistics-2017editted.pdf. Accessed 16 May 2020
22. Lagos State Government (2010) Lagos State urban and regional planning law 2010. The Lagos State Government, Alausa
23. Lagos State Government (2015) Demographic dividend in lagos state: the opportunity we must not Forgo. https://mepb.lagosstate.gov.ng/wp-content/uploads/sites/29/2017/08/Demogr aphic-Dividend-in-Lagos-State-2015-1.pdf. Accessed 16 May 2020
24. Lawanson T (2016) Urbanisation in Nigeria: the need for a paradigm shift in planning education and practice. Environ Technol Sci J 7(1):13–23
25. Lawanson T, Oduwaye L (2014) Socio-economic adaptation strategies of the urban poor in the Lagos Metropolis, Nigeria. African Rev Econ Fin 6(1):139–160
26. Marrengane N (2014) Co-production: AURI there are no shortcuts: the rocky road to enabling knowledge co-production in African cities. In: Presented at African urban research initiative (AURI) policy conference, Nairobi, Kenya
27. Ministry of Housing and Urban Development (2006) National Urban development policy. The Federal Ministry of Housing and Urban Development, Abuja
28. National Bureau of Statistics (2020) https://www.nigerianstat.gov.ng/. Accessed 16 May 2020
29. National Population Commission of Nigeria (2020) https://www.population.gov.ng/. Accessed 16 May 2020
30. Nigeria Institute of Town Planners (2016) Communiqué of 2016 Nigeria Institute of Town (NITP) annual conference. https://nitpng.com/communique-of-the-golden-jubilee-annive rsary-47th-annual-conference-of-the-nigerian-institute-of-town-planners-nitp-held-between-the-31st-october-and-4th-november-2016-at-the-nicon-luxury-hotel-a/. Accessed 16 May 2020
31. Nubi T, Oyalowo B, Muraina O (2018) Sustainable regeneration of waterfront slums in Lagos: a transdisciplinary approach. In: 13th UNILAG annual research conference, 28–30 Aug 2018

32. Oleribe O, Taylor-Robinson S (2016) Before sustainable development goals (SDG): Why Nigeria failed to achieve the millennium development goals (MDGs). Pan African Med J 24(1):156

33. Olunubi T, Oyalowo B (2010) Housing finance between social needs and economic realities: the dilemma of policy transfer under neo-liberalism. In: Proceedings of the comparative housing research conference. Delft University of Technology, Delft, pp 368–386

34. Osasona T (2015, July 24) Governance institutions: Local government in Lagos. In: CPPA's series on Nigerian institutions of governance. https://cpparesearch.org/nu-en-pl/governance-institutions-local-government-in-lagos/. Accessed 16 May 2020

35. Patel Z (2014) Univer-City partnerships:co-production for urban sustainable development in Cape Town. Presented at African Urban Research Initiative (AURI) policy conference, Nairobi, Kenya

36. Pieterse E (2008) City futures: confronting the crisis of urban development. Zed Books, London and New York

37. Pieterse E (2015) The contemporary African city: crises, potentials and limits. The Antipode foundation's 5th international institute for the geographies of justice. 23 June, University of the Witwatersrand, Johannesburg

38. Regions 4 Sustainable Development (2020) https://www.regions4.org/. Accessed 16 May 2020

39. Rose R (1993) Lesson-drawing in public policy: a guide to learning across time and space, vol 91. Chatham House Publishers, Chatham

40. Satterthwaite D (2016a) Editorial: a new urban agenda? Environ Urban

41. Satterthwaite D (2016b) Missing the millennium development goal targets for water and sanitation in urban areas. Environ Urban 28(1):99–118

42. Satterthwaite D (2016c) Successful, safe and sustainable cities: towards a new urban agenda. Commonwealth J Local Govern 19(1):3–18

43. Satterthwaite D (2016d) Where are the local indicators for the SDGs? https://www.environmentandurbanization.org/where-are-local-indicators-sdgs. Accessed 16 May 2020

44. The Lagos State Government (2013) Lagos state development plan 2012–2025. Main document. Alausa

45. UN-Habitat (2017) Resolution adopted by the general assembly on 23 December 2016. In: General assembly: seventy-first session. https://habitat3.org/wp-content/uploads/New-Urban-Agenda-GA-Adopted-68th-Plenary-N1646655-E.pdf. Accessed 16 May 2020

46. UN-Habitat (2018) Global municipal finance database. https://globalmunicipaldatabase.unhabitat.org/node/2312. Accessed 16 May 2020

47. United Cities and Local Governments (2020) www.uclg.org. Accessed 16 May 2020

48. United Nations (2020) Sustainable development goals. https://unstats.un.org/sdgs/metadata/. Accessed 16 May 2020

49. Watson V, Odendaal N (2013) Changing planning education in Africa: the role of the association of African planning schools. J Plann Educ Res 33(1):96–107

50. Word Bank (2016) https://projects.worldbank.org/en/projects-operations/project-detail/P071340?lang=en&tab=procurement. Accessed 16 May 2020

The Road not Taken: Policy and Politics of Housing Management in Africa

Abiodun Anthony Olowoyeye

Abstract Housing is central to meeting the Sustainable Development Goals in Africa. Housing holds everything together and reflects on social/economic inequality and injustice, poor infrastructure development, and poor health outcomes. To have a housing system that works for all, it is necessary to have a holistic housing policy. While the blame for Africa's housing woes has been laid on lack of funding, corruption, poor land management system and overpopulation for too long, the causes of Africa's housing problems go far much deeper. The aim of this chapter is to present a new approach to the problems of housing, identifying strategic as well as historical causes of the problems. The study notes that the foundation of the problem of housing was laid when, at independence, politicians and policymakers missed the opportunity to implement the recommendations of the expert committee of the Economic Commission for Africa at its 1963 meeting. This was the road not taken and, more than 50 years later, the consequences of this missed opportunity remain abiding. The chapter concludes that without robust policies to enable housing to become a key driver of economic, social and infrastructural development, meeting the SDG goals will remain a difficult goal to attain.

Keywords Housing · Policymakers · Historical · Sustainable development goals (SDG)

> Two roads diverged in a wood, and I—
> I took the one less traveled by,
> And that has made all the difference.
> (Frost, 1920)

A. A. Olowoyeye (✉)
ADIGO Limited, Answer House, Revenston Lane, Whitburn Bathgate EH478HJ, UK
e-mail: aolowoyeye98@gmail.com

© The Author(s), under exclusive license to Springer Nature Singapore Pte Ltd. 2021
T. G. Nubi et al. (eds.), *Housing and SDGs in Urban Africa*, Advances in 21st Century Human Settlements, https://doi.org/10.1007/978-981-33-4424-2_3

1 Introduction

For too long, Africa's housing problems have been blamed on the usual suspects of population growth, rapid urbanisation, poor land system and corruption. While these are certainly problems in the African context, Africa's housing problems go much deeper. In January 1963, experts and stakeholders, under the auspices of the Economic Commission for Africa, met in Addis Ababa, Ethiopia to consider housing problems and policies for Africa. Coming at a time when many African countries were gaining independence from their colonial masters, the Commission understood the fundamental benefits that better housing plans and delivery could bring to the development of nations across Africa. It therefore brought together experts with the purpose of [30]:

- exchanging views on the prevalent housing situation in Africa
- defining as clearly as possible housing problems, scope, and solution
- suggesting the general measures to facilitate the framing and implementation of housing policies
- proposing the organisational arrangements most suitable to ensure that housing problems shall be regularly and continuously studied within the Commission.

Although the conference had its shortcomings (the enumeration of which is beyond the scope of this paper), its purpose and outcomes were laudable in many ways. Most importantly, the conference placed housing at the centre of economic development, recommending practical solutions on how housing could support development across Africa. For instance, the conference stated that "good housing conditions contribute to economic progress; well-located dwellings and improved urban services. Better housing is a concrete proof of national progress and there is a close relationship between housing and industrialisation" [30].

The aim of this chapter is to provide a historical context on how the continent missed the opportunities to develop robust housing policies that would have enabled housing to become a key driver of economic, social and infrastructural development. The chapter also highlights housing problems in Africa and how tackling the housing problems can support the attainment of the sustainable development goals (SDGs).

The chapter is divided into three main sections. The first section provides a background on the subject. The second section provides historical and strategic perspectives on the problems of housing. The third section highlights the importance of housing to the delivery/attainment of the SDGs and concludes that without developing a housing system that cares and considers the needs of all, realising the SDGs will remain a mirage for Africa.

2 The Road That Should Have Been Taken

At the 1963 ECA meeting, experts set out a new development pathway that made housing a key driver for economic and infrastructural development for the newly independent African countries. This level of importance attached to housing was evidenced by a number of factors. First was the diversity of the stakeholders and experts who participated at the meeting. For example, in attendance were representatives from Burundi, Ethiopia, Ghana, Ivory Coast, Kenya, Liberia, Madagascar, Nigeria, Federation of Rhodesia and Nyasaland[1] [28], Sierra Leone, Republic of South Africa, Tanganyika[2] [27], France and the United Kingdom. In addition, there were observers from Belgium, the Netherlands, Poland, Sweden, the USA, the USSR[3] [9], the Secretariat des missions d'urbanisme et d'habitat (Google Translate: *Secretariat of urban planning and housing missions, France*) and the Centre d'information du batiment (Google Translate: Building Information Center, France). In addition, present at the meeting were representatives of the International Labour Office (ILO), the Food and Agriculture Organisation (FAO), the World Health Organisation (WHO) and the United Nations Children's Fund (UNICEF).

Second was the scope and robustness of the discussion. Notably, discussions focused on the following:

1. Understanding housing needs: the conference noted that it was essential to understand the extent of housing needs among African states and, until the needs were known, it was impossible to formulate long-term housing policies and programmes. The conference also stressed that housing needs should be robust, encompassing housing-related infrastructure such as roads, schools, health centres, and sites for various social activities. In addition, the conference noted that the quantity of housing supplied was meaningless if the dwellings did not meet minimum requirements on quality.
2. Securing investment in housing and financing: the conference noted that the main problem facing all developing countries was how to secure the required investments in order to bridge the gap between housing needs and the available financial resources. The conference considered a number of options to address

---

[1]Federation of Rhodesia and Nyasaland, also called Central African Federation, political unit created in 1953 and ended on Dec. 31, 1963, that embraced the British settler-dominated colony of Southern Rhodesia (Zimbabwe) and the territories of Northern Rhodesia (Zambia) and Nyasaland (Malawi), which were under the control of the British Colonial Office. Encyclopaedia Britannica, Inc.

[2]Tanganyika, historical eastern African state that in 1964 merged with Zanzibar to form the United Republic of Tanganyika and Zanzibar, later renamed the United Republic of Tanzania. Encyclopaedia Britannica, Inc.

[3]Soviet Union, in full Union of Soviet Socialist Republics (U.S.S.R.), Russian Soyuz Sovetskikh Sotsialisticheskikh Respublik or Sovetsky Soyuz, former northern Eurasian empire (1917/22–1991) stretching from the Baltic and Black seas to the Pacific Ocean and, in its final years, consisting of 15 Soviet Socialist Republics (S.S.R.'s): Armenia, Azerbaijan, Belorussia (now Belarus), Estonia, Georgia, Kazakhstan, Kirgiziya (now Kyrgyzstan), Latvia, Lithuania, Moldavia (now Moldova), Russia, Tajikistan, Turkmenistan, Ukraine, and Uzbekistan. The capital was Moscow, then and now the capital of Russia.

this problem, including setting the percentage of housing budget within total national investment, the utilisation of all financial means available at a national level, and supplementing national resources with external financial aid.

3. Identifying ways of reducing building costs: the conference examined the means of making most effective use of the available resources, with a focus on reducing "building costs. Along this line, the conference encouraged African countries, which had developed innovative solutions on reducing construction costs to share their good practice among other African countries.

4. Physical Planning: the conference recognised that physical planning has a crucial relationship with both national economic development plans and housing programmes, and it stressed the need to coordinate physical planning with national economic and social policies at all levels but particularly within the framework of regional planning development.

In addition, to support the development of housing and related systems across Africa and foster international collaboration in housing, building and. physical planning, the experts at the meeting made some recommendations including the following. That:

a. The secretariat should centralise and disseminate among member countries information on housing, building and physical planning problems, with partic- ular reference to the results and the application of research and, exchange of experience;

b. The ECA should encourage the establishment of regional or sub-regional centres on housing and building research and documentation in Africa; the secretariat should in particular and within its competence facilitate the co-ordination and dissemination of published information on housing and building research, in close co-operation with the International Council for Building Research Studies and Documentation (CIB) and its African member institutes;

c. The ECA, within the framework of comprehensive studies of the economic geog- raphy of African countries, should carry out or sponsor such surveys as will be necessary to permit the elaboration and the implementation of national and regional physical plan;

d. The secretariat should collaborate with the relevant international organisations in collecting and circulating suitable material and assist in the training of families (particularly housewives), concerning the maintenance and rational use of the house and its facilities.

Notably, these discussions and recommendations paved the road to the develop- ment of a robust housing system in Africa. This was the road that, should have been taken.

3 The Road not Taken: Problems of Housing in Africa

Housing is crucial to the development of any nation and there are consequences when a nation fails to put in place effective housing policies. Unfortunately, at independence, politicians and policymakers across Africa missed the opportunity to develop and implement policies that would have enabled housing to become a key driver of economic, social and infrastructural development on the continent. This was the road not taken and, more than 50 years later, the consequences of this missed opportunity remain with Africa even today.

3.1 Africa's Missed Opportunities

The following section explains how Africa missed this opportunity.

3.1.1 Internal Political Instability

Political stability is the key to the development of any society, as it impacts on macroeconomic stability, which is a necessary ingredient for the development of a stable housing system. When nations are stable, governments and the machinery of policymaking and implementation thrive, a situation that in turn enables the development of policies on economy, housing and infrastructure. Shepherd [25] puts this succinctly when he notes that while other types of politics are also important for growth, "political stability is no doubt key to growth". When governance is unstable, opportunities to harness internal resources for development are missed. Notably, most of the 12 African countries at the 1963 ECA housing conference on Africa were at the early stages of their independence. However, these countries' high hopes were soon dashed, as most of these countries began to experience political instability and as such were not in a position to take forward the recommendations of the experts at the conference.

Notably, across many African countries, the situation (political instability) at independence meant that it was either difficult or impossible to implement many of these recommendations by the experts. Consequently, the 12 African countries that attended the conference missed the opportunity to develop their capabilities to deliver good and quality housing and, become examples of 'good practice' to other African countries.

Chart 1 below provides a snapshot of political instability among the 12 African countries that attended the 1963 ECA meeting. Note that it is sometimes difficult to provide timelines for political events, as these tend to occur over time. For instance, by 1963, Burundi, Sierra Leone, and Liberia (which was not colonised) were embroiled in political unrest. Simultaneously, Ethiopia, Ivory Coast, and Madagascar fell under autocratic or one-party rule and soon afterward, civilian governments in Ghana,

Nigeria and the Central African Republic were displaced by war and/or military coups. Although Tanzania was relatively peaceful due to socialism as practised by Julius Nyerere's government, the overreliance of this policy on foreign aid was its undoing once foreign aid stopped.

This lack of political stability was worsened by prevalent inequality, poor economy, bitter struggle for power and inter-ethnic warring in the face of frequent military intervention. The consequences were the weakening of civil institutions, exposure to economic volatility and misalignment of the machinery of government, all of which prevented these African countries from putting in place efficient state apparatus to implement experts' recommendations on housing, building and. physical planning in Africa (Table 1).

3.1.2   Lack of Understanding of Housing Policy and Strategy

Across Africa, there is a prevalence of a misunderstanding or lack of understanding of the dynamics of housing among policymakers and politicians. This situation has prevented housing from being positioned as a driver of economic development. According to Kissick et al. [17] noted thus 'housing markets have significant macroeconomic impacts. Housing generates expenditures outside the housing sector itself, it is seen as a distinct asset class leading to greater market efficiency, stability, and liquidity, it is a basis for taxation for government taxation and is a significant contributor to the fiscal health of governments and their capacity to deliver basic urban services.' However, while on other continents, housing has, over the years, evolved from the traditional methods of 'bricks and mortar' into a system that is a major contributor to national GDPs, the approach to housing continues to be mundane and traditional across Africa. This means that housing continues to be on the periphery of policy and economic development.

In addition, there is a clear misunderstanding of market and affordable housing among politicians and policymakers. This misunderstanding is evidenced by a lack of clear definition of what constitutes affordable or market housing, as well as what qualifying criteria are used to determine affordability. This failure to define affordability, and to implement appropriate policies to support the development of a housing system that works for everyone, has by default created a 'housing class' problem across Africa, resulting in the 'haves' and the 'have-nots', with the 'have-nots' relying on informal or slum housing. Glossop [12] captured this summarily when he reports that while "housing matters to economic development, it can also lead to segregation and spatial concentrations of poverty." The result of this is the gradual emergence and growth of sub-standard and slum housing. Slum housing is not an epidemic that emerged suddenly but an endemic problem that has grown because of repeated failure by African states to develop a housing system that works for all. In 1963, the ECA Commission put this clearly when it stated that "the existing methods of housing policy were often found to be unsuitable for the solution of problems raised, particularly as low-income classes were not in a position to pay the rent for housing

Table 1 A Snapshot of Political Instability in African Countries at Independence

Countries (rows): Burundi, Ethiopia, Ghana, Ivory Coast, Kenya, Liberia Pre 1957, Madagascar, Nigeria, Central African Republic, Sierra Leone, South Africa, Tanzania

Years (columns): 1957, 1958, 1959, 1960, 1961, 1962, 1963, 1964, 1965, 1966, 1967, 1968, 1969, 1970, 1971, 1972, 1973, 1974, 1975, 1976, 1977, 1978, 1979, 1980, 1981, 1982, 1983, 1984, 1985, 1986, 1987, 1988, 1989, 1990, 1991, 1992, 1993, 1994, 1995, 1996, 1997, 1998, 1999, 2000, 2001, 2002, 2003, 2004, 2005, 2006, 2007, 2008, 2009, 2010, 2011, 2012, 2013, 2014, 2015, 2016, 2017, 2018, 2019

Colour / Format	Colour / Format Interpretation
	Democratically elected government
	Inequality/ Repression/ Political unrest /Violence
	Inequality/ War and/or Military intervention
	Autocratic rule/One Party with economic problem

Historical records of events are fluid in nature making strict periodic categorisation of events challenging. However, the author has categorised events into the nearest timescale as much as possible

Sources British Broadcasting Corporation (United Kingdom); Central Intelligence Agency World Factbook. The Security Council Report; African Journal of Political Science and International Relations

when this meant devoting 20 to 25% of their wages to rent". More than 50 years after this observation by ECA, the housing need in Africa remains as dire as ever!

3.1.3 Lack of Understanding of Housing Need and Demand

Housing need is an unconstrained assessment of the number of homes needed in an area. Assessing housing need is the first step in the process of deciding how many homes need to be planned for. It should be undertaken separately from assessing land availability, establishing a housing requirement figure and preparing policies to address this, such as site allocations [29]. Understanding housing need and demand involves carrying out periodic research and consultations while also recognising the unique regional differences and using this knowledge to inform housing policies and budgetary/delivery plans. Across most African countries, there is a lack of understanding by governments on the changing patterns of housing needs among the population. A related problem is the lack of structured data and information that would help in the robust understanding of national, regional and local housing needs and demands.

One of the key recommendations of the ECA meeting was the development of housing need and demand capabilities for African states, which is a crucial step in the process. Notably, it provides a realistic 'understanding of the number of housing units needed to meet existing and future housing need and demand. It thereby assists governments to develop robust policies to ensure continuous effective delivery of housing and housing-related infrastructure. It informs the delivery of the right housing in the places where people want to live.

Lack of understanding of housing need and demand among most African countries is indeed a serious problem, as it means that housing policies are frequently not informed by empirical evidence. The implication is that many governments are unable to plan for present and changing future housing and household needs. Taking Nigeria as an example, for about 10 years, Nigerian governments have based their housing policies on meeting an estimated 17 million accommodations shortage. However, it is not clear how the figure of 17 million was arrived at and what variables informed the housing need; the ratio of affordable and market housing between the urban area and the rural area is not indicated. Moreover, government has not specified what percentage has been met so far, just as it is not clear if backlogs have been calculated and recorded. By repeatedly quoting the 17 million figure, the government is essentially stating that while the population of Nigeria continues to grow, housing need has remained unchanged.

3.1.4 Misplaced Faith in the Market

There is a severe shortage of housing across Africa. According to the Sud, "African cities become the new home to over 40,000 people every day, many of whom find themselves without a roof over their heads". Bah et al. [1] in their research on housing deficit in African cities, observed, "the rapid urbanisation rates and lack of urban planning have resulted in very large housing deficits"

To meet the severe shortage, governments are increasingly delegating responsibilities to the private market for reducing housing pressures and meeting shortages. By market, it is meant housing development and sales/lease based solely on financial markets, such as mortgage instruments or setting rents based on market forces. Although the market can support the delivery of housing and a functioning market is indeed necessary for an effective housing system, it is only a part of the story.

Housing is both a social good as well as an investment tool and there is a significant difference between market and affordable housing. While market housing can enhance the delivery of social housing, it cannot replace it and African governments have to start investing in affordable housing alongside market housing. This is so because of a number of factors. For one, most Africans are unable to afford mortgage housing as such. Thus, relying solely on the market to meet housing needs means governments are not delivering housing and related benefits to their citizens. Essentially, without affordable housing, many citizens across Africa will continue to live in substandard homes with implications for their general well-being. For another thing, the housing market in Africa in its present form cannot deliver the much-needed level of investments required to meet housing shortages because of the challenging investment atmosphere. Bah et al. [1] noted that the 'weak legal and regulatory frameworks that result in high payoffs for rent-seeking behaviour, the dominance of the informal sector that affects the development of mortgage markets, and the shallow capital markets that are not a significant source of financing for long-term housing investments reflect the inadequacy of financial instruments and weak finance markets across many African countries'. Therefore, expecting the market to deliver housing for all is a misplacement of faith and responsibilities.

3.1.5 Lack of Adequate Collaboration Between Academia/Research Institutions and the Building/Housing Industry

There are 1685 universities in Africa, many of which offer courses in engineering, sustainability, building/construction, urban planning and technology [8]. Nevertheless, academic research has not sufficiently informed the practicality of housing development across Africa. Notably, across Europe, America and Asia, academic research and development (R&D) has informed the development of a housing system that continually adapts to changing household and population needs, government policies, financial markets and environmental sustainability. For instance, in the UK,

More than 100 university-owned science and enterprise parks bring together the best companies, personnel and ideas from academia and the private sector, driving productivity and innovation and offering further opportunities for graduates to bring their skills into the local jobs market [31].

Moreover, the UK is home to the Building Research Establishment (BRE), a leading multidisciplinary building science centre with a mission to improve buildings and infrastructure through research and knowledge generation (BRE). This has contributed immensely to housing development in the UK and across the world. However, R&D has not delivered similar values to housing development across Africa. Where formal academic research has been carried out, research outcomes are often presented in inaccessible formats that may not necessarily connect with the building industry and the wider population.

3.1.6   Failure to Aim Small, Miss Small

In the movie *The Patriot*, Benjamin Martin (Mel Gibson) reminded his sons as they prepared to ambush a unit of British soldiers, to "aim small, miss small." This means that if you aim at a man and miss, you miss the man, while if you aim at a button (for instance) and miss, you still hit the man [2]. The same lessons can be applied to the development of housing and infrastructure. Delivering housing in manageable quantities will allow African governments to develop relevant experience in house building, housing finance and risk management. Rather than focusing on understanding and delivering housing in manageable quantities as determined by robust housing needs, there is an ongoing obsession in many African countries to embark on large-scale new-city developments. Moreover, many of these large-scale projects are often advertised and amplified as robust solutions to the problems of affordable housing in these countries, but the reality could not be further from that.

Notably, from the Tatu and Konza Techno Cities in Nairobi (Kenya) to Hope City in Accra (Ghana), Kigamboni City in Dar es Salaam (Tanzania) and Eko Atlantic in Lagos (Nigeria), these cities are draped in the rhetoric of "smart cities" and "eco-cities", with the promise to modernise African cities and turn them into gateways for international investors and showpieces for ambitious politicians [34]. Watson therefore, identified the following common themes among these city development plans:

- they are large-scale and involve the re-planning of all or large parts of an existing city or restructuring a city through the creation of new but linked satellite cities;
- they consist of graphically represented and three-dimensional visions of future cities rather than detailed land use plans, influenced by cities such as Dubai, Shanghai or Singapore;
- there are clear attempts to link these physical visions to contemporary rhetoric on urban sustainability, risk and new technologies, underpinned by the ideal that through these cities Africa can be "modernised";
- they are either on the websites of the global companies that have developed them or are on government websites with references to their origins within private-sector companies;
- their location in the legal or governance structures of a country is not clear;

- there is no reference to any kind of participation or democratic debate that has taken place.

It is no wonder that these new-city development plans either fail or fall short of expectations. For instance, in 2015, the South African Government announced its plan to build the Modderfontein New City in Johannesburg, with the help of a Chinese developer. Touted as the 'Manhattan of Africa', it was pitched as the next big development in South Africa but despite the release of futuristic computer-generated images that led to significant publicity for the project, it was never built. Instead, the land was eventually sold off [4].

4 United Kingdom Housing Delivery Plan: An Example of the Road That Was Taken

A case study of how the UK was able to deal with the perennial housing shortage after World War I demonstrates that Africa can also overcome the challenges posed by housing. During the nineteenth Century, Britain's cities expanded due to workers and labourers migrating to the cities for work during the years of industrialisation and growing opportunity for employment [32]. At this time, house building was largely done by profit-seeking private builders and, because borrowing and mortgages had not become common, only the richest people could afford to own their own homes. As city populations increased, the problems arising from poor housing conditions started to grow. Malpass and Murie [20] observed the situation in the UK, noting that 'there was a serious decline in the level of housing production for most of the decade before 1914, and new buildings fell still further during the war itself. The result was that by 1918 there was a severe housing shortage which for economic reasons private enterprise could not tackle effectively, especially in the short term, and which for political reasons the state could not ignore.' Lund [19] noted that long before the Second World War was over, Government recognised that a large scale house-building programme would be needed in its aftermath. Around 458,000 houses were destroyed in the war and 250,000 badly damaged. Notably, in July 2019, Britain marked the centenary of the Housing Act 1919, which paved the way for large-scale council housing [32]. Although housing shortage continues to exist in the country, the success of housing in the UK is the result of a combination of factors that will now be discussed.

First was government's commitment to gain an understanding of housing need and put in place robust policies to increase access to housing and reduce the massive housing deficits. For example, in 1919, the Housing and Town Planning Act of 1919 (otherwise known as the Addison Act) was enacted to build 500,000 homes over three years. Moreover, in 1980 the government of Margaret Thatcher introduced the Right to Buy (RTB) policy, which allowed council tenants to buy their homes at huge discounts [6].

Second was the introduction of subsidies and grants by the government to fund the development of new housing at national and local levels. For instance, the 1919 Act, which introduced subsidies for councils to solve the blight of slum estates, is credited with establishing the principle of large-scale, state-funded provision of council housing at low rents. Although the 1919 Act was later replaced, the appetite for large-scale housing development had been born [16]. For instance, in Scotland alone, between 2013 and 2018, the Scottish Government, based on agreed criteria, earmarked £5 billion, with £3 billion allocated to supporting Councils and Housing Associations' new build and £2 billion to fund housing-related infrastructure. In addition to social and market housing, a wide range of housing policies were introduced to provide households with opportunities to access affordable rent or home ownership. These included initiatives such as 'Mid-Market Rent' which, provides quality, affordable homes for low to moderate income households i.e. tenants who would not qualify for social housing but cannot afford to pay market rent or buy a property (Link Housing); 'Home Owners' Support Fund' which, helps people at risk of having their home repossessed [21]; 'Help to Buy (Scotland)' scheme which, helps people purchase a new-build home without the need for a large deposit (Scottish Government); and Low-cost Initiative for First Time Buyers which, helps people buy a home (Scottish Government).

Third was tackling housing and social inequalities. Following large-scale destruction during World War II, the welfare system was introduced in Britain. Under this system, government provided free welfare benefits such as education, health, housing and unemployment payments to the population at the point of use, although paid for by general taxation (British Broadcasting Corporation). In addition, across Britain there is evidence of government's commitment to providing housing and support to homeless households and disabled persons.

Fourth was the introduction of innovation funding. For example, the UK government is constantly reviewing and introducing new policies to help attract investment funding from housing providers, including banks, private investors, councils, and pension funds.

5   2030 Sustainable Development Goals: The Housing Connection

Housing has an impact on attainment of the Sustainable Development Goals. For instance, housing has an impact on poverty and hunger, economic and infrastructure development, health and well-being, social and economic inequality, energy use and sustainability, good quality education, and human safety and resilience. According to Enterprise Community Partners [10]

Access to decent, affordable housing provides stability for vulnerable families, helps prevent homelessness and helps increase the discretionary income that low-income families have available to meet important family needs or save for the future.

Severe forms of housing instability can seriously jeopardise children's performance and success in school and housing instability (including high housing costs in proportion to income, poor housing quality, overcrowding and multiple moves) has serious negative impacts on child and adult health.

On his part, Collier [7] noted that 'housing policy is the single most important factor in Africa's economic development. It needs to be elevated to the highest political level". Moreover, the Habitat for Humanity has described the impact of housing on the SDG goals by submitting that

> housing is a driver, catalyst and contributor for 13 of the 17 SDGs. It is a platform for household resilience and sustainability, driving the Human Development Index and Multi-Poverty Index outcomes in health, education and standard of living, including indicators in nutrition, child mortality, school enrolment, energy, water, as well as sanitation and durable, healthy construction. In addition, housing as a process can create a sense of place and dignity, building community cohesion as well as one's social and financial network and assets. It improves household income and financial stability, often providing home-based industry opportunities. Adequate and affordable housing is a multiplier of community jobs and income, and housing is a prerequisite for inclusive, equitable, safe, resilient and sustainable cities [15].

5.1 Recommendations

If the governments and institutions in Africa would approach housing not as a 'brick and mortar' issue, but as a robust system (which affects the wellbeing of the individual, households, community and the wider society), then the housing hopes of Africans will not only be realised but will also support attainment of the SDGs. And in order to ensure good quality housing are accessible to Africans irrespective of their economic strata, African governments will have to put in place a number of actions, as discussed below.

5.1.1 Develop an Understanding of the Dynamics of Housing Need and Demands

This is a major challenge that Africa has to overcome in order to meet the continent's housing supply shortfall and alleviate the social and economic underdevelopment prevalent in the country. This is because housing has the capability to drive economic development and promote job creation in other sectors of the economy, including energy, construction, engineering, finance and transportation. Overcoming this challenge goes beyond investing only in bricks and mortar. It will entail doing the following:

- embracing a new system thinking and a complete 'housing re-engineering' that would include producing holistic policies and strategies that focus not only on bricks and mortar but also on educational and employment opportunities, transportation, health and well-being;

- making better use of available resources and embracing innovative building methods, which have been proven to work in other parts of the world along with traditional methods, and adapting such methods to housing development in Nigeria;
- developing smarter houses, which rely less on the use of conventional energy, and building houses that enhance the economic development of our communities;
- having a robust understanding of housing equality that the housing needs of people in cities across Africa differ from the housing needs of people living in the rural areas and working not only with fund managers and financial institutions but also the populace to meet their housing needs; and
- supporting academic/research institutions to become the bastion of data that can be used by policymakers and politicians to evidence housing needs and enable informed decision-making.

Governments and relevant institutions across Africa should as a matter of high priority set up strategic housing as part of their housing institutions to set the strategic direction for housing. Along this line, governments should commit to investments in gathering/analysing data for housing and infrastructure planning. This will not only inform policymaking and investment decisions but will also ensure that governments are able to plot how housing can help in attaining the SDGs.

5.1.2 Develop a Robust Legislation and Planning System that Delivers Good Quality Housing

African governments need to continually develop contemporary legislation and planning systems that reflect changing households' needs. Also, sustainable housing quality standards should be developed, enshrined and enforced to ensure suitable housing for all.

5.1.3 Implement Sound Robust Housing Financial Policies

African governments need to develop and implement stringent housing financial policies to ensure that governments and developers adhere to housing delivery agreements. To ensure that the housing budget is spent on delivering the type of housing that works for all, governments should provide funding streams to support:

- affordable home ownership and rentals, which involves setting out clear policies and assessment/eligibility criteria for developers and households;
- the private sector in delivering and growing the housing market;
- regeneration schemes that target, redevelop and improve slum housing;
- specialist housing that provides accessible housing at adaptable costs for people with physical impairment, including wheelchair users.

In addition, governments and their institutions should ensure all future housing developments do not encroach on or damage valuable public or private green and open spaces, in addition to ensuring that they are supported by adequate provision of, or planning for, infrastructure.

5.1.4   Tackle Housing Inequality

Housing is about promoting equality; as such, tackling housing inequality will enable governments to meet the SDGs. This means developing an understanding that the housing needs and demands of people and communities vary based on age, physical ability and other social factors. It means developing bespoke policies and support to ensure the housing needs of all are considered and that households have opportunities to access housing in line with their circumstances.

5.1.5   Develop Innovative Funding

Developing innovative funding can be done in two ways. First, the funding/loan parameters used by government and international lenders/donor's bodies for assessing funding applications and awards for housing delivery across Africa should be reviewed and widened to include factors beyond bricks and mortar in their assessment. This comprehensive housing 'package' assessment will support the delivery of the SDGs. Second, countries in Africa cannot continue to rely heavily on foreign donors, aid or foreign loans to supply housing across the continent, as FDI can never provide the level of investment needed for housing on the continent. It would mean unlocking the nation's wealth and newer sources of investments within the African continent and making better use of available resources.

5.1.6   Prioritise Regeneration

Cities and towns are living organisms; they are born, they live, they age and they die and regeneration helps cities and towns to continue to thrive. Given the scarcity of land especially in the urban areas, governments across Africa should urgently prioritise regeneration and strategic planning. This is particularly important especially in the urban centres because of the rapid rate of urbanisation that 'often results in the urbanisation of poverty and manifests itself in mushrooming urban informal settlements (slums) [33]. Regeneration is not about yanking people away from their land or place of abode, a practice that only shifts the gear of slum formation elsewhere. On the contrary, regeneration involves developing an understanding of the needs of people (housing, education, health and employment) and their environment and working with them to develop meaningful and sustainable solutions that put people at the centre of decision-making. Simply, regeneration is about [24]:

- stabilising neighbourhoods in decline through investments in housing projects and improving employment opportunities through economic strategies;
- consulting and involving local communities through effective partnerships and participation; and
- maintaining the viability of existing stock and increasing the value of other infrastructure investments.

5.2   Conclusion

Housing is a system and, like other systems, it has an impact on the economy, infrastructure development, environmental sustainability, the financial market, living standards, and the health as well as well-being of a nation. Housing is crucial to alleviation of poverty and eradication of disease. In addition, housing stores and distributes a nation's wealth and is a major driver of the economy. Perhaps, more than any other factor, housing (where one lives) determines one's life chances. Governments across Africa need to understand that meeting the housing needs of Africans is a very possible task. It is also a common denominator for the delivery/attainment of the SDGs, since, without good quality housing that works for all irrespective of social, economic or tribal class, realising the SDGs will remain a mirage.

References

1. Bah E, Faye I, Geh ZF (2018). Housing market dynamics in Africa, Springer eBook collection: economics and finance. Palgrave Macmillan, London. https://doi.org/10.1057/978-1-137-597 92-2
2. Basboll T (2010, September 27) Aim small, miss small. Research as a second language. http://secondlanguage.blogspot.com/2010/09/aim-small-miss-small.html
3. BRE https://www.bregroup.com/
4. Brill F, Reboredo R (2019) Failed fantasies in a South African context: the case of Modderfontein, Johannesburg. Urban Forum 30:171–189. https://doi.org/10.1007/s12132-018-9348-1
5. British Broadcasting Corporation (BBC) Responses to social and economic inequality. https://www.bbc.co.uk/bitesize/guides/zqr3y4j/revision/1#glossary
6. Brown C (2019, July 22) A history of council housing: a timeline. Inside Housing. https://www.insidehousing.co.uk/insight/a-history-of-council-housing-a-timeline-62359
7. Collier P (2014, December 11) Housing is the single most important factor in Africa's economic development. NewStatesman. https://www.citymetric.com/skylines/housing-single-most-important-factor-africas-economic-development-573
8. Cybermetrics Lab (2020, January) Webometrics ranking of World Universities. http://www.webometrics.info/en/Ranking_africa?page=16
9. Dewdney JC, Conquest R, Pipes RE, McCauley M (2018, December 20) Soviet Union, in full Union of Soviet Socialist Republics (U.S.S.R.), Encyclopaedia Britannica, Inc. https://www.britannica.com/place/Soviet-Union
10. Enterprise Community Partners (2014) Impact of affordable housing on families and communities: a review of the evidence base. https://homeforallsmc.org/wp-content/uploads/2017/05/Impact-of-Affordable-Housing-on-Families-and-Communities.pdf

11. Frost R (1920) The road not taken. https://www.poetryfoundation.org/poems/44272/the-road-not-taken
12. Glossop C (2008, November) Housing and economic development: moving forward together. Centre for Cities. https://www.centreforcities.org/wp-content/uploads/2014/09/08-11-06-Housing-and-economic-development.pdf
13. Google Translate. Building Information Centre, France. https://translate.google.com/translate?hl=en&sl=fr&u=https://data.bnf.fr/fr/12179218/centre_d_information_et_de_documentation_du_batiment_paris/&prev=search
14. Google Translate. Secretariat of urban planning and housing missions, France. https://translate.google.com/translate?hl=en&sl=fr&u=https://data.bnf.fr/fr/11878192/secretariat_des_missions_d_urbanisme_et_d_habitat_france/&prev=search
15. Habitat for Humanity, Trinidad and Tobago (2020, March 13) Housing and the UNSDGS. https://www.habitat-tt.org/housing-and-the-unsdgs/
16. Inside Housing (2019, July 31) Happy 100th anniversary Addison Act—celebrating 100 years of council housing. Inside Housing. https://www.insidehousing.co.uk/insight/the-addison-act–celebrating-100-years-of-council-housing-61980
17. Kissick D, Leibson D, Kogul M, Bachmann J, Anderson J, Eckert J (2006, June) Housing for all: essential to economic, social, and civic development. International Housing Coalition. World Urban Forum III Vancouver, Canada. http://www.habitat.org/lc/housing_finance/pdf/housing_for_all.pdf
18. Link Housing. Link2Let. https://linkhousing.org.uk/find-a-home/renting-options/link2let/
19. Lund B (2006) Understanding housing policy. Policy Press, Bristol
20. Malpass P, Murie A (1999, 5th Edition) Housing policy and practice. MacMillan Press Ltd. Basingstoke: Palgrave
21. Scottish Government (2019, September 4) Home Owners' Support Fund. https://www.mygov.scot/home-owners-support-fund/
22. Scottish Government https://www.gov.scot/policies/homeowners/help-to-buy/
23. Scottish Government, Housing and Social Justice Directorate. Low-cost Initiative for First Time Buyers. https://www.gov.scot/policies/homeowners/low-cost-initiative-for-first-time-buyers/
24. Scottish Government (2010).Regeneration. Housing and Social Justice Directorate, Local Government and Communities Directorate. https://www.gov.scot/policies/regeneration/
25. Shepherd B (2010) Political stability: crucial for growth? http://www.lse.ac.uk/internationalRelations/dinamfellow/docs/dinamshepherdpoliticalstability.pdf
26. Sud, N, Rapid urbanization is pushing up demand for housing in Sub-Saharan Africa. International Finance Corporation. https://www.ifc.org/wps/wcm/connect/news_ext_content/ifc_ext_ernal_corporate_site/news+and+events/news/trp_featurestory_africahousing
27. The Editors of Encyclopaedia Britannica (2010, March 15) Tanganyika. https://www.britannica.com/place/Tanganyika
28. The Editors of Encyclopaedia Britannica (2011, January 06) Federation of Rhodesia and Nyasaland. https://www.britannica.com/place/Federation-of-Rhodesia-and-Nyasaland
29. United Kingdom, Ministry of Housing, Communities and Local Government (2015, March 20) Housing and economic needs assessment. https://www.gov.uk/guidance/housing-and-economic-development-needs-assessments
30. United Nations Economic and Social Council (1963, January 9–18) Housing in Africa problems and policies. Economic commission for Africa meeting of experts on housing problems in Africa (Addis Ababa, Ethiopia). http://repository.uneca.org/bitstream/handle/10855/7119/Bib-47214_add%202.pdf?sequence=5
31. Universities UK (2015, June) The economic role of UK universities. https://www.universitiesuk.ac.uk/policy-and-analysis/reports/Documents/2015/the-economic-role-of-uk-universities.pdf
32. University of the West of England, Bristol (2008) The history of council housing. https://fet.uwe.ac.uk/conweb/house_ages/council_housing/print.htm

33. Wall RS, Maseland J, Rochell K, Spaliviero M (2018) The state of African cities 2018: the geography of African investment. United Nations Human Settlements Programme (UN-Habitat) and IHS-Erasmus University Rotterdam. https://unhabitat.org/books/the-state-of-african-cities-2018-the-geography-of-african-investment/
34. Watson V (2013, December) African urban fantasies: dreams or nightmares? SAGE J Environ Urbanization. https://doi.org/10.1177/0956247813513705

Learning from Experience: An Exposition of Singapore's Home Ownership Scheme and Imperatives for Nigeria

Hikmot A. Koleoso and Basirat Oyalowo

Abstract Sustainable Development Goal (SDG) 11.1 is a testimony that poor access to housing has been a serious cause of poverty in both developed and developing countries. Using a review of literature and Singapore as an example, this chapter presents how the government transformed the country into one of the most sophisticated economic hubs in the Asian continent within less than four decades. This study attempts to provide answers to the questions: How did Singapore achieve this feat in its public home ownership sector? What are the key features of these reforms and interventions? How were these reforms and interventions used for achieving SDG 11.1 and what can Nigeria adapt from these in the development of a sustainable, robust and effective housing policy? The study found that the Singaporean government was deeply committed to home ownership and that extensive resources were committed to researches that uncovered the precise socio-economic, cultural and financial needs of the people, therefore influencing the holistic, comprehensive and poverty eradicating nature of the scheme. The study identified key success features that could guide Nigeria in developing a successful housing policy that will provide sustainable solution to her persistent housing problem, while positioning her strategically to attain SDG11.1.

Keywords Housing in Nigeria · Sustainable development goal 11 · Singaporean public housing · Sustainable home ownership

H. A. Koleoso (✉) · B. Oyalowo
Department of Estate Management /Centre for Housing and Sustainable Development, University of Lagos, Akoka Yaba, Lagos, Nigeria
e-mail: hkoleosho@unilag.edu.ng

B. Oyalowo
e-mail: boyalowo@unilag.edu.ng

© The Author(s), under exclusive license to Springer Nature Singapore Pte Ltd. 2021
T. G. Nubi et al. (eds.), *Housing and SDGs in Urban Africa*, Advances in 21st Century Human Settlements, https://doi.org/10.1007/978-981-33-4424-2_4

1   Introduction

SDG 11 is regarded as the urban SDG because it crystallises the global aspiration for urban sustainability into its goals and targets. In particular, its first target—to ensure access to adequate, safe, affordable housing and basic services and to upgrade slums—typifies the ideal urban policy of cities in developing countries. In a recent progress report on SDG attainment, the UN recognised that inadequate housing impacts negatively on urban equity and inclusion, urban safety, livelihood opportunities and health. The UN also extended the indicators for measurement of the target to include housing inadequacy, in addition to traditional measures of deprivation in slums and informal settlements.

In Nigeria, housing quality and quantity is an urban problem [8]. The shortage of housing is one of the most serious and widespread consequences and causes of poverty in the country and also a major cause of poor environmental quality. To address associated problems of poor housing in Nigeria, the National Housing Policy was promulgated in 1991. Close to 30 years after, millions of Nigerians are still homeless while many others live in indecent houses. Housing is a national problem that the country must focus on if it is to achieve substantial growth and a reduction in the level of pervasive poverty. This makes it necessary to evaluate housing policies in other countries that have successfully overcome their housing problems, in order to assess the feasibility of policy transfer. Singapore offers particularly compelling leadership in housing provision globally, to the extent that it regularly receives observers from other countries [14] who seek to understudy how it has created a policy environment that has resulted in over 70% of its residents living in affordable decent homes of their own as at 2014 [6].

In 2015, Singapore's population was 5.54 million, of which 3.38 million were citizens, 0.53 million were permanent residents, and 1.63 million were foreigners. One-fifth of its land area comprises of reclaimed land with a high population density of over 7,600 persons per km^2 [14]. Like Nigeria, Singapore is a multi-ethnic, multi-religious, multilingual and multicultural country. However, Nigeria has a much larger population—approximately 200 million—and its over 250 ethnic groups are spread across a land area of 923,768 km^2, compared to Singapore's 719 km^2. Furthermore, both countries share a common British colonial history, having a system that Yang and Zhang [16] described as 'differentiating treatment amongst the greater ethnic groups'. While Singapore has been shown to have utilised its housing policy to ameliorate inter-ethnic conflicts [1, 16], Nigeria is yet to achieve this.

Socially, there are also vast differences between the two countries. Singapore like most urbanized East Asian Societies has low birth rates, high proportions of unmarried adults and late marriages with large number of people remaining single in their 30s, and less than 2% birth occurring outside of wedlock [15]. The implication of this is reduced household formation and reduced family housing demand, which enabled the state's use of housing policy as a tool to stimulate marriage and family formation. In addition, Singapore operates a unitary dominant party parliamentary republic system, typified by central control with its People's Action Party

Government being in power since after independence in 1965. This enabled a heavily state controlled housing policy, though with infusion of market driven institutions that provide subsidized financial services to eligible citizens. Nigeria on the other hand, has since independence in 1960, witnessed various governance ideals and structures, from regional governance, to military dictatorship and democratic governance in recent times. In Singapore, land and social policies as shown in subsequent sections, are inextricably linked up to meet with state's objective. Phang and Helble [14] observe that Singaporean land scarcity and high population density of over 7600 persons per km^2 provided additional justification for the dominance of state intervention in land ownership and the housing market. Although, Nigeria also adopted a land nationalization policy in 1976, there were no integrated policy for land re-allocation and the policy itself has been the subject of much criticisms, the most overarching of which are the constraints to land transfer and the dual land markets that it inevitably created. Nigeria's housing policy has also shifted from direct government provision to private sector provision and a mixture of both in recent times. Unlike in Singapore, land scarcity is not in general, a pressing problem in Nigeria except in primate cities and commercial centres. Thus, a case to case comparison is not considered a feasible methodology in this discourse so this is not a comparative study. Rather, the preferred approach is to ascertain how Singapore has achieved the feat of bringing home-ownership to over 90% of its residents and to examine the key policy tools and features that have been used to accomplish this.

In doing this, a second objective is to highlight aspects of the Singaporean housing system that are amenable to sustainable practices. Borrowing from the policy transfer tools of public policy research, areas of best fit both for policy action and for achieving SDG 11.1 that can be adapted for the Nigerian situation are presented.

However, it is not the objective of this chapter to examine ideological arguments about the usefulness of policy transfer or to discuss the various differences between policy transfer, lesson learning, policy diffusion and other concepts associated with the term in public policy studies. While acknowledging the limitations of policy transfer and lesson drawing, as argued by James and Lodge [9], and the constraints to policy transfer identified by Benson [3] as well as contextual differences between the two countries, this chapter analysed the Singaporean housing policy as a source of policy transfer and lesson drawing (used synonymously) for Nigeria. An adaptation of Dolowitz's conceptualization, as cited in Benson's [3] is used to define what policy transfer is. It is seen as the process whereby knowledge about policies, administrative arrangements and institutions, etc., in one time and or place (here, Post-Independence Singapore) are used (with due recognition for contextual differences) in the development of policies, administrative arrangements and institutions in another time and/or place (Nigeria, post-2019). Associated research questions are: How did Singapore achieve this feat in its public home ownership sector? What are the key features of these reforms and interventions? How were these reforms and interventions used as a platform for achieving SDG 11.1 and what can Nigeria adapt from these in the development of a sustainable, robust and effective housing policy?

2   The Success Story of the Singaporean Public Home-ownership Sector: Policy Structure

The State of Singapore operates a national public housing programme, accounting for over 70% of its residential real estate [1]. Yang and Zhang [16] attribute the rapid development of Singapore to the successful experience of the Government in addressing the nation's housing problems in its early post-independence years whilst also handling ethnic relations between the principal ethnic groups (74.1% Chinese, 13.4% Malays and 9.2% Indians with 3.3% from other nationalities).

In the wake of Singapore's political evolution from self-government, merger with Malaysia and unexpected independence in 1965, the housing challenge in the politically and ethnically turbulent state was one of massive shortage, lack of long-term investment, largely migrant population and lack of private sector resources (local and foreign). All of these were compelling factors for large-scale public housing provision by the newly elected government [14]. At the time, Singapore consisted of villages made up of shanties, squatter huts, single-storey sheds constructed with cardboards, zinc sheets, sticks and poles with unhygienic conditions [10], such as we have in today's Lagos slums, e.g., Makoko and Maroko.

The Singaporean public housing system is a combined system that provides public housing and low-cost rental housing funded by the Singapore Government and operated through the Housing Development Board (HDB). The system evolved as a result of the country's housing need in the early post-independence period, when, as at 1959, only 9% of the country had access to decent housing and nearly 40% of its two million people lived in shacks [16].

As at 2015, homeownership rate was above 90%, with 75% of the housing stock classified as public housing [14]. The Housing Development Board flats are in high-rise apartments of usually between 12 and 40 stories tall [15]. The median house type is a four-room (approximately 90 m^2) flat sold by the government and held on a 99-year leasehold basis [14].

There are discernible sustainability moments in the history of the Singaporean housing policy as summarized in Table 1. HDB started by building large numbers of rented apartments for the poorest people, who could not afford to pay rent. It was not until four years later that its mandate changed to homeownership through the establishment of the government's Home Ownership for the People Scheme (HOPS).

The scheme enabled sales of housing units on 99-year leasehold basis. Affordability was assured for low-income households, as the HDB offered loans that enabled them to pay less in monthly mortgage payments than they would have paid in rents. A significant proportion of the houses were also reserved for the really needy, who had no other viable option for having a home. As at 2012 the HDB had succeeded in constructing over 900,000 housing units [13]. This housed over 80% of the resident population [6]. The outcome of this was that housing units increased by about 50% in each decade from 1970 to 2000, with homeownership also doubling from 29% in 1970 to 59% in 1980 and 88% by 1990. HDB housing displaced private housing

Table 1 Key sustainability policy moments in Singapore's housing history

1959 and before: Public housing was available to only 8.8% of the population, with majority living in overcrowded rent-controlled apartments lacking access to water and modern sanitation
1960: The Housing Development Board was established in Singapore with the charge of planning, construction and management of public housing. The major thrust of the Government housing policy was to provide low-cost rental public housing
1964: Singapore announced implementation of the Home Ownership Scheme, so that low- and middle-income people could buy HDB flats by instalments, thus combining the policy of low-cost rental public housing with sales of public housing. (Affordability)
1966: Enactment of the Land Acquisition Act, which enabled massive land acquisition and reconstruction into flats that are then allocated to former landowners
1968: An expansion of the role of the Central Provident Fund into a Housing Finance Institution
1980: Since 1980, a series of intervention projects have been embarked upon to upscale the HDB flats. Examples are the renewal plan for HDB flats in the Central Zone and the Neighbourhoods Rehabilitation plan. These interventions were to comprehensively improve the living environment and sense of community among residents. (Adequacy)
1984: A Social Development Unit (SDU) was created in the Ministry of Community Development, Youth and Sports to stimulate 'love matches' between unmarried singles as a result of low marriage rates and this was tied to the housing policy, such that engaged couples could access a fast-track option of homeownership
1989: Government promulgated the policy of ethnic proportionality in HDB flat allocations by stipulating the proportion of ethnic groups in each HDB property. It also introduced balloting for selection and demolished old residential areas. (Inclusion)

Source Benabbou et al. [1], McCarthy et al. [12], Phang and Helble [14], Strijbosch [15], Yang and Zhang [16]

as low-density shop-houses, squatter settlements and villages were acquired by the government and demolished to make way for high-rise flats [14].

Phang and Helble [14] observe that the pillars of the Singaporean housing policy are Direct Government Provision through the establishment of a Housing and Development Board (HDB), State Control on land (through enactment of the Land Acquisition Act in 1966) and a complementary housing finance policy (through the expansion of the state's pension fund Central Provident Fund [CPF] into a housing finance institution. It would be useful to identify how these actions address issues related to SDG 11.1 (inclusion, urban equity, affordability, adequacy and access to basic services).

3 Land Assembly and Regeneration Strategy

The processes for public housing construction, land acquisition, slum clearance, resettlement and urban renewal in Singapore have been closely interwoven and the country's land policy has been described as "land reform in an urban setting," which involves massive transfer of scarce land resources from minority wealthy family landowners and colonial British private organisations to public use [14]. This was achieved in the first two decades after the country's independence. Like its housing policy, the land reform has evolved from its origination with enactment of the Land Acquisition Act of 1966. Upon acquisition, land resource was redistributed among the "landless majority" through the development of industrial estates, financial districts, commercial developments, large public housing programmes and public-sector infrastructure development [14]. The Land Acquisition Act enabled government to provide land on lease to government agencies, who then developed high-rise flats that are sold on 99-year lease to households. For this purpose, the government had to embark on massive clearance of existing villages, slums and low-density neighbourhoods.

Through its urban redevelopment programme, Singapore makes land available to the private sector for real-estate investments. Under the Act, the government can compulsorily acquire any land for public purposes and public utility as well as for any residential, commercial or industrial purposes. While landowners are entitled to compensation, they cannot object to terms set by the state and have recourse to an Appeal Board and not to a Court [14]. Compensation only takes into account the existing use or zoned use of the land, whichever is lower; as such, it disregards any potential value for more intensive uses. Inevitably, the compensations paid are usually much lower than the market price. The compensation policy has been reviewed from the statutory date method (payment made at land prices of 1973 which remained in force for 14 years up till 1987 without market valuation or landowners' purchase price, then 1986, 1992 and 1995) to compensation based on full market value by 2007. This is to take care of the widening gap between increasing market land prices and compensation paid for them.

Phang and Helble [14] observed that land owned by the State increased from 44% in 1960 to 76% in 1985 and 90% in 2005. The equitable redistribution of wealth that promotes social sustainability is evident in the actions taken by the Singaporean government after land acquisition. There has not been strong resistance by land owners due to the redistribution actions which included initiatives such as providing better quality alternatives for all affected by the land acquisition, transparency and observance of rule of law, market compensation for even squatters and effective resettlement planning. It is also notable that political resistance to land acquisition in the nation was minimal because in addition to transparent processes and redistribution initiatives only a minority owned land. This was because at the time the legislation was enacted, Singaporean land was largely held by a few land holding families and British post-colonial land owners.

4   Housing Design and Construction Strategy

At the commencement of the scheme, the housing designs were simple and utilitarian. Beng [2] recounts that it included blocks with variations of 1, 2 and 3-bedroom flats which came in cost effective sizes of at least 23 m^2 with basic amenities such as

Table 2 Accommodation types and their typical sizes at inception of scheme

Type	Typical size
1-Room Emergency	23 m^2 (250 sq. ft)
1-Room Improved	33 m^2 (360 sq. ft)
2-Room Emergency	37 m^2 (400 sq. ft)
2-Room Improved	45 m^2 (480 sq. ft)
2-Room Standard	41 m^2 (440 sq. ft)
3-Room Improved	60 m^2 (650 sq. ft)
3-Room Model 'A'	75 m^2 (810 sq. ft)
3-Room New Generation	69 m^2 (740 sq. ft)
3-Room Simplified	65 m^2 (700 sq. ft)
3-Room Standard	54 m^2 (580 sq. ft)

Source HDB [7]

pipe-borne water and electricity that are quite basic by today's standards, but met the needs of the people at that time.

Table 2 provides further details on the initial designs and their space provisions. Although most of these sizes are typically no longer being built, it is important to mention that they continued to be built for as long as they were in demand. In addition, in response to the criticisms of shoddy workmanship and building defects of the early quickly-built projects such as cracks in walls and ceilings, inferior fittings, frequent lift breakdown etc., HDB housing evolved through research and innovation to quickly rectify these problems and modify the next cycle of constructions to include significant quality considerations [2]. The scheme also adopted effective technology in construction. For example, it started to use prefabrication technology in its construction since 1980 when the first contract for this type of development was awarded to build 3 and 4 room flats in Hougang, Tampines and Yishun. A development known as The Pinnacle@Duxton, achieved engineering breakthroughs with the construction of almost the entire 50-storey building using modularised and off-site prefabricated components [7].

Public housing was also designed in mixed-use layouts, meaning that every new housing layout was designed as sustainable, self-contained new satellite towns that contained—alongside mixed-type residential apartments—educational, social, institutional, industrial and community facilities. These developments had a hawker centre, a court where inexpensive cooked foods were sold. Residents were therefore able to meet their social and economic needs within the new town. Queens Town and Toa Payoh were the pioneer New Towns, followed by others such as Ang Mo Kio and Bedok in the 1960s to 1970's [5]. These new towns were all carefully located and planned within the country's macro-spatial planning [17], with considerations for access to infrastructure. Each public housing block has a common area known as void decks built into the design to promote social interaction.

5 Inclusiveness Adequacy and Strategy for Affordability

To promote inclusiveness, the HDB has continued to respond to the changing demography through designs and in multiple other ways. It has introduced new designs to suit its rapidly ageing population, alongside schemes such as the Home Office Scheme that permits owners to conduct their businesses from home, thus saving cost of business start-up, commuting time and expenses, and enabling residents to spend more time with their families. This is critical for integrating social and economic sustainability.

Along with citizenship and household income, marital status (with children or without children) is a strong condition for eligibility for housing in Singapore. Eligibility conditions are however constantly revised to ensure that allocation is equitable. Phang and Helble [14] cite an instance in the 1960s and 1970s, in which there were long waiting lists for flats and the HDB allocated flats with priority given to households affected by resettlement and on a first-come-first-served basis for other households.

The ability of the state to respond to the lifestyles of Singaporeans has ensured that the public housing sector now provides high quality affordable housing. Through macro-economic policies, the apartments have continued to be functional and affordable. This significantly aligns with the sustainable communities' goal of SDG 11 and specifically addresses the drive for adequate housing as encapsulated in SDG11.1, while also responding to inclusiveness and affordability—the core indicators of sustainable communities. However, the allocation policy, which is quite restrictive to non-citizens, has been described in some quarters as not being inclusive for all categories of residents.

6 Financial Strategy Using the Pension Funds

Singapore is one of the few countries that have successfully linked their housing policy with their pension fund and social security system and then provided mechanisms to limit the risks of macro-economic externalities arising from a heavily subsidised housing market. This pension fund, which is also the Singaporean social security programme, is known as the Central Provident Fund. Through the Fund, the Singaporean government executes its affordable, subsidised housing policy, offering residents substantial grants and awards of up to S$60,000 for first-time buyers towards the purchase of an HDB apartment.

The CPF was initially established as a forced retirement savings system for employed citizens of Singapore as well as for permanent residents. However, like most policies in Singapore, it has evolved over time and as at 2001 now also offers life and health insurance, extensive home-purchase support and educational accounts [14, 12]. The contribution rates in 2016 are 20% of wages for employee and 17% of wages for employers, up to a monthly salary ceiling of $6,000 and this has not changed in nominal terms since 1984 [12, 14]. McCarthy et al. [12] also noted that as at 2001, the CPF covered more than 2.5 million wage-and-salary employees, and in 2014 housing mortgage loans to GDP ratio was 55.5%, from only 4% in 1970, with total contributions to the pension funds varying with the age of the worker, from 40% payable by those under age 55 to 7.5% for older workers above 65 years. Once deducted, the contributions are allocated to the three different accounts within the Fund (the Medisave that allows withdrawals for medical care, the Ordinary Account which funds could be used to finance housing, insurance premiums and education expenses, etc.). The special account is the third account and it holds the old-age retirement savings, even though it can also be invested in "approved" assets. Just as there are variances across age bands in the savings scheme, there are also variances in the re-distribution into these three accounts, with a ratio of 6% to 30% to 4% for a young worker into the Medisave, ordinary and special accounts respectively [12]. In Singapore, an eligible worker may use up to 100% of his Ordinary Account CPF savings as a down payment for their house or flat. In addition, up to their entire monthly CPF Ordinary Account contributions may be used to service mortgage payments [4].

HDB enjoys government loans to finance its mortgage lending activities, upon which interests are paid at prevailing CPF savings rates. These loans are issued to buyers of new and resale leasehold flats as long-term (25 year) mortgage loans, made at 80% of selling price at 0.1 percentage point above the CPF ordinary account savings interest rate. Subsidies for homes were deliberately significant, sometimes more than 50% particularly for the smaller flats, this helped to keep monthly mortgage repayment (rent) and purchase price at no higher than 20% of monthly income and twice the annual income respectively. This promoted, accessibility, affordability and hence inclusiveness [7].

While many nations see housing subsidy as a means of redistributing income, Singapore has been able to use it as a way to grow the economy through its cyclical and indirect contributions to national income and GDP resulting from the volume of constructions. There are also positive spill-overs to the commercial banking sector. When there was a change of policy to allow commercial banks have first claim on a property should a borrower default, they entered into the HDB mortgage market and with the ensuing competition, commercial banks have been able to offer loans even below HDB mortgage loans at a 2.6% interest floor [14].

7 Security of Tenure

The scheme ensured that the tenure of the average citizen in his or her home is well secured. In 1981, the Home Protection Scheme, administered by the CPF Board, was implemented to ensure that dependents of HDB flat owners would not lose their homes because of a default in loan repayments, in the event of death or permanent incapacity of the sole breadwinner [7]. Although it is provided that the flats would be recovered after a 3-month continuous default in mortgage repayment, recovery has hardly ever occurred. In a number of instances mortgage loan rescheduling schemes have been adopted to reduce the financial burden of mortgage instalments in the case of owners with financial difficulties. This is a conscious attempt to secure the tenures of homeowners in the face of financial difficulties has further encouraged citizens to participate in the scheme. This has promoted tenure security and reduced fears of forced evictions which could destabilize mortgage markets in affordable housing which pervades other countries.

The HDB-CPF framework has been noted to have transformed urban Singapore, especially as government has been consistent in this policy by maintaining it for five decades. In fact, between 1960 and 2013, the ratio of housing investment to gross domestic product (GDP) averaged an internationally competitive 7%. Moreover, the ratio of housing investment to total investment averaged 23% [14]. Such policy consistency is important for ensuring investment in a relatively stable market and in assuring affordability and accessibility buoyed by trust in the system. This has ensured the continuous popularity of the HDB flats in Singapore.

8 Maintenance Strategy

For Singapore, significant planning and financing had been provided to ensure proper and adequate maintenance activities in public housing developments. On a longer-term basis, public flats and estates have been progressively upgraded with resident participation. It is said that on average about 10-13% of HDB annual operating expenditure is spent on flat upgrading and improvements. According to Yuen [18], the then Minister of Finance, the spending was justified as a means of redistributing economic growth and increasing the housing assets of Singapore citizens.

In 2005 a new warranty for Defects Liability Period for new flats was introduced, involving a five-year period for ceiling leakage and external water seepage and a 10-year period for spalling concrete. To enhance affordability of these improvements, public housing residents pay only a small fraction of the costs to guarantee this warranty and at times nothing at all. The general maintenance of common areas, which would include the common corridors, void decks, lifts, water tanks, external lighting and the open spaces surrounding the estates, are carried out by Town Council Authorities, which are similar to Local Governments in Nigeria. Rental flats, however, are maintained directly by the HDB to ensure that they are made serviceable for the next occupant. Where homeowners engage third-party contractors to carry out renovations, the HDB has direct authority to oversee such works. It therefore imposes strict renovation rules to ensure such maintenance works do not result in structural damage and that noise control provisions are observed during renovation works.

However, large-scale improvement works to existing public housing developments and improvements to older estates such as modernising the town centres, adding or upgrading community facilities are carried out in the form of various programmes under the Estate Renewal Strategy [7]. All of these works ensure that the developments continue to be available to the citizens, meet contemporary needs and is sustainable.

9   Limitations to Lesson Learning

The Singapore model has attracted considerable interest from other Asian countries, even as it contains key policy elements that could be transferred to Nigeria and other African countries. However, the system is not, without its criticism and risks [14], a major one of which is its allocation policy. Researchers such as Strijbosch [15] have undertaken ethnographic studies on the state's policy to allocate flats to only married couples thereby excluding the singles. The study documents that only singles above the age of 35 can apply to purchase an HDB flat. In addition, the prevailing strong cultural norms of the Singaporean nuclear family ideal, which discourages children from leaving their parents' home before marriage, as well as the lack of affordable housing in the private housing market, all ensure that unmarried young people are at a disadvantage in the housing allocation system.

On the contrary, there are strong pro-marriage (and hence pro-childbearing) incentives in the housing policy, as under a fiancée-fiancé scheme, couples can apply for a home and have their waiting time on HDB flats shortened considerably, with allocation completed once they are married. However, owing to strong cultural affiliations, Strijbosch [15] showed that young, single and well-educated women who fall between the ages of 25 and 30 are not particularly bothered about the discriminatory allocation practices, since they also desire to get married at some point. This brings a cultural dimension to the discourse, thereby serving as a reminder of the need to contextualise and localise international policies such as the SDG, since what is measured as inclusiveness would vary based on cultural affiliations.

Literature review shows that Singapore uses its housing allocation system to promote ethnic diversity among its four major ethnic groups: Chinese, Malay, India, and others. This is done by creating a quota system whereby each ethnic group cannot own more than a certain percentage in a housing project. The major objective of this is to have every neighborhood contain residents from all ethnic groups thereby building the Singaporean multicultural nationhood [10]. This is certainly instructive; although research by Benabbou et al. [1] indicate that there are certainly externalities, such as reducing access to people from minority groups while increasing that of majorities like Chinese, since the quota is based on absolute population figures. The allocation policy, which came on the heels of the introduction of the Country's Ethnic Integration Policy *'effectively separated individuals from homogenous ethnic groups and pushed them into individual heterogeneous group public HDB flat communities, thereby breaking the cultural foundation of ethnic communities, effectively promoting the disintegration of traditional settlement patterns and for two generations, fostering a fresh sense of shared national history.'* ([16]: 250).

Moreover, the ethnic integration policy produced heavy regulations in the housing resale market. Where a household desired to sell its HDB flat, it was not allowed to sell to an ethnic group whose quota had been filled at the block or neighbourhood level. The seller household would therefore have to put the property on the market for as long as it would take to find a buyer from an ethnic group whose quota was yet to be filled. However, while the ethnic integration policy was meant to address ethnic inclusiveness, it has been admitted to be a most intrusive yet important Singaporean social policy. Being based on the experience of ethnic profiling and unrest, it is acknowledged to be the most important policy for achieving inclusive communities. If Singapore abandons this policy, it risks creating ethnic communities that will activate slum enclaves and social isolation. To this extent, a thorough understanding of the contextual background of Singapore has been a mitigating factor in the enactment of this policy. The rationale for transferring such a quota-based system to countries with more nuanced ethnic problems remains a critical consideration.

The joined-up nature of Singaporean financial and social policy is a strength, though with profound risks. The mandatory nature of the CPF, together with the dominance of the HDB, could have resulted in over allocation of resources to housing, since the CPF collects from members more than what is required for housing [14]. The illiquid nature of the CPF savings and the risk of house price declines particularly place households at risk of being "asset rich and cash poor" [11]. The focus on homeownership has also led to a rather undeveloped rental-housing sector, which is usually a first step to affordable housing for the foreign nationals in rapidly expanding economies and in the promotion of economic growth. All of these considerations suggest the need to exercise caution in promoting adoption of the Singaporean housing system in Nigeria, even though there are still pathways for policy transfer in Nigeria.

10   Conclusions and Recommendations for Nigeria

As a newly independent country in 1965, Singapore was a nation with limited resources and poor infrastructure just as with Nigeria. However, in just over four decades it evolved from villages made up of shanties, squatter huts, ill constructed single storey sheds in unhygienic neighbourhoods into one of the world's most sophisticated cities and business hubs.

It is widely recognized that its unique and progressive home ownership scheme was responsible for a large part of this growth. The Singaporean government rightfully recognized the housing sector as a tangible asset in the country, a means of financial security and a hedge against inflation. It also recognized affordable housing as a right of the people. The foregoing initiated the commitment of government to its public housing scheme leading to the development of reforms and innovation which ensured that the home ownership scheme is properly planned and made contextual to the need of the people from the planning to the maintenance and re-improvement stages. In this, there are lessons for Nigeria.

It is important to understand that some identifiable features of the HDB contributed to its success. This is the direction that we propose for policy transfer in Nigeria. The first is that HDB is the sole agency in charge of public housing from Independence to date. This provided it with the opportunity to adopt effective resource planning and housing allocation approaches in its activities. The agency is provided with the authority to secure land, raw materials and manpower for large-scale construction in a way that optimised resources and achieved economies of scale. This centralization has circumvented the common problems of lack of policy continuity, duplication and fragmentation of duties, and bureaucratic rivalries associated with multi-agency implementation [18].

The second is that the organization enjoyed strong political and financial commitment from the Government in the form of (i) annual grants from the current budget to cover deficits it incurs on development, maintenance, and upgrading of estates; (ii) loans for mortgage lending and long-term development purposes; and (iii) land allocation for comprehensive HDB housing and town planning [14]. This enabled HDB to acquire and utilize the best resources in its operations. The Singaporean government did not take housing as a social problem to be addressed after the achievement of economic progress, rather it saw both as problems with equal dimensions, needing to be tackled together in view of their symbiotic relationship [18].

The third and probably most important feature is the integrated approach that the organisation adopts in its scheme [2, 18]. Its strategies and activities are holistic and innovative, thus engendering proper planning which makes it possible to meet the contextual needs of the people. The strategy provided for various perspectives of the scheme including design, finance, legislation, regulation, operation, allocation and physical management and re-generation of the properties.

In the understanding that strategies that are applicable elsewhere may not be appropriate for the Singaporean context, deep seated researches were embarked upon to develop strategies that were closely knitted with the precise research indicated needs of the people in terms of financial ability, socio-economic, cultural and political inclinations. It is therefore necessary to understand these strategies and the features of the citizenry that make the strategies compatible with the peoples' needs.

Policy transfer or lesson learning between nations have been facilitated by globalization and access to information across boundaries. Also, institutional interest in global governance and environmental protection has ensured that there are institutions who actively promote international policy adoption and transfer of best practices to other countries and their localization within these countries. The achievement of the goals and targets of the SDGs across multiple countries that have signed on for it is arguably innately based on the process of policy transfer or lesson learning. This has been the foundation for this chapter. The Singaporean policy environment provides opportunities for lesson learning on core actions and initiatives that could catalyse the attainment of the SDG11.1 in Nigeria particularly the aspects of the policy that relates to how the urban poor might be helped through housing delivery. The Nigerian government has to be more dedicated to affordable housing for low-income earners, which must be based on intentional policies that run through the housing development value chain. Second, inclusiveness must be strongly considered as an end, and this should be based on continuous learning and refinement which allowed the precise needs of the citizens to be identified and provided as necessary. Nigeria should be deliberate about learning from other countries and then localizing for change in its own environment.

References

1. Benabbou, N., Chakraborty, M., Ho, X.V., Sliwinski, J., Zick, Y. (2018). Diversity constraints in public housing allocation. *In Proceedings of the 17th International Conference on Autonomous Agents and Multiagent Systems* (pp. 973-981). International Foundation for Autonomous Agents and Multiagent Systems. Stockholm, Sweden
2. Beng, Y.C. (2007). Homes for a Nation — Public Housing in Singapore. https://www.csc.gov.sg/articles/homes-for-a-nation-public-housing-in-singapore. Accessed 1 October 2019
3. Benson, D. (2009). Constraints on Policy Transfer, CSERGE Working Paper EDM. *University of East Anglia, The Centre for Social and Economic Research on the Global Environment (CSERGE), Norwich,* 9–13
4. Central Provident Fund Board (1999) Residential Properties Scheme. Central Provident Fund Board, Singapore
5. Chua, B.H. (2007). Emerging Issues in Developmental Welfarism in Singapore. *The Crisis of Welfare in East Asia,* 27–42
6. Housing Development Board (2014) Annual report 2013/2014 Singapore Housing and Development Board. Housing and Development Board, Singapore
7. Housing Development Board (2019). Housing Development Board History and Towns. https://www.hdb.gov.sg/cs/infoweb/hdb-message. Accessed 1 Dec 2019
8. Ibimilua AF (2011) The Nigerian National Housing Policy in Perspective: A Critical Analysis. Journal of Social Development in Africa. 26(2):165–168
9. James O, Lodge M (2003) The Limitations of 'Policy Transfer' and 'Lesson Drawing' For Public Policy Research. Political Studies Review 1:179–193
10. Jensen, S.S. & Jochumsen S.C. (2012). Singapore's Successful Long-Term Public Housing Strategies. International Federation for Housing and Planning. http://www.ifhp.org/ifhp-blog/singapore%E2%80%99s-successful-long-term-public-housing-strategies#.VFTSCcVdVEM. Accessed 1 Dec 2019
11. McCarthy D, Mitchell OS, Piggott J (2003) Asset Rich and Cash Poor: Retirement Provision and Housing Policy in Singapore. Journal of Pension Economics and Finance. 1(3):197–222. https://doi.org/10.1017/S1474747202001130
12. McCarthy, D., Mitchell, O.S., Piggott, J. (2001). Asset Rich and Cash Poor: Retirement Provision and Housing Policy in Singapore. In: Pension Research Council Working Paper, The Wharton School, University of Pennsylvania. Accessed October 2019 from. https://www.researchgate.net%2Fprofile%2FJohn_Piggott2%2Fpublication%2F4772693_Asset_Rich%5Fand%5FCash%5FPoor%5FRetirement%5FProvision%5Fand%5FHousing%5FPolicy%5Fin%5FSingapore%2Flinks%2F55bcf7a108ae092e966382d1%2FAsset%2DRich%2Dand%2DCash%2DPoor%2DRetirement%2DProvision%2Dand%2DHousing%2DPolicy%2Din%2DSingapore.pdf
13. Miller, W., Stenton, J., Worsley, H., Wuersching, T. (2014). Strategies and Solutions for Housing Sustainability: Building Information Files and Performance Certificates. *Program 1: Greening the built environment.* http://eprints.qut.edu.au/67271/1/140113 Strategies and Solutions for Sustainable Housing Final report.pdf. Accessed 1 Dec 2019
14. Phang, S.Y. & Helble, M. (2016). Housing Policies in Singapore. 1–27. *Research Collection School of Economics.* https://ink.library.smu.edu.sg/soe_research/1802. Accessed 1 Dec 2019
15. Strijbosch K (2015) Single and the City: State Influences on Intimate Relationships of Young, Single, Well-Educated Women in Singapore. Journal of Marriage and Family 77(5):1108–1125. https://doi.org/10.1111/jomf.12221
16. Yang, Y. & Zhang, N. (2019). Research on Urban Ethnic Inter-Embedded Community Construction from the Perspective of ''Singapore Experience'. *2019 4th International Social Science and Education Conference.* https://webofproceedings.org/proceedings_series/ESSP/ISSEC%202019/ISSEC19052.pdf. Accessed 1 Dec 2019

17. Yuen B (2004) Safety and Dwelling in Singapore. Cities 21(1):19–28
18. Yuen, B. (2009). Guiding Spatial Changes: Singapore Urban Planning. In: Lall S.V., Freire M., Yuen B., Rajack R., Helluin JJ. (eds) Urban Land Markets. Springer, Dordrecht. https://doi.org/10.1007/978-1-4020-8862-9_14

Housing and Possible Health Implications in Upgraded Informal Settlements: Evidence from Mangaung Township, South Africa

John Ntema, Isobel Anderson, and Lochner Marais

Abstract Both international and South African researches on informal settlement upgrading indicate that a possible 'direct' relationship with the health of residents is connected more to neighbourhood access to social amenities than to physical housing conditions. Despite South African government attempts to align the National Development Plan (NDP) with Agenda 2030, and policy references to health and social amenities as integral to socially and economically integrated communities, the Department of Housing's programme is remarkably silent in respect of health outcomes. This chapter assesses the knowledge gap on health outcomes in upgraded informal settlements, within the wider context of Agenda 2030 and SDGs 3, 6, and 11. Through a survey, in-depth qualitative interviews and focus group discussions in an upgraded informal settlement, it contributes new evidence from analysis of households' perceptions of health. It then makes policy recommendations such as ensuring improved integration of health services and boosting sanitation and housing in upgrading programmes in order to achieve both a more preventive approach to health locally and to contribute more effectively to achieving the SDGs on health, water and sanitation, and housing.

Keywords Informal settlement upgrading · Health · Service infrastructure

J. Ntema (✉)
Department of Human Settlements, University of Fort Hare, Alice, South Africa
e-mail: lntema@ufh.ac.za; lejone12345@gmail.com

I. Anderson
Faculty of Social Sciences, University of Stirling, Stirling FK9 4LA, UK
e-mail: isobel.anderson@stir.ac.uk

L. Marais
Centre for Development Support, University of the Free State, Bloemfontein, South Africa
e-mail: MaraisJGL@ufs.ac.za

© The Author(s), under exclusive license to Springer Nature Singapore Pte Ltd. 2021   71
T. G. Nubi et al. (eds.), *Housing and SDGs in Urban Africa*, Advances in 21st Century
Human Settlements, https://doi.org/10.1007/978-981-33-4424-2_5

1   Introduction

Health is an integral part of housing policy, in addition to promoting operational-isation of the Agenda 2030 Sustainable Development Goals (SDGs) and the New Urban Agenda. Housing and neighbourhoods influence health and well-being (SDG 3), even as they provide shelter and a secure home (SDG 11). This chapter shows a possible 'direct' relationship between informal settlement upgrading and the health of residents by discussing infrastructure and access to social amenities rather than the physical aspects of housing unit(s). In an earlier South African study, Marais and Cloete [32] found that "health outcomes are more impacted by service-related factors than by housing structure." Other South African research has shown that improving the top structure, especially in upgraded informal settlements, has in some cases led to a decrease in the provision of infrastructure and social amenities, a situation that might have adverse health consequences [1, 34, 59]. The literature indicates that the majority of informal settlers suffer disproportionately from health-related issues, although there is limited research evidence on the health impacts of broader settlement environment on upgraded informal settlements [1, 23, 32, 49, 59].

This chapter draws on new empirical evidence of informal settlement upgrading in the South African context to examine the complex relationship between health and the immediate neighbourhood as prioritised in SDG 3 (Good Health and Well-being) and SDG 6 (Clean Water and Sanitation). The evidence reveals the health implications of the growing and long-standing backlog in the provision of water, sanitation and health facilities in an upgraded Phase 6 area in Mangaung Township (Bloemfontein)—all of which factors constrain effective achievement of the SDGs.

2   Housing, Health and the Sustainable Development Goals

Notwithstanding the existing infrastructural development backlog, the South African government sought to align its National Development Plan (NDP) with the United Nation's 2030 Agenda and its Sustainable Development Goals (SDGs). Indeed, in 2000, South Africa, along with other countries, had committed herself to the earlier United Nation's Millennium Development Goal 7 Target 7D on environ-mental sustainability and improving the lives of slum dwellers [24]. In light of the growing number of signatories, especially in developing countries, the research ques-tion emerges as to how far historical policies have determined the national and local implementation of Agenda 2030 and the New Urban Agenda (NUA) in different countries. In particular, the year 2000 commitment by South Africa is linked directly to the new Sustainable Development Goals, especially their 2030 targets on the provi-sion of water, sanitation and health facilities, although a global housing agenda can be traced from the 1970s [53]. No doubt, housing is central to the New Urban Agenda [53–55] and integral to the 2030 Agenda for Sustainable Development [57], which set the 17 integrated Sustainable Development Goals, with targets to be achieved by

2030. The United Nations [56] SDG knowledge platform summarises progress so far on these 17 goals, concluding as follows for the two SDGs most relevant to this chapter:

- SDG3 (Good Health and Well-being) recorded progress in increasing life expectancy, reducing maternal and child mortality and fighting against leading communicable diseases. However, progress was stalled or slow with regard to addressing significant diseases, such as malaria and tuberculosis. At least half the global population still did not have access to essential health services (often linked to financial hardship), potentially pushing them into extreme poverty.
- SDG6 (Clean Water and Sanitation) recorded limited progress, with billions of people still lacking safe water, sanitation and handwashing facilities. Data suggested the need to double the current annual rate of progress.

Related both to the two SDGs above and housing issues is SDG 11, which seeks to "make cities and human settlements inclusive, safe, resilient and sustainable." Its related target 11.1 seeks to ensure access for all to adequate, safe and affordable housing and basic services and to upgrade 'slums' by 2030 [57]. Consequently, the Implementation Plan (paragraph 32) of the New Urban Agenda, adopted at the HABITAT III congress in Quito, 2016, committed signatories (South Africa included) to promoting the development of age- and gender-responsive housing policies and to integrating healthcare and social integration sectors. The upgrading of informal settlements, as explored in this chapter, is a crucial benchmark for both the New Urban Agenda and SDG 11 in Agenda 2030. For SDG 3 on health, the targets most relevant to this chapter are the pursuit of universal health coverage and access to healthcare services (3.8), and substantially reducing the number of deaths attributed to unsafe water, unsafe sanitation and lack of hygiene (3.9.2). SDG 6 seeks to achieve universal and equitable access to safe and affordable drinking water for all by 2030, as well as to increase the proportion of the population using safely managed sanitation services. As stated in SDG 11 target 11, the prevalence of backlogs on all these basic services is particularly high among communities in slums or informal settlements.

The analysis of empirical data in this chapter (see Sects. 1.5–1.7) is underpinned by a consideration of housing as a social determinant of health [6, 36]. Globally, home and housing are at the heart of communities and represent a vital sphere in which municipalities and their partners can contribute to the health and well-being of residents, including the most marginalised groups in society [18, 19, 52]. In line with housing and health in South Africa (see discussion below), evidence from the global north reveals continuing challenges of proving cause and effect in the housing-health relationship. Nonetheless, potential health benefits have been identified concerning a range of housing and homelessness initiatives [14]. Broadly, research evidence indicated a positive relationship between being well-housed and being well, while experiences such as homelessness or other housing problems had adverse health effects, often compounded by poverty and inequality. Garnham and Rolfe [14] found reasonably clear evidence of the negative impact of poor physical housing quality on both physical and mental health [5, 13, 29, 37, 50, 63]. Evidence was also reasonably clear on the negative health consequences of acute (street) homelessness [2, 40, 45,

62]. Insecurity of tenure appeared to negatively affect mental and physical health [12, 27], while housing that 'felt like home' was found to generate psychosocial benefits [14, 20, 28]. World Health Organisation guidance on housing and health also recognises the importance of affordability, security and the surrounding environment in achieving overall well-being [63].

While there may be no single conceptual approach that best explains the intersection of health and housing at the community level, evaluation of programmes such as the upgrading of informal settlements can shed new light on evolving debates. This chapter presents a neighbourhood-level case study where a range of health and housing factors may come into play in shaping well-being in informal settlements.

3 Health, Housing and Informal Settlement Upgrading in South Africa

A growing body of work has emerged in South Africa on the relationship between health and housing, including informal settlement upgrading. Neither the original White Paper on Housing nor the revised policy proposals in the "Breaking New Ground" policy document extensively addresses health-related issues. The informal settlement upgrading policy of 2004 and subsequent documents were more vocal in emphasising the importance of the health benefits of basic infrastructure. Furthermore, the informal settlement upgrading policy stressed the role of informal settlement upgrading as a poverty alleviation strategy [23].

The evidence between housing and health in South Africa remains scant. Often other factors such as age, food and income have more direct implications for health [39, 58]. Several studies suggest that housing units funded by the state contribute to better physical health outcomes than the health outcomes in informal settlements [8, 32, 42, 49, 61]. However, some findings are rather mixed, indicating that factors other than housing are influencing outcomes. Controlling for age, Shortt and Hammond [49] and de Wet et al. [8] found no difference between informal and formal housing before or after the upgrading of an informal settlement. However, residents in informal settlements showed higher HIV-infection and TB rates than people in formal houses [15, 31, 48, 51, 59, 61]. Studies also found significantly better health outcomes in residents of houses provided by the state subsidy programme than in informal settlements for the following conditions and attributes: blood pressure [61], yellow eyes [32], skin disorders [32], rashes [32], diarrhoea [15–17, 32, 39, 51], child and infant mortality [32], and adult and child physical growth [32, 61]. However, these studies reported higher levels of asthma among residents of subsidised houses compared to residents in informal settlements [32]. One study reported that an urban renewal programme resulted in a decrease in pneumonia cases [51]. At least one other study linked improved children's health with improved infrastructure and housing conditions [38, 59].

Findings in respect of mental health also are contradictory. In two studies, the residents of houses obtained under the housing subsidy programme had improved mental health outcomes compared to residents in informal housing [39, 49]. Three other studies found either no relationships [35, 61] or only modest relationships between the built environment and health [47]. Crowding, irrespective of the housing type, but often more severe in informal settlements, was found to lead to mental health problems [32, 42] and was closely related to sexual abuse of children [11].

Access to health facilities is a crucial aspect of housing. One of the main criticisms of the housing subsidy programme has been the poor location concerning jobs and social infrastructure. The empirical evidence shows that new housing locations do not necessarily result in better access to healthcare facilities compared to the access in informal settlements [30, 32, 42]. This finding also suggests that informal settlement in situ upgrading, rather than greenfield developments, would ensure better access to healthcare infrastructure.

There is very little evidence that housing typologies are the main contributing factor to housing and health in South Africa. Access to infrastructure is more critical [3]. The two main factors are distance from water and having to share toilet facilities with other households [32]. Longer distances between water and housing units contribute to lower levels of water use and, thus, more inadequate hygiene. The distance to water also increases the dependence on water containers. Often the problem lies in the bacteria in the containers. The more dependent households are on these containers, the higher the likelihood of bacterial contamination. Shared toilet facilities probably have a bearing on the cleanliness of such facilities. Other research has pointed out the health benefits of the construction of self-build housing [60]. Higher temperatures related to global warming mean that indoor housing temperatures will have implications for health. South African research shows that informal settlements and government-funded houses are at risk in this respect [41]. One study also investigated indoor pollution and found higher levels of pollution in informal housing structures than in formal housing structures [26, 30].

Historically, health concerns provided a rationale for urban planning. Colonial and apartheid governments used health reasons to enforce modern town planning, as well as to encourage racial segregation and marginalise urban black people [7]. With the demise of apartheid planning and the development of post-apartheid housing policy, health motivations became less critical. The housing policy focused on the backlog and the segregated and fragmented nature of urban areas. The South African post-apartheid housing policy is a one-off, targeted and capital subsidy. Although the government provided large numbers of houses in the first ten years using the housing subsidy, it did not make much progress with informal settlement upgrading [21, 22]. Breaking New Ground thus emphasised the development of assets for the poor and provided a new informal settlement upgrading policy. The new policy on the upgrading of informal settlements highlighted the role of upgrading in dealing with poverty [22]. Based on poverty concerns in upgrading, health concerns re-emerged in the policy. The policy proposes three mechanisms to assist in this respect: location and access, secure tenure, and infrastructure [22, 25]. Good locations should provide adequate access to health services and employment and therefore reduce access costs, just as secure tenure should help build assets and support stability (which has proven health and well-being benefits) and infrastructure should ensure access to water, sanitation and electricity, which policymakers link with improving health outcomes [32]. As shown in the previous discussion, the new upgrading policy has placed health-related concerns back on the planning and housing agenda in South Africa.

4   Method

Methodologically, the chapter followed a case study approach, focusing on an upgraded area known as Phase 6, located in Mangaung Township (Bloemfontein). The strategic planning documents and interviews with municipal officials and ward councillors reflected a certain level of commitment on the part of Mangaung Metropolitan Municipality to continue undertaking informal settlement upgrading. Bloemfontein, the core city of the Mangaung Metropolitan Municipality, continues to experience an influx of migrants and the growth of informal settlements. This continued in-migration comes against the background of a history of influx control and displaced urbanisation. There are currently 34 informal settlement areas with an estimated population of 29,000 households in the city. Prior evaluations of the upgrading process [4, 33, 44] failed to pay detailed attention to health outcomes and this limitation reflects a knowledge gap in South African research on informal settlement upgrading [1, 25]. Further, not enough has been learned from city-based (city of Bloemfontein included) responses in most cities in the country [46].

We gathered data through mixed methods. A household survey of 200 was undertaken in 2016. One key criterion in selecting participants was that such individuals should be adults and (one way or the other), owners of dwellings. Participants were identified through a systematic random sampling method. The survey was complemented with two focus group discussions—each group comprising nine participants and involving five in-depth interviews with heads of households, as well as further in-depth interviews with five senior municipal officials responsible for informal settlement upgrading and a ward councillor.

5   Health Profile of Households in an Upgraded Phase 6 Area (Mangaung Township)

The health profile is broken into two categories: self-reported symptoms and self-reported diagnoses of selected illnesses (Table 1). Despite several differences, the prevalence of some of the diseases in our case study area corresponds, in part, with some research results and literature [32]. The work by Marais and Cloete [32] is based on data collected in 2008 during the National Income Dynamics Study (NIDS). The study was conducted by the Southern Africa Labour Development and Research Unit

Table 1 Health profile of respondents in phase 6 area (Mangaung Township), 2016

In the last 30 days-respondents with symptoms of	%
Flu	33.8
Fever	29.9
Persistent cough	9.0
Rash	5.0
Skin disorder	4.5
Cough with blood	2.0
Eye infection	2.0
Diarrhoea	2.0
Weight loss	1.5
Respondents diagnosed with:	%
High blood pressure	17
Arthritis	9.5
Tuberculosis (TB)	3.0
Diabetes	2.5

Source Authors (2016, Household Survey)

at the University of Cape Town. This was a panel study which consisted of approximately 16,500 adults (aged 15 and older) and 9500 children (under 15 years of age) in 7300 households. Data was collected using a stratified two-stage cluster sample design. The Statistical Package for the Social Sciences (SPSS), Version 20 was used for modelling. Where similar data exist between our 2016 research findings and the data used by Marais and Cloete [32], the chapter uses the latter for benchmarking purposes.

The information about the health profile of respondents in Phase 6 area is vital for two reasons. First, it assists in gaining a better understanding of the local context in terms of the prevalence of certain illnesses in the area at the time of the study, filling the existing knowledge gap on health outcomes in an upgraded informal settlement. Second, it also assists with benchmarking with similar prior studies.

When compared with specifically selected indicators in the study by Marais and Cloete [32], our 2016 survey findings in Table 1 show some interesting trends and patterns. Our survey in Phase 6 has recorded the highest level of respondents with symptoms of flu in any of the four housing typologies in the national study. For instance, Phase 6 area has 33.8% of respondents with symptoms of flu as compared with Marais and Cloete's study in which they found flu symptoms in 28.5% in subsidised housing, 31.4% in informal settlements, 27.8% in informal housing and 23.2% in formal settlements. Similarly, our study shows a higher level of fever cases (29.9%) compared to 19.9% in subsidised housing, 20.7% in informal settlements, 16.9% in informal housing and 13.4% in informal settlements [32]. The levels of skin disorder were also higher in Phase 6 (4.5%) compared to 1.9% in subsidised housing, 3.6% in informal settlements, 3.4% in informal housing and 2.2% in formal housing. Skin disorder is often directly related to the lack of water.

There are also levels of self-reported symptoms and diagnosed illnesses where our study findings in Phase 6 show levels lower than those recorded by Marais and Cloete [32]. Our study recorded 9.0% for persistent cough, 2.0% for diarrhoea, 2.0% for eye infection, 1.5% for weight loss and 3.0% for tuberculosis in Phase 6. Marais and Cloete [32] recorded slightly higher figures across the various typologies they considered. Finally, although slightly lower than levels in the four housing typologies investigated by Marais and Cloete [32], the 3% of respondents in Phase 6 diagnosed with TB should be a matter of concern. TB is a curable disease, so this figure should generally be lower.

6   Access to Sanitation, On-Site Water Supply and Clinics

Our focus now shifts to a discussion of access to basic infrastructure. Our findings show that where services exist, they are mostly either inadequate or poorly maintained. We highlight three main issues. First, interviewees complained that a significant number of households are still without adequate sanitary system or flushing toilets. The household survey shows that 68% of the households use a pit latrine. Approximately two-thirds of the respondents said that they are dissatisfied with the sewerage in the area. This is confirmed by remarks such as, "There are sections in our community that are still without flushing toilets and using pit latrine toilet… this system is a big challenge." Second, 85% of the houses are without provision of water on the stand and are dependent on communal taps. Third, the area is still without a permanently built clinic. This is despite the fact that the area is home to over 3000 households. It is therefore not surprising that almost 70% of respondents expressed dissatisfaction about the state of, and access, to health facilities in their neighbourhood.

Evidence from our research shows that the current lack of sanitation, on-site water and clinics in the area is a matter of concern. The latest policy on informal settlement upgrading emphasises the principle of "progressive access to adequate housing" [9] and "incremental provision of basic services and social amenities" [10]. After over ten years, the question is whether incrementalism has become a way not to provide these services. For example, the growing backlogs in provision of these basic services (water, sanitation, public health and education facilities), in upgraded areas may create a danger that incrementalism ultimately implies almost no or limited progress for the community in accessing basic services. Consequently, policy principle on 'incrementalism' may be used post-relocation phase especially by politicians in government to endlessly avoid accountability for some of unrealistic service infrastructure development promises they might have made during their elections campaigns in un-serviced informal settlement communities.

7 Perceived Impact of Lack of Adequate Sanitation, On-Site Water Supply and Clinics on Health

This section is an analysis of the perceptions of project beneficiaries, ward councillors and municipal officials. The focus is on opinions on the possible health risks associated with lack of adequate sanitation, on-site water supply and a clinic in an upgraded Phase 6 area since 1999.

7.1 Absence of Clinics and Possible Health Implications

The twin factors of lack of a clinic in a neighbourhood of more than 3000 households and an unemployment rate of 45.8% have left the community with two options: a planned mobile clinic or travelling 5 km to the nearest clinic. There was an initial agreement on a mobile clinic, scheduled to visit the area every second week. However, this did not always materialise. The lack of access has two main implications for residents in Phase 6. First, respondents cited the high travel costs to visit clinics. Given the high rate of unemployment (45.8%), it was not a surprise that most interviewees complained about unaffordable taxi fares. A round trip costs about R20 while walking to the clinic will take almost two hours. One interviewee complained:

> Majority of us are unemployed and cannot afford taxi fee … so an almost two-hour walk to the nearest clinic is a common practice although impossible for the elderly and the weak who cannot afford either transport costs or walk.

There is little doubt that the community has not been appropriately treated about accessing health services.

Second, interviewees insisted that the mobile clinic was unreliable. The clinic either fails to arrive on scheduled dates or does arrive but without enough or required medication. The inability to keep to the planned times was also influenced by the amount of rain and the quality of the road infrastructure. While both the unaffordable transport costs to nearest clinics and the unreliable and under-resourced mobile clinic could potentially affect the health of all patients in the area, specific services need urgent attention.

First, considering the large percentage of females who participated in the household survey and the qualitative interviews, it does not come as a surprise that "access to family planning methods" was prioritised. The unreliability of the mobile clinic means that pregnancy prevention medication is not available and that the women are not sure that they will have access to the medicine in the next month. Consequently, interviewees complained about the high rates of teenage pregnancies. Expressing their collective frustration, one woman said:

> The mobile clinic is a serious problem here—sometimes it comes, sometimes it does not. Imagine if you are on a family planning and the mobile clinic fails to come, and you have to wait for another two weeks, or your child falls sick … this is bad indeed.

Another respondent noted:

More and more girls are falling pregnant … we do not have exact numbers, but we do see these things.

Second, the immunisation or vaccination of children is also widely raised as a concern. Childrens' vaccinations are critical to achieving the SDGs through ensuring longevity but also improve learning and development outcomes.

Third, the unaffordable transport cost to clinics, as well as the unreliable and under resourced mobile clinic, impacts on scheduled treatment for tuberculosis (TB) patients, as well as on availability of chronic medication for high blood pressure, diabetes and HIV. Some interviewees knew of family members or neighbours who were forced to unintentionally default on their TB treatment and chronic medication due to their inability to access clinics. One interviewee reported:

There are people who are unemployed and on chronic medication for TB and other ailments. If they do not have the R20 taxi fee, they usually default on their medication, and although difficult, we are busy trying to profile our chronic patients for a possible intervention where necessary.

Fourth, the absence of a clinic also has negative implications for any campaign or programme that seeks to encourage and promote a healthy lifestyle in the community. A ward councillor expressed deep frustration over the lack of a local clinic. For the councillor, such a clinic should undertake joint programmes on health screening for lifestyle-related diseases such as high blood pressure, diabetes and HIV. The absence of a clinic contributes to ignoring the preventative nature of health planning and practice.

7.2 Lack of Adequate Sanitation, on-Site Water Supply and Possible Health Implications

Despite the area being upgraded in 1999, 68% of respondents were still using a pit latrine toilet system and 66% expressed dissatisfaction with the sewerage system. This situation fuelled a perception that their health and well-being was being compromised to a certain degree. The current pit latrine toilet system caused a threefold health risk. First, the toilets were stinky and attracted swarms of flies, especially during hot summer days. Second, toilets were always full due to failure by the municipality to regularly dispatch a 'sewerage truck' to drain them. The respondents said that the municipal trucks were either broken down or without a diesel supply. Residents were left on their own to deal with full toilets (some) for more than a year. Third, most of these toilets were poorly built and prone to collapse. Once again, the failure of the state to perform essential functions suggests that the problem will remain.

Most interviewees reported that bad odour, particularly from full toilets, compromised their health and hygiene while making them vulnerable to germs and airborne diseases such as TB (although this is not the way the TB virus is spread). One interviewee noted:

Toilets are a crisis in the area ... there are households who have been over a year now waiting for the municipal truck to drain their full toilets. This poses a serious health risk, especially now that people are unable to do regular checkups for TB.

Furthermore, respondents saw the high number of collapsing toilet structures as posing a danger to children and the elderly. Most interviewees claimed to have now sought an alternative. It is possible that the reason why 8.5% of respondents is currently sharing toilet facilities with neighbours might be due to their toilets that are forever full or have collapsed. It is important to stress the worrying matter of sharing toilet facilities, which has potential to cause the skin infection called 'scabies', contracted from a burrowing mite [43]. Indeed, sharing toilets compromises hygiene and health in general [32].

In sections of Phase 6 area where there is no on-site water supply, communal taps are the only source of water supply. An interviewee said: "In some streets, there is only one communal water tap and with regular interruptions in supply ... we always have to ensure that we collect and store enough water." With inference drawn in the literature between water storage in containers and bacteria [32], it is little wonder that 4.5% of interviewees complained about skin disorder (Table 1).

8 Conclusion

The following concluding remarks are worth making. First, while the study on Phase 6 area in Mangaung Township (Bloemfontein) did not explicitly set out to test implementation of Agenda 2030, its findings demonstrate the continuing need to integrate interventions better to achieve SDGs 3, 6 and 11. Second, it appears that lack of access to adequate sanitation and health facilities (clinics) has largely deprived a significant number of respondents in Phase 6 area of their constitutional rights to "adequate sanitary and health facilities including waste disposal." There is a perception that absence of a permanent clinic and unreliable services from a mobile clinic compromises health, especially collection of scheduled chronic medication, family planning and vaccination of toddlers. This has raised concerns about a significant number of patients who are possibly defaulting unintentionally on their chronic medication, very likely with dire health consequences. An example in this regard could be the 3.0% of respondents diagnosed with TB for more than a year, despite this being a curable disease within six months provided a patient does not default on their treatment. The pit latrine toilets that are full and remain uncollected for several months and the tendency by some households to keep water in containers may also have severe health consequences. It is thus possible that the rate of self-reported diseases shown in Table 1 might slightly change if more people have easy access to well-resourced clinics for regular and voluntary medical checkups and health screening. The current level of access to health facilities in Phase 6 area does not make it easy for the community to voluntarily subscribe to the notion that prevention is better than cure. Overall, our findings suggest that the anticipated health benefits of housing improvements in upgraded informal settlements may be undermined by the growing backlog in the provision of clinics, water and sanitation. Backlogs in provision of these basic services in upgraded areas may create a danger that incrementalism ultimately implies almost no or limited progress for the community, in addition to impeding wider achievement of the UN Sustainable Development Goals. To ensure that 'no-one is left behind' in terms of access to health facilities as well as sanitation and water in Phase 6 area, responsible government departments may have to consider adopting a joint and integrated project planning underpinned by the principle of 'sound intergovernmental relations'. This needs to include a commitment to provision of adequate healthcare, water and sanitation, which are core to the upgrading of informal settlements.

Acknowledgements We acknowledge the financial contribution of the National Research Foundation (South Africa) towards funding the fieldwork for this chapter (Grant no: 89788).

References

1. Ambert C (2006) An HIV and Aids lens for informal settlement policy and practice in South Africa. In: Huchzermeyer M, Karam A (eds) Informal settlements: a perpetual challenge?. UCT Press, Cape Town, pp 146–164
2. Anderson I, Barclay A (2003) Housing and health. In: Watterson A (ed) Public health in practice. Palgrave, London, pp 158–183
3. Ataguba J, Day C, McIntyre D et al (2015) Explaining the role of the social determinants of health on health inequality in South Africa. Glob Health Action 8(1):1–11
4. Botes L, Krige S, Wessels J et al (1991) Informal settlements in Bloemfontein: a case study of migration patterns, socio-economic profile, living conditions and future housing expectations. Research report, Urban Foundation, Bloemfontein
5. Braubach M, Jacobs DE, Ormandy D et al (2011) Environmental burden of disease associated with inadequate housing. Research paper, WHO, Europe, Copenhagen
6. Dahlgren G, Whitehead M (1991) Policies and strategies to promote social equity in health. Stockholm: Research paper, Institute for Futures Studies
7. Davies RJ (1981) The spatial formation of the South African city. GeoJournal 2:59–72
8. De Wet T, Plagerson S, Harpham T et al (2011) Poor housing, good health: a comparison of formal and informal housing in Johannesburg, South Africa. Int J Public Health 56(6):625–633
9. Department of Housing (1994) White paper on housing: a new housing policy and strategy for South Africa. Government Gazette No. 16178, Pretoria
10. Department of Housing (2004) "Breaking new ground": a comprehensive plan for the development of sustainable human settlements. Department of Housing, Pretoria
11. Department of Social Development (2012) Violence against children in South Africa. Department of Social Development, Pretoria
12. Downing J (2016) The health effects of the foreclosure crisis and unaffordable housing: a systematic review and explanation of evidence. Soc Sci Med 16:88–96
13. Fisk WJ, Eliseeva EA, Mendell MJ et al (2010) Association of residential dampness and mould with respiratory tract infections and bronchitis: a meta-analysis. Environ Health 9:72. https://doi.org/10.1186/1476-069X-9-72
14. Garnham L, Rolfe S (2019) Housing as a social determinant of health: evidence from the Housing through Social Enterprise study. Research paper, University of Stirling and Glasgow Centre for Population Health, UK

15. Govender T (2011) The health and sanitation status of specific low-cost housing communities as contrasted with those occupying backyard dwellings in the City of Cape Town, South Africa. Unpublished Doctoral thesis, University of Stellenbosch, Stellenbosch
16. Govender T, Barnes J, Pieper C et al (2011) Contribution of water pollution from inadequate sanitation and housing quality to diarrheal disease in low-cost housing settlements of Cape Town, South Africa. Am J Public Health 10(7):4–9
17. Govender T, Barnes J, Pieper C et al (2011) Housing conditions, sanitation status and associated health risks in selected subsidized low-cost housing settlements in Cape Town, South Africa. Habitat Int 35(2):335–342
18. Heatherington K, Hamlet N (2015) Restoring the Public Health response to Homelessness in Scotland. Research paper, Scottish Public Health Network, Scotland
19. Heatherington K, Hamlet N (2019) Health and homelessness. In: Bonner AB (ed) Social determinants of health: an interdisciplinary perspective on social inequality and wellbeing. Policy Press, Bristol, pp 195–210
20. Hiscock R, Kearns A, MacIntyre S, Ellaway A et al (2001) Ontological security and psychosocial benefits from the home: qualitative evidence on issues of tenure. Hous Theor Soc 18(1–2):50–66
21. Huchzermeyer M (2001) Concent and contradiction: scholarly responses to the capital subsidy model for informal settlement intervention in South Africa. Urban Forum 21(1):71–106
22. Huchzermeyer M (2004) From "contravention of laws" to "lack of rights": redefining the problem of informal settlements in South Africa. Habitat Int 28:333–347
23. Huchzermeyer M (2006) The new instrument for upgrading informal settlements in South Africa: contributions and constraints. In: Huchzermeyer M, Karam A (eds) Informal settlements: a perpetual challenge. UCT Press, Cape Town, pp 41–61
24. Huchzermeyer M (2010) A legacy of control? The capital subsidy for housing and informal settlement intervention in South Africa. Int J Urban Reg Res 27(3):591–612
25. Huchzermeyer M, Karam A (2006) Informal settlements: a perpetual challenge. University of Cape Town Press, Cape Town
26. Jafta N, Barregard L, Jeena P, Naidoo R et al (2017) Indoor air quality of low and middle-income urban households in Durban, South Africa. Environ Res 156:47–56
27. Jelleyman T, Spencer N (2008) Residential mobility in childhood and health outcomes: a systematic review. J Epidemiol Community Health 62(7):584–592
28. Kearns A, Hiscock R, Ellaway A, Macintyre S et al (2000) 'Beyond Four Walls'. The Psychosocial benefits of home: evidence from West Central Scotland. Hous Stud 15(3):387
29. Liddell C, Guiney C (2015) Living in a cold and damp home: frameworks for understanding impacts on mental well-being. Public Health 129(3):191–199
30. Makene C (2008) Housing-related risk factors for respiratory disease in low cost housing settlements in Johannesburg, South Africa. Unpublished masters dissertation, University of the Witwatersrand, Johannesburg
31. Marais H (2007) The uneven impact of AIDS in a polarized society. AIDS 21(3):21–29
32. Marais L, Cloete J (2014) "Dying to get a house?" The health outcomes of the South African low-income housing programme. Habitat Int 43(1):48–60
33. Marais L, Krige S (1997) The upgrading of Freedom Square informal settlement in Bloemfontein: lessons for future low-income housing. Urban Forum 8(2):176–193
34. Marais L, Ntema J (2013) The upgrading of an informal settlement in South Africa: two decades onwards. Habitat Int 39:85–95
35. Marais L, Sharp C, Pappin M, Cloete J, Lenka M, Skinner D, Serekoane J et al (2013) Housing conditions and mental health of orphans in South Africa. Health Place 24:23–29
36. Marmot M (2010) Fairer society, healthy lives. The Marmot Review 2010, University College London, London
37. Marsh A, Gordon D, Heslop P, Pantazis C et al (2000) Housing deprivation and health: a longitudinal analysis. Hous Stud 15(3):411–428
38. Mathee A, Barnes B, Naidoo S, Swart A, Rother H et al (2017) Development for children's environmental health in South Africa: past gains and future opportunities. Dev South Afr 35(2):283–293

39. Mathee A, Harpham T, Barnes B, Swart A, Naidoo S, De Wet T, Becker P et al (2009) Inequity in poverty: the emerging public health challenge in Johannesburg. Dev South Afr 26(5):721–732
40. Munoz M, Crespo M, Perez-Santos E et al (2005) Homelessness effects on men's and women's health. Int J Mental Health 34(2):47–61
41. Naicker N, Teare J, Balakrishna Y, Wright C, Mathee A et al (2017) Indoor temperatures in low-cost housing in Johannesburg, South Africa. Int J Environ Res Public Health 14(11):1–18
42. Narsai P, Taylor M, Jinabhai C, Stevens F et al (2013) Variations in housing satisfaction and health status in four lower socio-economic housing typologies in the eThekwini Municipality in KwaZulu-Natal. Dev South Afr 30(3):367–385
43. Ngcukana L (2019) Desperate for decent toilets. City Press, p 13
44. Ntema LJ (2011) Self-help housing in South Africa: paradigms, policy and practice. Unpublished Doctoral thesis, University of the Free State, Bloemfontein
45. O'Connell J (2005) Premature mortality in homeless populations: a review of the literature. Research paper, National Healthcare for the Homeless Council, Nashville
46. Oldfield S (2002) Partial formalization and its implications for community governance in an informal settlement. Urban Forum 13(2):89–102
47. Ralston M (2018) The role of older persons' environment in aging well: quality of life, illness, and community context in South Africa. Gerontologist 58(1):111–120
48. Shisana O, Rehle T, Simbayi L, Parker W, Zuma K, Bhana A, Connolly C, Jooste S, Pillay V et al (2005) South African national HIV prevalence, hiv incidence, behaviour and communication survey, 2005. Research report, HSRC, Cape Town
49. Shortt N, Hammond D (2013) Housing and health in an informal settlement upgrade in Cape Town, South Africa. J Hous Built Environ 28(4):615–627
50. Thomson H, Thomas S, Sellstrom E, Petticrew M et al (2013) Housing improvements for health and associated socio-economic outcomes. Cochrane Database Syst Rev https://doi.org/10.1002/14651858.CD008657..pub2
51. Tlhabanelo M (2011) The impact of urban renewal on the health status of the community of Evaton. Unpublished Masters dissertation, University of Stellenbosch, Stellenbosch
52. Tweed E (2017) Foundations for well-being: reconnecting public health and housing. A Practical Guide to Improving Health and Reducing Inequalities. Research paper, Scottish Public Health Network (ScotPHN), Scotland
53. UN-Habitat (2015) Housing at the centre of the new urban agenda. UN-HABITAT, Nairobi
54. UN-Habitat (2017) The new urban agenda. HABITAT III, Quito
55. UN-Habitat (2017) Action framework for implementation of the new urban agenda. HABITAT III, Quito
56. United Nations (2019) UN sustainable development goals, knowledge platform. https://sustainabledevelopment.un.org/.pdf. Accessed 22 Nov 2019
57. United Nations (2015) The world population prospects: 2015 revision. United Nations, New York
58. Vearey J (2008) Migration, access to ART, and survivalist livelihood strategies in Johannesburg. Afr J AIDS Res 7(3):361–374
59. Vearey J (2011) Challenging urban health: towards an improved local government response to migration, informal settlements, and HIV in Johannesburg, South Africa. Glob Health Action 4:1–9
60. Venter A, Marais L, Morgan H et al (2019) Informal settlement upgrading in South Africa: a preliminary regenerative perspective. Sustainability 11:1–15
61. Vorster H, Venter C, Kruger H, Kruger A, Malan N, Wissing M, de Ridder J, Veldman F, Steyn H, Margetts B, MacIntyre U et al (2000) The impact of urbanization on physical, physiological and mental health of Africans in the North West Province of South Africa: the THUSA study. S Afr J Sci 96(9/10):505–514

62. Wolf J, Anderson I, Van den Dries L, Filipovic-Hrast MF et al (2016) The health of homeless women. In: Mayock P, Bretherton J (eds) Women's homelessness in Europe. Palgrave, London, pp 155–178
63. World Health Organisation (2018) Global strategy and action plan on ageing and health (2016–2020). World Health Organisation, Geneva

Housing, Health and Well-Being of Slum Dwellers in Nigeria: Case Studies of Six Cities

Johnson Bade Falade

Abstract This study analyses the contributions of slum housing to satisfying the health and well-being needs of residents. Using data collected from the slum characterisation studies conducted in the six Nigeria cities of Lagos, Kano, Port Harcourt, Onitsha, Gombe, and Karu. The results of the analyses indicate that only 48.1% of the total households in the six slums have their health and well-being needs met for physiological needs and psychological needs, as well as for protection from diseases and protection from injury. The results also show that only 12 out of the 26 housing indicators investigated were significantly provided for in the six slums. It is noteworthy that key health indicators such as access to safe water, safe waste disposal methods and provision for adequate health facilities contributed less than 80%, coupled with high occurrences of malaria, cholera and typhoid, which all show the disease burden facing slum dwellers. Also discussed are the implications of the findings of this study for promoting sound health and well-being among urban residents, as well as for urban planning, slum upgrading and attainment of the Sustainable Development Goals and the New Urban Agenda.

Keywords Health · Housing · Poverty · Slum upgrading · Well-being

1 Introduction

Research interests in the nexus between housing and health date to Florence Nightingale's time, when she postulated that "the connection between health and dwelling is one of the most important factors that exist" [11, 31, 33, 36, 47, 85]. However, in 1987, the World Health Organisation (WHO) and United Nations Environment Programme (UNEP) used the term "adequate shelter" to describe good housing that provides not only for shelter but also caters to the health and well-being needs of residents. Over the years, research interest in the nexus between housing, the built environment and health conditions of the people has continued to attract scholarly

J. B. Falade (✉)
Gotosearch.com Ltd, Ajah, Lagos, Nigeria
e-mail: go2search@yahoo.co.uk

attention. Many of these research interests have focused on the various diseases associated with housing and the urban environment, such as malaria, typhoid, respiratory infections, asthma, lead poisoning, injuries, mental health, noise exposure, thermal conditions, ventilation, mould, and home accidents [3, 7, 10, 13, 15, 25, 31, 36, 37, 39, 43, 44, 47, 56–58, 62, 83, 85].

The literature review indicates that studies of slums have been undertaken for some Nigerian cities. Many of these studies focused on formation of slums, the demographic and socio-economic characteristics of slum dwellers, housing conditions, strategies for urban renewal, poverty, income and informal activities of the slums in selected cities, as well as the role of urban planners and architects [1–6, 14, 22, 30, 41, 42, 45, 48, 52, 53, 55, 61, 68, 69, 71].

Among existing research works on slums is the slum characterisation studies of six cities in Nigeria as undertaken by the Foundation for Development and Environmental Initiatives [30] namely, Lagos, Port Harcourt, Karu, Onitsha, Bauchi, and Kano. This research provided valuable data on many aspects of the slums, such as (i) location and types of slums (ii) socio-economic and demographic data on slum dwellers including sex, religion, age, marital status, household population, occupation of household heads, education, income, affluence, engagement in informal activities, reasons for living in slums etc. (iii) housing conditions and building characteristics (iii) access to basic infrastructure (iv) environmental conditions and diseases and access to health facilities and social problems (v) land tenure Structure (vi) governance and management structure for the slums and (vii) improvements undertaken in the slum by government, organised private sector and donor agencies and (viii) demands for infrastructure for the improvement of the slums.

Only a few of these studies have focused on the health and disease burdens of living in slums [3, 30]. Aliyu and Amadu [3] have identified urbanisation and bad housing as major public health challenges:

> …urbanization, inadequacies of water supply, people living in squalor and shanty settlements, poor sanitation, inadequate solid waste management, and the inefficient, congested, and risky transport system, all these have increased the risk and disease burdens of living in towns and cities [3].

Although studies on the nexus of housing and health are on the increase, there are growing concerns in some quarters about the appropriateness of their methodology, scope and value for urban planning and housing design. Existing studies have been unduly focused on diseases while neglecting their value for urban planning, housing provision and architectural design [37, 57]. Consequently, it is impossible to tease out specific housing indicators and community characteristics that are related to diseases [37]. Thus, Rauh et al. [57] advocated more studies focusing on diseases as well as the physical and social features of housing and neighbourhoods. Furthermore, Hood [37] suggested the need to study housing from a healthy planning perspective and environmental factors. This study has been undertaken from this perspective to shed some light on the nexus between slum housing, health and the well-being of residents that will provide answers to urban planners, housing and health experts and policymakers.

2 Goals and Objectives of the Study

With an estimated 200 million population and being Africa's most populous country, Nigeria is on the fast track of urbanisation but with serious housing shortages and slum proliferation that pose health problems to the more than 50.2% of the country's total urban population living in slums [28]. With the country's urbanisation rate of 50%, this figure translates to more than 50 million urban residents living in slums and paints a grim picture of the country's housing situation.

Therefore, the goal of this study is to explore the nexus between slum housing, health and well-being, utilising data collected from households living in slums in six Nigerian cities. The objectives are to: (i) define housing indicators and health and well-being goals/roles for analysing the relationship between housing, health and well-being (ii) generate and analyse the data on selected housing indicators to determine which indicators are provided and their significant contributions to meeting the health and well-being goals (iii) discuss the health and well-being implications of the housing indicators and (iv) based on research findings, to make appropriate recommendations for improving the slums and to discuss their implications for urban planning, slum upgrading and realisation of the relevant targets and goals of Agenda 2030 and the New Urban Agenda (NUA) 2036.

3 Definitions of Terms and Research Questions

3.1 Definition of Terms

(i) *Health and well-being*

The definition of health advanced by WHO and UNEP [83] is the one adopted for this study, viz: "a state of complete physical, mental and social well-being and not merely the absence of disease or infirmity".

Health and well-being are related terms. To be in a state of well-being involves:

> feelings of happiness, curiosity and engagement, which are characteristic of someone having a positive sense of themselves. Having positive relationships, control over one's own life and having a sense of purpose are all attributes of functioning well [38].

Being in a state of health and well-being has also been linked to socio-economic and demographic factors such as a good education, a well-paid job that can afford the salary earner to pay for good healthcare and good housing [49, 83].

(ii) *Adequate Shelter*

According to WHO and UNEP [83], the term adequate shelter means:

more than a roof over one's head. It means to have a home, a place which protects privacy, contributes to physical and psychological well-being, and supports the development and social integration of its inhabitants – a central place for human life.

From this definition, the role of housing is more than providing shelter. It includes meeting the various health and wealth-being needs of the people to lead a normal life.

In 1996, *The Habitat Agenda,* which was adopted at the 2nd Global Conference on Human settlements that was held in Istanbul, Turkey, defined 'adequate shelter' as follows:

Adequate shelter means more than a roof over one's head. It also means adequate privacy; adequate space; physical accessibility; adequate security; security of tenure; structural stability and durability; adequate lighting, heating and ventilation; adequate basic infrastructure, such as water-supply, sanitation and waste-management facilities; suitable environmental quality and health-related factors; and adequate and accessible location with regard to work and basic facilities: all of which should be available at an affordable cost. Adequacy should be determined together with the people concerned, bearing in mind the prospect for gradual development. Adequacy often varies from country to country, since it depends on specific cultural, social, environmental, and economic factors. Gender-specific and age-specific factors, such as the exposure of children and women to toxic substances, should be considered in this context ...' [64 para. 60 p. 35, 7].

From this definition, the roles of housing include not only shelter but also other health, environmental, social and economic needs of the people. The roles of housing include providing privacy, space, security (safety, tenure and protection from toxic pollution), lighting and energy, provision of basic infrastructure (water, sanitation and waste management), accessibility, cultural, social, environmental and economic factors and responding to the needs of the people, women and children. The use of the word 'adequate' in relation to these housing roles implies that such a house must be complete and lack nothing to fulfil its several roles. However, it is noteworthy that the authors of this definition also recognize that what constitutes adequate shelter can vary from country to country. Thus, what is adequate in Nigeria might not be adequate in the UK and USA and vice versa.

(iii)   *Slums*

The definition of slum adopted for the slum characterisation studies in six Nigerian cities is also the one adhered to in this study, as advanced by the Expert Group Meeting (EGM). Thus, a slum is housing that has one or more of five issues: (i) inadequate access to safe water (ii) inadequate access to sanitation and other infrastructure such as access roads (iii) poor structural quality of housing (iv) overcrowding and (v) insecure residential status tenure [78].

3.2   Research Questions

Many studies assume that good housing will positively impact good health and well-being of residents [7, 8, 11, 13, 16, 32, 33, 37, 40, 60, 72, 73, 79, 80, 82, 83, 86]. Given this assertion, it is therefore inconceivable that slum housing, which represents the worst form of housing, can significantly contribute to meeting the health and well-being needs of residents. Since a high percent of the urban population in developing countries like Nigeria live in slums, it is necessary to explore the nexus between slum housing, health and well-being of the residents. The results of this kind of investigation can help to identify the ways of improving slum housing to provide healthy living.

The research questions are:

(i)   What are the elements/indicators of adequate shelter that can significantly contribute to achieving the health and well-being needs of the people?
(ii) What are the elements/indicators of good housing that are provided for in existing slums in Nigeria? What are their contributions to meeting the health and well-being needs of residents?
(iii) What can be done to further improve the existing slums to meet the health and well-being needs of the people?

4   Housing, Health and Well-Being: A Framework for Analysis

4.1   The Need for a Framework for Analysing Housing, Health and Well-Being

To successfully carry out this study, it was deemed necessary to evolve a robust framework for exploring the nexus between housing and health and the well-being of residents. This study follows the framework provided in the second chapter of *Healthy Housing Reference Manual* (2016), a publication by the Centers for Disease Control and Prevention (CDC) and US Department of Housing and Urban Development. The chapter specifies six health and well-being goals and the roles of housing for their realisation, viz: (i) physiological needs (ii) psychological needs (iii) protection from diseases (iv) protection from injury (v) protection from fire and (vi) protection from toxic pollution. Table 1 is a summary of the six goals and the corresponding roles of housing for realising them.

Table 1 Basic health and well-being goals and the role of housing

Health and well-being goals	Roles of housing
1. Meeting physiological needs	1. Ensuring protection from the elements (rain, sun snow, etc.)
	2. Providing a thermal environment that will avoid undue heat loss
	3. Providing a thermal environment that will permit adequate heat loss from the body
	4. Providing an atmosphere of reasonable chemical purity
	5. Ensuring adequate daylight illumination and avoidance of undue daylight glare
	6. Ensuring direct sunlight
	7. Providing adequate artificial illumination and avoidance of glare
	8. Protection from excessive noise
	9. Providing adequate space for exercise and for children to play
2. Meeting psychological needs	1. Providing adequate privacy for the individual
	2. Providing opportunities for normal family life
	3. Providing opportunities for normal community life
	4. Providing facilities for possible performance of household tasks without undue physical and mental fatigue
	5. Providing facilities for maintaining cleanliness of the dwelling and of the person
	6. Providing for aesthetic satisfaction in the home and its surroundings
	7. Ensuring concordance with prevailing social standards of the local community
3. Protection against diseases	1. Providing a safe and sanitary water supply
	2. Protecting water supply system against pollution
	3. Providing toilet facilities that minimize the danger of transmitting disease
	4. Protecting against sewage contamination of the interior surfaces of the dwelling
	5. Avoiding insanitary conditions near the dwelling
	6. Excluding vermin from the dwelling, which may play a part in transmitting disease
	7. Providing facilities for keeping milk and food fresh
	8. Providing sufficient space in sleeping rooms to minimise the danger of contact infection

(continued)

Table 1   (continued)

Health and well-being goals	Roles of housing
4. Protection against injury	Providing minimum standards for the protection of life, limb, property, environment, and for the safety and welfare of the consumer, the public and the owners as well as occupants of residential building
5. Protection against fire	Controlling conditions that promote the initiation and spread of fire
6. Protection against toxic wastes	Protecting against gas poisoning

Source Centers for Disease Control and Prevention and US Department of Housing and Urban Development (2006): *Healthy Housing Reference Manual*

4.2   Selection of Health and Well-Being Goals and Related Housing Indicators

Ideally, the six health and well-being goals of housing listed in Table 1 can be adopted for the study. However, as this study relied on secondary data, it became necessary to select only the goals for which data are available. Thus, four out of the six goals were selected. Table 2 is a summary of the selected goals and the 26 indicators on good housing that have been selected for investigation.

4.3   Method of Study

4.3.1   Overview of the Slum Characterisation Studies and Methods

This study relied on data collected during the slum characterisation studies; thus, a detailed account of the methodology used for the study has been presented in a national report [30]. Therefore, only highlights of the study's background, goal and methodology are provided here.

Study background and Goal: The slum characterisation study was conducted in 2014, as initiated by the Federal Ministry of Lands and Housing (FMLHUD), the apex national organ for policy formulation and coordination on housing, urban and regional development issues in the country. The FMLHUD commissioned the Foundation for Development and Environmental Initiatives (FDI) to execute the research project. The study was aimed at collecting relevant data on the chosen slums that will be used to develop urban renewal strategies and for the preparation of the National Report submitted for the Habitat III Conference in 2015.

Methods of data collection: Data collection involved specifying the criteria for selecting the slums and cities, employing a combination of methods for data collection in each slum and putting in place a robust management structure for the study.

Criteria for selecting cities and slums: The criteria for the selection of the slums and the cities included geopolitical considerations as well as type of slum and population. The selected slums and cities are (i) Abesan in Lagos, Southwest zone (ii) Dorayi in Kano, Northeast zone (iii) Okpoko in Onitsha, Southeast zone (iv) Elechi

Table 2 Selected goals, housing roles and indicators for the study

Health and well-being housing goals (1)	Role of housing (2)	Selected housing indicators (3)	Remarks (4)
Goal 1: Meeting physiological needs of residents	1.1 Protection from excessive noise	1. Households not experiencing noise pollution	Many of the roles of housing to fulfil this goal are covered by building regulations and land use standards
	1.2 Providing adequate space for exercise and for children to play	2. Households with access to open space for exercise and for children to play	
Goal 2: Meeting the psychological needs of residents	2.1 Providing adequate privacy for the individual	3. Providing adequate privacy	Two indicators are defined for 2 of the roles of housing
	2.2 Providing facilities for possible performance of household tasks without undue physical and mental fatigue	4. Households provided with kitchen	
	2.3 Providing facilities for maintaining cleanliness of the dwelling and of the person	5. Households provided with bathrooms 6. Households provided with toilets	
	2.4 Providing for aesthetic satisfaction in the home and its surroundings	7. Households who are satisfied with estate 8. Household with improved housing in the past five years	
Goal 3: Protection of residents from diseases	3.1 Providing for safe and sanitary water supply	9. Households with access to safe water	

(continued)

Table 2 (continued)

Health and well-being housing goals (1)	Role of housing (2)	Selected housing indicators (3)	Remarks (4)
	3.2 Providing toilet facilities that minimize the danger of transmitting disease	10. Households provided with Toilet 11. Households with no incidences of water borne diseases like cholera 12. Households with no incidences of water borne diseases like typhoid 13 Households with no incidence of a child in the past year	Collection of data on the incidence of diseases is considered necessary to know if the slums are contributing to the health and well-being of the resident
	3.3. Protecting against sewage contamination of the interior surfaces of the dwelling	14. Household with access to safe waste disposal methods	
	3.4 Providing for aesthetic satisfaction in the home and its surroundings	15. Households with good housing environment 16. Households that experienced no environmental disasters 17. Households with no incidence of disease related to environmental disaster 18. Households that has suffered from malaria in the previous year.	

(continued)

Table 2 (continued)

Health and well-being housing goals (1)	Role of housing (2)	Selected housing indicators (3)	Remarks (4)
	3.5 Providing sufficient space in sleeping rooms to minimise the danger of contact infection	19. Household with adequate sleeping space 20. Household with no incidence of communicable diseases like TB 21. Household with no incidence of communicable diseases like HIV/AIDS	Collection of data on the incidence of diseases is considered necessary to know if the slums are contributing to the health and well-being of the residents
Goal 4: Protection against injury	4.1 Providing minimum standards for the protection of life, limb, property, environment, and for the safety and welfare of the consumer, general public, and the owners and occupants of residential buildings	22. Houses constructed with durable floor materials 23. Houses constructed with durable roof material 24. Houses constructed with durable 25. Household with street lights 26. Households with access to good roads	Building with durable materials is an index of structural stability and safety

Source Columns 1 and 2 from Centers for Disease Control and Prevention and US Department of Housing and Urban Development (2006): *Healthy Housing Reference Manual*. Columns 3 and 4 compiled by the Author in December 2019

in Port Harcourt, South-south zone (v) Masaka-Angwin in Karu, North-central zone and (vi) Makama/Sarkin in Bauchi, North-east zone (Fig. 1). The six slums typify the two main categories of slums found in Nigerian cities. While Abesan, Dorayi and Makama are planned housing estates that degenerated into slums, Masaka-Angwan Police, Okpoko and Elechi are unplanned/informal slums. The six cities selected for study are among the most populous in each geopolitical zone of the country. It is noteworthy that Bauchi was chosen in preference to either Maiduguri or Yola because of the Boko Haram insurgency, which could affect safe and timely delivery of the project [30].

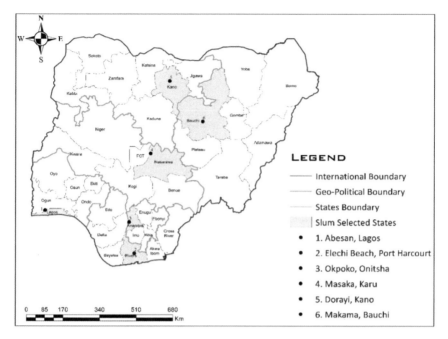

Fig. 1 Map of Nigeria showing the states and cities selected for the study. *Source* FDI (2014). National Report on Slum Characterisation Study

Methods used for data collection at the field level: The data collection methods included (i) land use survey in each slum to prepare the land use map (ii) questionnaire design for household survey (iii) field data collection (iv) organisation of public consultation meeting in each slum to discuss several thematic issues, including secure tenure, reactions to evictions, satisfaction with living conditions, health, safety and governance (see Plate 1).

Generally, the field data collection in the various cities was very successful, with relevant data and information collected on several aspects of slum dwellers, some of which have been used in this study. The residents of the slums appreciated the participatory approach adopted for the study, which they described as very different from the top-down approach often adopted by government agencies. Names like Maroko in Lagos and Elechi Beach in Port Harcourt are reminders of worst-case scenarios of improper urban renewal undertaken in the country [30].

Project Management: FDI adopted a decentralised and well-coordinated management approach to execute the project, as it engaged five city experts to conduct and report on the fieldwork. FDI coordinated the project and reported to FMLHUD (see Fig. 2, Plate 1).

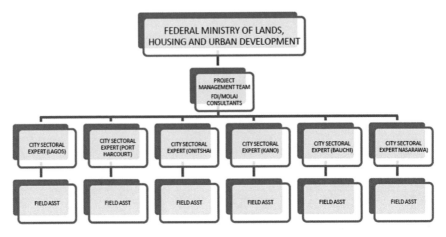

Fig. 2 Project management structure adopted for the study. *Source* FDI (2014)

Plate 1 Pictures of public consultation meetings held with slum dwellers. Left picture: Cross-section view of participants at Okpoko Slum Consultation. Right picture: Facilitators at the Consultation held in Abesan, Lagos. Source: FDI (2014)

4.3.2 Methods of Data Generation and Analysis for This Study

The data used for the study were generated by running a frequency analysis of the computerised data collected for the slum characterisation studies. This was done for each indicator in order to obtain the total number of houses that provided for the selected 26 indicators out of the total sample. This process is undertaken for the six slums for all the selected indicators. For instance, in a slum estate, if 45 out of 100 sampled houses provided for bath facilities, the computed percent total household with bath is 45%. Thereafter, the results were tabulated with the weighted averages computed. The results of the analyses for each set goal is undertaken and presented in tables and charts.

The significant contributions of the selected indicators to achieve the four set goals were analysed, using the Pareto Statistical Analysis Graph. The Pareto Analysis Graph is a statistical technique in decision-making used for selecting a limited number of tasks that produce significant results. In interpreting the results, scores of 80% and above are regarded as significant contributions while scores less than the pass mark are not significant.

5 Research Findings and Discussions

5.1 Analysis of Slum Housing in Meeting the Set Health and Well-Being Goals

5.1.1 Meeting Physiological Needs

The roles of housing in meeting the physiological needs of the residents are numerous, including (i) protection from weather elements, rain, sun, snow, etc. (ii) providing a conducive thermal environment and an atmosphere free from pollution (iii) ensuring direct day lighting into homes but preventing sun glare (iv) offering protection from excessive noise and (v) providing adequate space for exercise for adults and children (See Tables 1 and 2). For reasons of data availability only two of these roles were investigated. The two indicators chosen to analyse the roles of housing in meeting physiological needs are (i) access to open spaces and (ii) protection from noise. From Table 3, nine out of 20 houses (45.5%) in the six slums provided facilities for meeting the physiological needs of residents. Individually, the results varied from 26.2% recorded in Okpoko to 65.2% for Abesan. In satisfying this goal, the performances of Abesan and Dorayi are highest when compared with the scores of other slums. The three planned estates comprising Abesan, Dorayi and Makama performed better than the three other slums, which are unplanned.

5.1.2 Meeting Psychological Needs

From Table 4, only 36.5% of the total houses in the six slums provided for the psychological needs of residents. The high figure of 60.1%, which was recorded in Abesan, crashed to 16% in Elechi and 18.9% in Okpoko. On a comparative scale, this goal is better satisfied in Abesan, Dorayi and Makama when compared with the scores for the other three slums. The low scores recorded for privacy and an aesthetically pleasing environment only show that slum residents live in overcrowded houses surrounded by poor and unsightly living environments. These results show that slum dwellers are exposed to many health risks associated with overcrowding and an insanitary environment.

Table 3 Percent households providing for meeting physiological needs of residents

S/N	Indicator	Percent total of households by slum						
		Okpoko	Abesan	Dorayi	Elechi	Makama	Masaka	All slums
1	Protection against excessive noise	12.1	**71.4**	**55.2**	32.5	34.5	**72.6**	46.4
2	Access to organized open spaces	40.3	**58.9**	**53.3**	**68.5**	42.9	4.0	44.7
	Column average	26.2	**65.2**	**54.3**	**50.5**	38.7	38.3	45.5

Source FDI (2014). National report on slum characterisation study. NB. Scores above 50% are in bold text

Table 4 Percent households providing for meeting psychological needs

S/N	Indicator	Percent total of households by slums						
		Okpoko	Abesan	Dorayi	Elechi	Makama	Masaka	All slums
1	Adequate privacy	11.8	**55.7**	**52.6**	42.4	4.9	25.5	14.7
2	Households with bath	22.5	**98.0**	**79.9**	2.0	**74.6**	**85.5**	**60.4**
3	Households with kitchen	24.0	**99.0**	**92.5**	11.0	**93.3**	43.5	**60.6**
4	Good and pleasing environment	45.5	**61.7**	**78.0**	24.5	**61.0**	52.3	**53.8**
5	Improved housing environment	9.5	45.9	41.0	16.1	**59.4**	6.6	29.7
	Column average	18.9	**60.1**	**57.3**	16.0	**48.9**	35.6	36.5

Source FDI (2014). National report on slum characterisation study. NB. Scores above 50% are in bold text

5.1.3   Protection from Diseases

As shown on Table 5, more than half (56.4%) of the total households sampled provided facilities for protecting residents from diseases, with the highest score of 72.4% being recorded in Masaka, while Elechi (32.7%) and Okpoko (43%) recorded the least scores. The result also shows the prevalence of diseases in the slums. On average, almost a third of the total households (29.4%) in the six slums had not suffered from cholera, while as low as 6.2% of the total households had not suffered from malaria fever in the previous year. Less than half of the total households (48.8%) in the six slums have access to adequate health faclities. Households that have access to safe waste disposal methods and do not experience environmental disasters are (33.3%) and 44.1% respectively. These results are indicative of insanitary living, high prevalence of diseases, lack of health facilities and the disease burden in the slums.

5.1.4   Protection of Residents from Injury

On the whole, 54% of all houses in the six slums provided for the protection of residents from injury. The high-scoring estates are Abesan (70.1%), Okpoko (59.6%) and Makama (55.6%). The figure for the remaining slums varied from 34.3% for Elechi, which has the least, to 48.7% for Masaka. On a comparative scale, this goal is better satisfied in Abesan, Okpoko, and Makama than in the remaining three slums. Generally, there is a high premium for the use of durable materials for wall, roofs and floors in the six slums, which averaged 78.5%, 84.7% and 86.1% respectively. In addition, the result shows low provision for access to good roads, streetlights and security posts, which averaged 14.8%, 22.8% and 36.9% respectively. This result shows the low premium placed on providing essential infrastructure that might contribute to protection of residents from injury and by government (Table 6).

5.2   *Overall Scorecard of Slum Housing in Meeting the Set Four Goals*

The results of the significant contributions of the selected indicators in meeting the set goals in each slum and for all the six slums are summarised in Table 7 and shown on the radar chart (Fig. 3). With regard to the satisfaction of the four set goals, less than half (48.1%) of the houses in the six slums significantly provided the required facilities for meeting the four goals. Across the six slums, the score is highest in Abesan, which had 65.8%, and lowest in Elechi, which had 33.3%. The provisions for all the four set goals are better met in Abesan, Dorayi and Masaka when compared with the other slums (see Fig. 3; Table 7).

Table 5 Total households providing for protection from diseases

S/N	Indicator	Percent total households by slum						
		Okpoko	Abesan	Dorayi	Elechi	Makama	Masaka	All slums
1	Access to safe drinking water	**91.5**	**64.5**	**75.4**	41.4	**54.9**	**89.5**	**69.5**
2	Access to safe waste disposal methods	25.1	**67.8**	**50.9**	20.5	22.1	13.7	33.3
3	Adequate sleeping space	**59.7**	41.1	46.7	31.5	**57.1**	**96.0**	**55.1**
4	Households with toilet	24.0	**99.5**	**92.5**	2.5	**95.8**	**84.5**	**66.5**
5	Households not experiencing environmental disasters	8.0	32.8	46.2	35.2	47.2	**96.0**	44.1
6	Households not suffering from environment-related diseases	**86.5**	40.1	**62.3**	**72.6**	**73.9**	31.8	**61.2**
7	Households not suffering from cholera	34.0	**98.5**	**66.7**	34.7	44.6	**97.8**	**62.7**
8	Households not suffering from typhoid	3.5	**75.0**	1.2	13.0	20.6	**53.0**	29.4
9	Households not suffering from malaria	3.0	23.0	1.5	1.0	7.3	1.5	6.2
10	Households not suffering from TB	**88.5**	**99.0**	**95.2**	**61.0**	**69.3**	**100.0**	**85.5**
11	Households with no cases of HIV/AIDS	**62.9**	**100.0**	**99.4**	19.4	**88.0**	**99.5**	**90.2**
12	Households not experiencing death of a child in the previous year	49.2	**97.9**	**80.4**	**88.2**	**75.8**	**96.5**	**81.3**
13	Access to adequate health facilities	22.8	40.2	**85.6**	4.0	**58.8**	**81.8**	48.8
	Column average	3.0	**67.6**	**61.8**	32.7	**55.0**	**72.4**	**56.4**

Source FDI (2014). National report on slum characterisation study. NB. Scores above 50% are in bold text

Table 6 Percent total households providing for protection from injury

S/N	Indicator	Percent total of households by slums						
		Okpoko	Abesan	Dorayi	Elechi	Makama	Masaka	All slums
1	Durable wall materials	**97.0**	**99.0**	**65.1**	**57.8**	**54.3**	**97.5**	**78.5**
2	Durable floor materials	**96.0**	**89.9**	**67.0**	**90.4**	**72.1**	**92.5**	**84.7**
3	Durable roof materials	**88.9**	**62.4**	**74.6**	32.8	**83.9**	**100.0**	**86.1**
4	Access to good roads	14.1	6.1	22.8	3.0	42.3	0.5	14.8
5	Street lights	4.0	**75.4**	18.1	0.5	38.1	0.5	22.8
6	Access to security	**57.5**	**87.6**	11.6	21.0	42.8	1.0	36.9
	Column average	**59.6**	**70.1**	43.2	34.3	**55.6**	48.7	**54.0**

Source FDI (2014). National report on slum characterisation study. NB. Scores above 50% are in bold text

Table 7 Percent houses that provided for physiological, psychological needs and protection from diseases

Health and well-being goals/need of residents	Overall weighted scores (%)						
	Okpoko	Abesan	Dorayi	Elechi	Makama	Masaka	Six slums
Meeting physiological needs	26.2	**65.2**	**54.3**	**50.5**	38.7	38.3	45.5
Meeting psychological needs	18.9	**60.1**	**57.3**	16.0	48.9	35.6	36.5
Protection from diseases	43.0	**67.6**	**61.8**	32.7	**55.0**	**72.4**	**56.4**
Protection from injury	**59.6**	**70.1**	43.2	34.3	**55.6**	48.7	**54.0**
Meeting the four goals	36.9	**65.8**	**54.2**	33.4	49.6	48.8	48.1

Source Analysis done by author in 2019. NB: Score above 50% is in bold

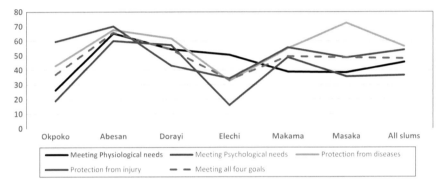

Fig. 3 Overall performances of the six slums in meeting the four goals

5.3 *Analysis of Significance of the Contributions of Selected Housing Indicators to Meeting the Set Goals*

One of the goals of this study is to assess the contributions of the selected 26 indicators to meeting the health and well-being needs of residents. This helps in identifying which indicators are well provided for and those which are not. Therefore, analysis of the significant contributions of the housing indicators was undertaken using the Pareto statistical analytical technique and adopting an 80% and above level of contribution as significant. The result of the analysis is summarised in Table 8. Figures 4 and 5 show the rankings of indicators that significantly contributed to meeting the set goals for the six slums and for the individual slums respectively.

The results of the analysis of the selected 26 indicators for the six slum as a whole show 12 indicators that contributed at an 80% and above level of significance.

The 12 indicators that scored 80% and above are: (i) household with toilet (ii) protection from noise (iii) access to open space (iv) no case of HIV/AIDS (v) household with kitchen (vi) durable floor material (vii) durable roof material (viii) no

Table 8 Summary of the levels of significant contributions of the 26 selected indicators by six slums and total

S/N	Indicators	Percent level of significant contribution	Ranking
1	Households with toilet	95	1
2	Protection against excessive noise	90	2
3	Access to organized open spaces	90	2
4	Households not reporting cases of HIV/AIDS	90	2
5	Households with kitchen	85	3
6	Durable floor materials	85	3
7	Durable roof materials	85	3
8	Households not suffering from tuberculosis	85	3
9	Adequate sleeping space	80	4
10	Households with bath	80	4
11	Durable wall materials	80	4
12	Households not experiencing death of a child in the previous year	80	4
13	Access to safe drinking water	70	5
14	Households not suffering from environment-related diseases	60	6
15	Good and pleasing environment	54	7
16	Access to adequate health facilities	50	8
17	Households not experiencing environmental disasters	44	9
18	Improved housing environment	40	10
19	Street lights	35	11
20	Access to safe waste disposal methods	35	11
21	Households not suffering from cholera	30	12
22	Households not suffering from typhoid	30	12
23	Adequate privacy	20	13
24	Access to good roads	20	13
25	Access to security	20	13
26	Households not suffering from malaria	5	14

Source Analysis by author
Note The prevalence of red colour in this table shows that these slums failed to adequately provide for these indicators

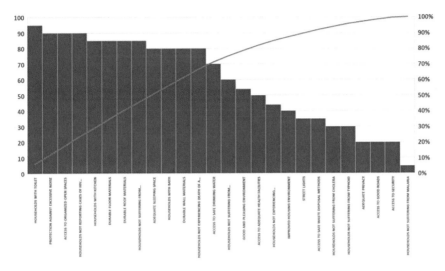

Fig. 4 Ranking of significant contributions of the selected indicators for all slums. NB: While provisons for toilet topped the list, households not protected from malaria was the least

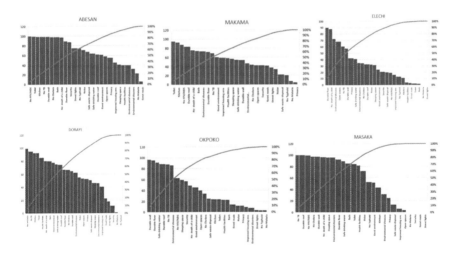

Fig. 5 Ranking of levels of contribution of the selected indicators by the six slums

incidence of TB (ix) adequate sleeping space (x) household with bath (xi) durable wall material (xii) no death of a child (see Fig. 4). It is noteworthy that among these indicators are important building components such as bath, toilet, kitchen, wall, roof and floor; there are also indicators such as "no occurrences of HIV/AIDS and TB". This shows that households valued these facilities/factors and they significantly provided for them to meet their health and well-being needs. The indicators that meet the pass mark of 80% contributions represent critical areas to build upon in devising city-specific upgrading plans and programmes for improving the lives of people living in slums.

Conversely, the 14 indicators that failed to make the pass mark of 80%, which constitute weak areas, are (i) access to safe drinking water (ii) no incidence of environmental disasters (iii) good and pleasing environment (iv) access to adequate health facilities (v) no incidence of environmental disaster-related disease (vi) improved housing environment (vii) street lights (viii) access to safe waste disposal methods (ix) no incidence of cholera (x) no incidence of typhoid (xi) adequate privacy (xii) access to good roads (xiii) access to security post and (xiv) no incidence of malaria. The above list contains critical areas to focus on in developing robust actions for improving the lives of people living in slums.

Among the six slums, the number of indicators that contributed significantly at an 80% and above level across the six slums varied widely, as shown in Fig. 5. The scores vary from a minimum of four indicators in Elechi to a maximum of 9 in Abesan. These results show that some of the slums are more deprived than others. While Abesan stands out as the best, Elechi is simply the worst of them all.

5.4 Health and Well-Being Implications of Selected Housing Indicators

All the indicators selected for study have both positive and negative implications for healthy living. Some of these benefits and otherwise are highlighted below.

(a) *Protection from noise*

With the exception of Okpoko and Elechi slums, residents of the slums are exposed to noise pollution. This means that the slum dwellers are exposed to stress, poor concentration, productivity losses in the workplace, communication difficulties and fatigue from lack of sleep associated with noise pollution, which can cause cardiovascular disease, cognitive impairment, tinnitus, etc. [81, 84].

(b) *Access to Open space*

The study showed that the residents of Makama lacked access to open spaces. Lack of access to open spaces can deprive residents of the opportunity to engage in recreational activities such as walking, playing games, meeting diverse community members and visiting parks to explore nature benefits [17–21, 23, 24]. Access to recreational open spaces has several health benefits, such as improved mental health and cognitive function, reduced cardiovascular morbidity and mortality, reduction in the prevalence of obesity and Type 2 diabetes, improved pregnancy outcomes, reduced mortality and increased life span [9].

(c) *Adequate sleeping space*

Lack of adequate space is reflected in overcrowding as recorded in the slums, which ranges from three (3) to seven (7) persons per room [30]. Such overcrowding can be a major factor in the transmission of diseases such as acute respiratory infections, meningitis, typhus, cholera, scabies, etc. [29, 31, 35, 43, 58].

(d) *Provision for bath, toilet and kitchen facilities*

Generally, provision for bath, toilet and kitchen facilities in houses is high in all the slums but relatively higher in Abesan, Dorayi, and Makama when compared with the three other slums. The health and well-being benefits of providing suitable toilets in houses include preventing germs from getting into the environment, protecting the health of the household population, enhancing privacy, comfort, dignity, cleanliness and respect, as well as saving the time of residents [12, 35, 59].

(e) *Durable materials for roofing, walls and floors*

The study recorded high scores for the use of durable materials for wall, roof and floors in the six slums. It shows the premium placed on using durable materials and on the soundness and longevity of the buildings. It also shows that despite the various deprivations recorded, many of the existing houses can be rehabilitated instead of being demolished. The health and well-being benefits of using durable building materials include ensuring the safety of lives and property, providing shelter from inclement weather conditions, and having a peaceful mind and feeling secure living in these houses. However, it needs emphasising that the use of asbestos for roofing, which is recorded in the study, is harmful to the health of the people and should be banned.

(f) *Access to safe drinking water and safe disposal of wastes*

The study showed gross lack of access to safe drinking water, poor hygiene and using unsafe waste disposal methods. The main sources of water are streams, wells and boreholes (see Plate 2). According to the World Health Organisation, lacking these facilities exposes residents to the risk of contracting several tropical diseases, such as soil-transmitted parasitic helminth infections, schistosomiasis and trachoma; there is also the transmission of enteric pathogens that can cause diarrhoea and environmental enteropathy, which can also lead to chronic problems with absorbing nutrients. In turn, this can lead to stunting, wasting and being underweight. Improving WASH will thus have significant implications for poverty reduction and human development outcomes.

(g) *Non-occurrences of cholera, typhoid, malaria, tuberculosis, HIV/AIDS and the death of a child in the previous year*

The findings of the study, in terms of the high incidences of vector diseases in these slums, are indicative of the dreadful health risks of living in slums. Cases of households suffering from malaria, typhoid, TB and typhoid and the death of a child in the previous year are high. In particular, the relatively high incidence of

Plate 2 Borehole, Okpoko Slum, Onitisha: An example of water supply provided by private commercial enterprise. *Source* FDI (2014)

the death of children in these slums has also corroborated other findings on higher rates of childhood malnutrition, childhood morbidity, and mortality in slums when compared to other well-developed neighbourhoods in cities [34, 50]. Furthermore, the high incidences of malaria, typhoid and cholera recorded in these slums can be linked to poor sanitation and low provision for safe drinking water, waste disposal and overcrowding (Plate 3). It is therefore necessary to upgrade these slums to ensure access to safe water, safe waste disposal, general landscape improvement and public enlightenment on how to live clean and safe.

(h) *Environmental disasters and environment-related diseases*

The study showed that occurrence of environmental disasters and cases of suffering from related diseases are common in the slums. The disasters frequently mentioned by the households include flooding, uncollected solid wastes, soil erosion, as well as air

Plate 3 Okpoko Slum, Onitsha: A view of two three-storey tenement houses, constructed with durable materials but disfigured by refuse. *Source* FDI (2014)

and noise pollution. Many households have attributed their diseases to environmental problems such as unsightly housing environment, overcrowding and poor sanitation and lack of access to safe water [30]. These results show that environment-related diseases are common in the slums.

(i) *Access to adequate health facilities*

Generally, the provision for health facilities in the slums is grossly inadequate. Plates 4, 5 and 6 show some of the low-quality health and educational facilities provided in these slums [30]. The major factors inhibiting access to medicines the *shortage* of resources and the *lack* of skilled personnel [83, 84], thus making the slums

Plate 4 Two views of health facilities in Makama, Bauchi: 'Get Well' and 'Best Clinic' are some of the names of these ramshackle clinics. *Source* FDI (2014)

Plate 5 View of a nursery school in Okpoko Slum, Onitsha: An example of a school run in an unsafe building and environment. *Source* FDI (2014)

Plate 6 Informal trading activities in Makama Slum, Karu (Left: Selling on the alley. *Right*: Front yard converted to grinding spot for pepper, tomatoes, beans, etc. These indicate the level of poverty, noise pollution and encroachment common to many slums) *Source* FDI (2014)

unsafe. As such, people are more likely to lose their lives during bouts of serious illness, since they are mostly exposed to quacks. Consequently, it is imperative to provide adequate health facilities in accessible locations in these slums. Moreover, government needs to provide financial support to private-sector health operators.

6   Implications of the Research Findings

This section addresses the several implications of the findings of the study with respect to urban planning, slum upgrading, as well as implementation of the Agenda 2030 and the NUA.

6.1   Implications for Urban Planning and Slum Upgrading

(a)   *Implications for urban planning*

The six slums studied belong to the two types of slums found in Nigerian cities. Three of these areas—Abesan, Dorayi, and Makama—originally benefitted from planning but have now become slums owing to ineffective development control and lack of sustained estate management (Fig. 6). Moreover, three of the slums—Okpoko, Elechi, and Masaka – grew informally without benefitting from planning (Fig. 7).

   A sharp contrast may be observed on the performance of the two types of slums with respect to the selected indicators of good housing investigated in this study. The results of the analyses showed that Abesan, Dorayi, and Makama, which are the planned estates, performed better than Okpoko, Elechi, and Masaka, which did not benefit from planning. The sharp differences recorded in the performances of the six slums may be attributed partly to the importance attached to urban planning and estate management. This study thus confirms findings from other studies that

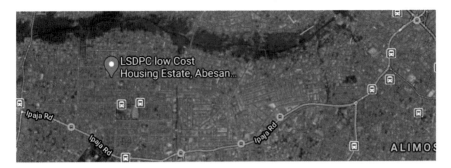

Fig. 6 Aerial view of Abesan, Lagos: An example of a planned estate that degenerated into a Slum *Sources* FDI (2014) and Google Map

Fig. 7 Aerial view of Elechi and other slums, Port Harcourt: Examples of unplanned informal housing developed on prime lands. *Sources* Google Map and FDI (2014)

have identified the absence of planning or the ineffective operation of the planning system, population growth and unplanned rapid urbanisation as contributing to the formation of slums in developing countries such as Nigeria [26, 54, 67, 70, 72, 74, 75]. According to FMLHUD [26], "while urban planning is an important tool for managing urban development, unfortunately most Nigerian cities continue to grow without adopting any form of physical plans to guide them, thus making them non-functional, disorderly, non-aesthetic and unappealing."

The Nigerian government is committed to promoting adequate shelter that is well-planned and managed in the adopted National Housing and National Urban Development policies. While the goal of the National Housing Policy (NHP) is to promote "adequate shelter for all Nigerians", that of the National Urban Development Policy (NUDP) is to ensure "total elimination of slums in Nigerian cities" [26, 27]. These provisions underscore the need to adopt and implement land use plans towards preventing the proliferation of slums and promoting orderly development and growth of towns and cities.

(b) *Implications for Slum Upgrading*

From the findings of the different strands of analyses already presented, the six slums failed to provide adequately for all selected indicators of good housing, with the types of deprivations varying with the different slums. Generally, each of the slums suffers a number of deprivations but the deprivations that are common to all the six slums include the following:

(i)   provision of safe drinking water and safe waste disposal
(ii)   prevention of environmental disasters and related diseases
(iii)   improvement of housing, the surrounding environment and open spaces
(iv)   provision of adequate health facilities
(v)   provision of access to safe waste disposal methods
(vi)   provision of adequate privacy in housing
(vii)   provision of good roads, drainages and street lights
(viii) provision of security post, and
(ix)   prevention of incidence of malaria, typhoid, cholera and TB.

These eye-opening findings have implications for upgrading these slums in order to improve the lives of residents. Special efforts must be made to provide for well-maintained open spaces and protection of residents from air, water and soil pollution, as well as to provide good, functional and aesthetic settings for the various housing units to enhance their beauty and function while boosting the well-being of residents of all ages (Falade, 1985) [20, 21]. The plan must provide for facilities for meeting the various health and well-being needs of the people.

The study found that many buildings in the six slums studied were built with durable materials for wall, roof and floor. This is a strong case to promote slum upgrading and shun total clearance, which characterised the past approach [71]. This recommendation is in furtherance of achieving the goals of the adopted National Urban Development Policy (NUDP) is aimed at promoting the adoption of slum upgrading plans in a participatory manner.

It is recommended that the slum upgrading programme for these slums embrace the following: (i) active involvement and participation of slum dwellers (ii) consideration of the socio-economic needs of slum dwellers and the local community (iii) strategic planning covering financial, institutional and regulatory decisions (iv) fund mobilisation for implementation of the programme and (v) political commitment at the highest level. The community should be empowered to form associations to whom the management of the estate can be transferred after upgrading activities have been completed.

6.2 Implications for Achieving the NUA and Agenda 2030

(a) Implications for implementing NUA

The findings of the study have implications for how Nigeria can realise her global commitment to implement the New Urban Agenda, which was adopted in Quito, Ecuador in October 2016. The goal of the NUA is to guide nations to achieve sustainable urban development by 2036. To achieve this goal the NUA has committed the country to ensuring the planning and management of urban spatial development and integrated urban spatial development strategies, which cover (i) urban and territorial planning (ii) housing and slum upgrading (iii) land and secure tenure (iv) transport and mobility (v) urban basic services and (vi) heritage and culture [76, 77].

It is interesting to note that Nigeria adopted a revised NUDP in 2012, which covers all the thematic areas. In 2019, Nigeria commenced a review of the extant policy. It must be emphasised that Nigeria has not lacked policies and legislations. However, the country has been slack in implementing these plans, policies and legislations. The country needs to be seen to be committed to implementing the adopted plans and policies. Nigeria will need to adopt a blueprint and a national programme for promoting urban planning and slum upgrading in the country's various towns and cities. More importantly, for successful realisation of the NUA, the country needs to implement the following:

- Revising institutional frameworks for the implementation of NUA in the country
- Creating spaces for participation, engagement and coordination at all tiers of government
- Establishing mechanisms for ensuring effective contribution from all stakeholders
- Putting in place a monitoring and evaluation framework to effectively track implementation of NUA in the country.

(b) Implications for Implementing Agenda 2030

(i) *Implications for Agenda 2030*

Nigeria is a signatory to Agenda 2030, which has committed the country to promoting sustainable development in all sectors of the economy. Many findings of this study have implications for successfully implementing the country's commitments to Agenda 2030.

(ii) *Implications for SDG 11 Target 1 on adequate shelter for all*

Agenda 2030, on SDG 11 on sustainable cities, which is in line with the national objectives for housing and urban development, has further committed the country for their realisation. SDG 11 of Agenda 2030 on making cities and human settlements inclusive, safe, resilient and sustainable by ensuring access to safe and affordable housing for slum dwellers/people living in informal settlements has also committed the country to implementing the NUA. Particularly, SDG 11 Target 1 has committed the country to total eradication of slums by 2030 as follows:

By 2030, ensure access for all to adequate, safe and affordable housing and basic services and upgrade.

Thus, the country must upstage its plan to promote aggressive urban planning and slum upgrading and find good results to report on by 2030.

(iii) *Implications for SDG 3 on Good health and well-being for people and SDG 6 on clean water and sanitation*

The findings on the role of housing in protecting residents from diseases are of great relevance for achieving many of the SDGs of Agenda 2030. Many of the findings of the study on the lack of access to adequate health facilities, safe drinking water and safe waste disposal methods, as well as the high figures recorded for malaria, typhoid, cholera, TB, HIV/AIDS and infant mortality in these slums, have implications for achieving SDG 3 and SDG 6. Specifically, SDG 3 has committed the country to promoting good health and well-being for people, especially to focus on achieving universal health coverage, including access to essential medicines and vaccines, preventing death of new-borns and children under 5, reducing maternal mortality and ending epidemics such as AIDS, TB, malaria, ending water-borne diseases by increasing access to safe drinking water and sanitation, prevention and treatment of substance abuse, deaths and injuries from traffic accidents and hazardous chemicals and prevention of air, water and soil pollution.

Similarly, **SDG 6** on clean water and sanitation has further committed the country to ensuring the availability and sustainable management of water and sanitation for all, as well as to addressing the problem of lack of access to safe water and safe waste disposal [63].

(iv) *Implications for SDG 16 on Peace and Security*

From the findings of the study, high figures have been recorded for heinous crimes such as rape, robbery, mob action and drug abuse [30]. These crimes tend to make the slums unsafe not only for residents but also the entire city population. There is the need to formulate and implement plans for promoting security and peace in these slums, which will positively impact realisation of the country's commitment to achieving SDG **16** on peace and security. It is also imperative to note that local prograslums studied did not significantly provide for all themmes for promoting peace in the slums should be supported under this SDG and others.

(v) *Implications for achieving other SDGs on Poverty, Gender, Energy, Employment, Popular participation and Partnership*

Apart from the areas already discussed, the six slums faced many socio-economic issues that other SDGs are focused on. The slum characterisation studies show that majority of the households living in these slums are poor and unemployed. They also have low educational attainment. There is equally a prevalence of gender discrimination against women, who mostly lack access to land and sustainable energy for cooking [30]. Beyond addressing the housing and well-being needs of slum dwellers, it is necessary to promote poverty reduction, gender empowerment, employment generation, access to sustainable energy for cooking, access to universal education, partnership building and popular participation. Implementing these programmes will substantially contribute to achieving the several goals and targets of Agenda 2030, which include (i) SDG 1 on poverty (ii) SDG 5 on Gender Equality (iii) SDG 7 on Energy (iv) SDG 8 on Decent Work and Economic Growth (v) SDG 10 on Reduced Inequalities and (vi) SDG 17 on Partnerships.

From the foregoing, the slum areas in Nigerian cities are obvious candidates for implementing any programme under Agenda 2030 and NUA. It is thus recommended that implementation of Agenda 2030 in Nigeria should target the many of developmental challenges facing slum dwellers.

7 Conclusions and Recommendations

Four major conclusions can be reached from the various results already presented. First, the framework adopted for this study is found to be helpful in ensuring that good housing indicators are provided for in slum houses in an objective manner. It helps to examine the performances of the six slums with respect to the 26 indicators. Second, the results of the analyses showed that housing in slums suffer multiple deprivations, as more than half (51.9%) of the total households in the six slums studied did not significantly provide for all the good housing indicators investigated. The implication of this is that it is either difficult or highly unlikely for slum housing to satisfy the various health and well-being needs of slum dwellers. Third, the study identified 12 housing indicators that significantly contributed to achieving the four set goals and the 14 indicators that did not. Fourth, across the six slums, the results indicate that Abesan, Dorayi, and Makama performed better than Elechi, Masaka, and Okpoko, which, as discussed in the paper, can be attributed to the degree of importance attached to land use planning and development control. The former three slums, which benefitted from planning, performed better than the latter three, which did not.

7.1 Recommendations

Based on the research findings the following actions are recommended for improving the lives of slum dwellers:

i.   Promote urban planning and slum upgrading as a two-pronged strategy to tackle and prevent slums in cities, as provided for in NHP and NUDP, towards achieving Nigeria's commitment to Agenda 2030 and the NUA towards achieving adequate shelter and sustainable human settlements development and management.

ii.   Adopt slum upgrading as the first choice in preference to total clearance in order to conserve and improve on existing buildings as well as to promote social cohesion and reduce cost of implementing plan.

iii.   Develop and implement a wide range of cross-cutting projects/programmes in the slums that will address many of the several development challenges such as poverty reduction, gender empowerment, employment generation, access to clean water, prevention of diseases such as TB, cholera, malaria and typhoid, as well as prevention of infant mortality, which can be supported under Agenda 2030 and the NUA towards improving the lives of slum dwellers.

iv.   Provide ample open spaces to address dire shortage in the slums and promote general landscape improvement of open spaces and the housing environment, including tree planting, drainage construction, paved walkways and tarred roads, etc., and provision of play equipment for children.

v.   Promote access to land and ensure security of tenure for slums dwellers.

vi.   Review existing building regulations and standards with respect to building materials, size of rooms, room density and sustainable estate maintenance standards that can be enforced and that focuses on enhancing the health and well-being of the people.

vii.   Establish Land Owners/Tenant Association in housing estates that will be charged with monitoring and evaluating sustained maintenance of the development in order to prevent these estates from degenerating into slums.

viii.   Adopt land use planning and urban design approaches that are focused on ensuring adequate provision for privacy, aesthetics and communal meetings halls in the slums.

ix.   Provide financial support to private-sector operators providing services such as schools, health facilities and water supply, refuse disposal and the need to regulate their services.

x.   Build the capacity of planning authorities to adopt robust, inclusive and dynamic land use plans to guide growth of cities.

xi.   Implement public awareness programmes to educate slum dwellers on the benefits of urban planning while addressing any resistance to change.

xii.   Build and strengthen the capacities of public agencies and community-based organisations for slum upgrading and urban planning.

xiii.   Provide loans to homeowners to upgrade their houses and facilitate access to housing finance for house owners for their structural upgrading through partnership with financial institutions.

xiv.   Establish proper development control and adequate management of the city's urban environment that can lead to a sustainable urban development process.

xv.   Promote partnership with the private sector for the provision of neighbourhood services such as schools, hospitals, markets and banks.

7.2 Conclusions

In concluding this study, it is necessary to comment on the value of the framework adopted for this study, which finding has clearly identified the nexus between housing, health and the well-being of residents. The results of the various strands of the analyses have helped to identify housing indicators that are significantly met and those not met. This method of assessing the housing provision in slums based on some good indicators of housing can be combined with the use of geographical information systems (GIS) to show the spatial distribution of the houses surveyed upon which slum upgrading plans can be based.

The recommendations made in this study are invaluable for promoting urban planning and slum upgrading in the country. Considering the pervasiveness of the proliferation of slums in Nigerian cities, the current efforts need to be doubled up and fast-tracked. This is especially because future projections for slum populations will continue to increase if the present situation persists. By 2050, Nigeria's urban population would have reached the 60% mark, by which time more slums would have been created [46, 51, 55, 65, 66].

Apart from the adopted national urban and housing policies that have committed the country to urban planning and slum upgrading, the commitment of Nigeria to Agenda 2030 and NUA provides further impetus to address the twin issues of urban planning and slum upgrading to improve the lives of slum residents. To this end, the tiers of government in Nigeria must rise to the occasion to upgrade existing slums, promote urban planning, implement grassroots programmes and project for improving the lives of slum dwellers and eradicating poverty in our cities. It must be stressed that achieving the various SDGs of Agenda 2030 and NUA for slum eradication in Nigeria is still within reach. The country needs to redouble its efforts to achieve all the sustainable goals of Agenda 2030. There must also be serious commitment to the principles and transformative agenda of the NUA via the implementation of slum-upgrading plans focused on housing improvement, poverty reduction, employment generation and provision of health facilities.

Acknowledgements The author is the Executive Director of FDI. He was the Principal Researcher/Coordinator of the slum characterisation studies. The contributions of City Researchers in the study are duly acknowledged: Mr. Bunmi Ajayi (Abesan, Lagos), Professor Yunusa Bello (Dorayi, Kano, and Makama, Bauchi), Mr. Barnabas Atiyaye (Masaka, Karu), Dr Azuka Onweluzo (Okpoko, Onitsha), Dr Stanley Okafor (Elechi, Port Harcourt). Also acknowledged is the support received from FDI staff Mr. Gabriel Adebajo, Mr. Bernard Idowu Oluwaseun and Mr. Lukman Rabe in the preparation of this paper.

References

1. Agbola T, Agunbiade EM (2009) Urbanization, slum development and security of tenure: the challenges of meeting millennium development goal 7 in Metropolitan Lagos, Nigeria. Urban

population-environment dynamics in the developing world: case studies and lessons learned. Committee for International Cooperation in National Research in Demography (CICRED), Paris. http://www.populationenvironmentresearch.org/workshops.jsp#W2007; http://www.cie sin.columbia.edu/repository/pern/papers/urban_pde_agbola_agunbiade.pdf

2. Akinwale O, Adeneye A, Musa A, Oyedeji K, Sulyman MA, Lekan JO, Adejoh P, Adeneye A (2014) Living conditions and public health status in three urban slums of Lagos, Nigeria. South East Asia J Pub Health 3(10):3329. https://doi.org/10.3329/seajph.v3i1.17709

3. Aliyu AA, Amadu L (2017) Urbanization, cities, and health: the challenges to Nigeria—A Review. Ann Afr Med 16:149–158

4. Ayoade AO (2019) Analysis of slum formation in core area of Ilesa, Nigeria. https://www. semanticscholar.org/paper/Analysis-of-Slum-Formation-in-Core-Area-of-Ilesa%2C-Yoade/ 5953acc203c4cdc428560307378a2f5cb1ac7ba1

5. Ayuba M (2019) Slums proliferation in Nigeria: exploring the spatial manifestations, formations and implications. https://www.researchgate.net/publication/334443295

6. Bobadoye S, Fakere Y (2013) Slum prevalence in Nigeria: what role for architects? J Environ 3:45–51 https://www.researchgate.net/publication/339080826_Slum_Prevalence_in_ Nigeria_What_Role_for_Architects

7. Bonnefoy X (2007) Inadequate housing and health: an overview. Int J Environ Pollut 30(3/4):411–429

8. Brakarz J (2013) Evaluation of slum upgrading programs literature review and methodological approaches. InterAmerican Development Bank, Washington, DC. http://idbdocs.iadb.org/wis doms/getdocument.aspx?docnum=38339263

9. Braubach M, Egorov A, Mudu P, Wolf T, Thompson CW, Martuzzi M (2017) Effects of urban green space on environmental health, equity and resilience. In Kabisch N et al (eds.) Nature-based solutions to climate change adaptation in urban areas, theory and practice of urban sustainability transitions. https://doi.org/10.1007/978-3-319-56091-5_11

10. Burridge R, Ormandy D (eds) (1993) Unhealthy housing: research, remedies and reform. E&FN Spon, London

11. Centers for Disease Control and Prevention (CDC) and US Department of Housing and Urban Development (2006) Healthy housing reference manual. US Department of Health and Human Services, Atlanta

12. Centre for Civic Education (2020) Lesson 6: what are the possible consequences of privacy? https://www.civiced.org/resources/curriculum/lesson-plans/1782. Visited in 21 Jan 2020

13. Cohen DA, Mason K, Bedimo A, Scribner R, Basolo V, Farley TA (2003) Neighborhood physical conditions and health. Am J Public Health 93(3):467–471

14. Daniel M, Wapwera S, Akande E, Choji C, Aliyu A (2015) Slum housing conditions and eradication practices in some selected Nigerian cities. J Sustain Dev 8:230–241

15. Dunn JR, Hayes MV (2000) Social inequality, population health, and housing: a study of two Vancouver neighborhoods. Soc Sci Med 51:563–587

16. Ellaway A, Macintyre S, Kearns A (2001) Perceptions of place and health in socially contrasting neighbourhoods. Urban Stud 38(12):2299–2316. Establishment Report 289, BRE, Watford

17. Falade JB (1981) Glasgow parks system: a case study. M. Phil. Thesis, University of Edinburgh, Department of Architecture, Edinburgh, Scotland

18. Falade JB (1985a) Nigeria's urban open spaces: an inquiry into their evolution, planning and landscape qualities. Ph.D., Thesis, Department of Architecture, Edinburgh University, Edinburgh, Scotland

19. Falade JB (1985) Nigeria's urban open spaces: an assessment of the current provision on outdoor recreational pursuits. Edinburgh Arch Res (U.K.) 12:89–103

20. Falade JB (1988) Amenity and Open Space Planning in Nigeria. Land Use Policy (U.K.) 6(2):162–171

21. Falade JB (1989) Open space planning in Nigeria: a comparative study of Ile-Ife and Jos. Ife Res Publ Geogr 3:75–93

22. Falade JB (2007) Housing in National Planning Commission (2009). National Economic Development Strategies (NEEDS) II, Abuja, Nigeria

23. Falade JB (2011) Tourism and recreation. In Ogunleye M, Alo (eds) State of the environment report—Lagos 2010. Tomprints Production in Lagos State, Lagos, pp. 188–221
24. Falade JB, Bakare AA (1990) An assessment of the recreational potential of Lekki Beach, Lagos State, Nigeria. In: Studies in Environmental Design in West Africa, vol 11, pp 22
25. Fiedler K (1997) Alles uber gesundes Wohnen Wohnmedizin in Alltig Munchen, Beck
26. FMLHUD (2012a) National urban development policy, Abuja
27. FMLHUD (2012b) National housing policy, Abuja
28. Foreign Commonwealth Office and UN-Habitat (2018) Prosperity fund global future cities programme for Lagos—terms of reference for development of guidelines for urban renewal programmes in Lagos, Nigeria December 2018
29. Foster HD (1992) Health, disease and environment. Florida, London
30. Foundation for Development and Environmental Initiatives (FDI) (2014) National report on slum characterisation study
31. Fuller-Thomson E, Hulchanski JD, Hwang S (2000) The housing/health relationship: what do we know? Rev Environ Health 15(1–2):109–133
32. Gilbert A (2007) The return of the slum: does language matter? Int J Urban Regional Res 31(4). http://abahlali.org/files/Return%20of%20the%20Slum_0.pdf
33. Goldstein G, Novick R, Schaefer M (1990) Housing, health and well-being: An international perspective. J Sociol Soc Welfare 17 (1) Article No 10
34. Gomez C (2012) Morar Carioca: reducing infrastructure gaps in the favelas. http://metx01ret rofit.tumblr.com/post/68990623847/ morar-carioca-reducing-infrastructure-gaps-in-the
35. Gomez-Jacinto L, Hombrados-Mendieta I (2002) Multiple effects of community and household crowding. J Environ Psychol 22:233–246
36. Hayward E, Ibe C, Gudzune KA (2015) Linking social and built environmental factors to the health of public housing residents: a focus group study. BMC Public Health 2015(15):351–368
37. Hood E (2005) Dwelling disparities: how poor housing leads to poor health. Environ Health Perspect 113(6):A310–A317
38. Huppert F, So T (2013) Flourishing across Europe: Application of a new conceptional framework for defining well-being. Soc Ind Res, pp 837–861
39. Ineichen B (1993) Homes and health: how housing and health interact. E&FN Spon, London
40. Jackson RJ (2003) The impact of the built environment on health: An emerging field. Am J Public Health 93(9):1382–1384
41. Kabiru CW (2012) "Making It": understanding adolescent resilience in two informal settlements (Slums) in Nairobi, Kenya. Child Youth Serv 33(1):12–32. https://doi.org/10.1080/014 5935x.2012.665321. pmc 3874576. pmid 24382935
42. Lagos State Urban Renewal Agency (2013) Handbook of slums in Lagos. Ikeja, Lagos Nigeria
43. Lawrence RJ (2000) Urban health: a new agenda? Rev Environ Health 15(1–2):1–12
44. Lowry S (1991) Housing. BMJ 303:838–840
45. Lukeman Y, Bako AI, Omole K, Nwokoro I, Akinbogun SO (2014) Socio-economic attributes of residents of slum and shanty areas of Lagos state, Nigeria. https://www.researchgate. net/publication/293245411_Socio-economic_attributes_of_residents_of_slum_and_shanty_ areas_of_Lagos_state_Nigeria
46. Mitlin D, Satterthwaite D (2013) Urban poverty in the Global South. Routledge, London
47. Monteiro LA (1985) Florence nightingale on public health nursing. Am J Public Health 75(2):181–185
48. Morakinyo KO, Ogunrayewa MO, Olalekan KB, Adenubi OO (2012) Urban slums as spatial manifestations of urbanization in Sub-Saharan Africa: a case study of Ajegunle slum settlement, Lagos, Nigeria. https://www.semanticscholar.org/paper/Urban-Slums-as-Spatial-Manifesta tions-of-in-Sub-A-MorakinyoOgunrayewa/445b529e121eb8687858dd8084c88c4e31748833
49. NBS (2017) Social statistics report 2017. Abuja
50. Ndukwu CI, Egbuonu I, Ulasi TO, Ebenebe JC (2013) Determinants of undernutrition among primary school children residing in slum areas of a Nigerian city. Niger J Clin Pract 16(2):178–183

51. Nicolai S, Bhatkal T, Hoy C, Aedy T (2015) Projecting progress: the SDGs in Latin America and the Caribbean. Overseas Development Institute, London
52. Obemebe T, Odebunmi K, Olalemi A (2018) Determinants of family size among men in slums of Ibadan, Nigeria. Ann Ibadan Postgraduate Med 16:2–22. https://www.ncbi.nlm.nih.gov/pmc/articles/PMC6143888/
53. Okoye P, Ezeokonkwo J, Mbakwe C (2017) Survey of housing conditions and improvement strategies in Okpoko peri-urban settlement of Anambra State, Nigeria. https://www.researchgate.net/publication/320034440_Survey_of_Housing_Conditions_and_Improvement_Strategies_in_Okpoko_Peri-Urban_Settlement_of_Anambra_State_Nigeria
54. Pat-Mbano E, Nwadiaro ECC (2012) The rise of urban slum in Nigeria: Implications on the urban landscape. Int J Dev Manage Rev. file:///C:/Users/Bade/Downloads/79315-Article%20Text-186083-1-10-20120720%20(1).pdf
55. Potts D (2012) Challenging the myths of urban dynamics in sub-Saharan Africa: the evidence from Nigeria. World Dev 40(7)
56. Ranson R (1991) Healthy housing—a practical guide. E&FN Spon on behalf of the WHO Regional Office for Europe, Chapman & Hall, London
57. Rauh VA, Landrigan PJ, Claudio L (2008) Housing and health. Ann N Y Acad Sci 1136:276–288. https://doi.org/10.1196/nyas.2008.1136.issue
58. Raw GJ, Hamilton RM (1995) Building regulation and health. Building Research
59. Residential Landlord Association (2020) Minimum national amenity standards for licensable houses for multiple occupation (HMOS). https://www.rla.org.uk/landlord/guides/housing_act/docs/all/minimum_national_amenity_standards_for_licensable_hmos.shtml
60. Rojas E (ed) (2010) Building cities, neighbourhood upgrading and urban quality of life. Inter-American Development Bank, Washington, DC
61. Simon RF, Adegoke AK, Adewale B (2013). Slum settlement regeneration in Lagos Mega-city: an overview of a Waterfront Makoko Community. Int J Educ Res 1:1–16. https://www.researchgate.net/publication/312740981_Slum_settlement_regeneration_in_Lagos_Mega-city_an_overview_of_a_Waterfront_Makoko_Community
62. Steemers K (2020) Architecture for well-being and health. http://thedaylightsite.com/architecture-forwell-being-and-health/
63. The World Bank (2017) A wake-up call nigeria water supply, sanitation, and hygiene poverty diagnostic. Washington DC
64. UNCHS (1997) The Istanbul declaration and the habitat agenda with subject index, Nairobi
65. UN-DESA (2013) World economic and social survey: sustainable development challenges. United Nations Department of Economics and Social Affairs, New York. www.un.org/en/development/desa/policy/wess/wess_current/wess2013/WESS2013.pdf
66. UN-DESA (2014) World urbanization prospects. The 2014 revision. Highlights. United Nations Department of Economics and Social Affairs, New York. http://esa.un.org/unpd/wup/highlights/wup2014-highlights.pdf
67. UN-Habitat (2003) The challenges of slums: global report of human settlement, Earthscan, London. UN
68. UN-Habitat (2005) Financing urban shelter: global report on human settlements 2005. Earthscan, Abingdon and New York
69. UN-Habitat (2006) State of the world's cities report 2006/7. Earthscan and United Nations Human Settlements Programme
70. UN-Habitat (2008) A framework for addressing urban challenges in Africa. State of African Cities. UN-Habitat, Nairobi
71. UN-Habitat (2009) Evictions and demolitions in port Harcourt report of fact-finding mission to Port Harcourt city, Federal Republic of Nigeria 12–16 Mar 2009, Nairobi and Abuja
72. UN-Habitat (2010) Cities for all: bridging the urban divide. State of the World's Cities 2010/11. UN-Habitat, Nairobi
73. UN-Habitat (2011) Building urban safety through slum upgrading. UN-Habitat, Nairobi. http://unhabitat.org/books/building-urban-safety-through-slum-upgrading/

74. UN-Habitat (2014) A practical guide to designing, planning, and executing citywide slum upgrading programmes. UN-Habitat, Nairobi
75. UN-Habitat (2014) Global urban indicators database 2014. UN-Habitat, Nairobi
76. UN-Habitat (2015a) A practical guide to designing, planning, and executing citywide slum upgrading programmes. UN Human Settlements Programme, Nairobi. http://unhabitat.org/books/a-practical-guide-to-designing-planning-andexecuting-citywide-slum-upgradingprogrammes
77. UN-Habitat (2015b) Sustainable urban development and agenda 2030: UN-Habitat's Programme Framework. https://www.mypsup.org/library_files/downloads/SDGs%20and%20PSUP.pdf
78. UN-Habitat (2002) Report of the expert group meeting on slums, Nairobi. ISBN 92-1-131322-8
79. Wakely P (2014) Urban public housing strategies in developing countries: whence and whither paradigms, policies, programmes and projects. DPU60 Working Paper Series: Reflections No. 163/6. https://www.bartlett.ucl.ac.uk/dpu/latest/publications/dpu-reflections/WP163.pdf
80. Wakely P, Riley E (2011) The case for incremental housing. Cities Alliance Policy Research and Working Paper Series No. 1, Cities Alliance, Washington, DC
81. WHO (2011) Burden of disease from noise pollution. https://www.science.org.au/curious/earth-environment/health-effects-environmental-noise-pollution
82. WHO (2018) How air pollution is destroying our health. https://www.who.int/airpollution/ambient/health-impacts/en/
83. WHO & UNEP (1987a) Housing—the implications for health. Report of a WHO Consultant, Geneva, June 1987. WHO/EHE/RUD/87.2
84. WHO & UNEP (1987b) Global pollution and health. Global Environment Monitoring System (GEMS). Results of health-related environmental monitoring, Yale Press, sUK
85. Woodham-Smith C (1950) Florence Nightingale (1920–1910). Constable, London, p 460
86. Xavier HF, Magalhães F (2003) Understanding slums: case studies for the global report on human settlements. The case of Rio de Janeiro. www.ucl.ac.uk/dpu-projects/Global_Report/pdfs/Rio.pdf

Adverse Impact of Human Activities on Aquatic Ecosystems: Investigating the Environmental Sustainability Perception of Stakeholders in Lagos and Ogun States, Nigeria

Temitope Olawunmi Sogbanmu, Opeyemi Anne Ogunkoya, Esther Iyanuoluwa Olaniran, Adedoyin Kehinde Lasisi, and Thomas-Benjamin Seiler

Abstract Environmental risk perception of stakeholders for various human activities is germane to the sustainable development of a society. In urban Africa, rising population rates pose environmental challenges for the management of aquatic resources. Examples of two metropolitan cities in urban Africa are Lagos and Ogun states, Nigeria, with their teeming populations. The proximity of the Lagos lagoon and Ogun River to human settlements predisposes their use as sinks for disposal of wastewaters from potentially polluting activities such as sawmills and livestock processing (abattoirs). This chapter evaluates the environmental risk perception of specific stakeholders in the states whose activities result in potential adverse impact on aquatic ecosystems and associated ecosystem services. Copies of a structured questionnaire were administered to stakeholders at selected anthropogenic sites, i.e. Okobaba Sawmills and Kara Cow Market, in Lagos and Ogun states respectively. At the Okobaba sawmills adjoining the Lagos lagoon, respondents stated that sawdust is the major (84%) waste generated, most (90%) of which is burnt. Over half (51–90%) of the respondents noted that air quality, water quality and aquatic animals are adversely impacted by the sawmill activities. At Kara Cow Market, which adjoins Ogun River, respondents reported that they utilise the river for domestic activities, such as source of drinking water and for washing of cows. Most respondents (70–82%) acknowledged that the river is polluted, wastes are deliberately disposed into the river and wastewater from the abattoir is never treated. It is suggested

T. O. Sogbanmu (✉) · O. A. Ogunkoya · E. I. Olaniran
Ecotoxicology and Conservation Unit, Department of Zoology, Faculty of Science, University of Lagos, Lagos, Nigeria
e-mail: tsogbanmu@unilag.edu.ng

A. K. Lasisi
Environmental Management Department, Lagos State Ministry of the Environment and Water Resources, Lagos, Nigeria

T.-B. Seiler
Aachen Biology and Biotechnology, Department of Ecosystem Analysis (ESA), Institute for Environmental Research (Biology V), RWTH Aachen University, Worringerweg 1, 52074 Aachen, Germany

that stakeholder environmental education, advocacy, risk communication, as well as demand for, and implementation of, evidence-based policies for the management of these ecosystems are crucial steps to be taken in addressing the situation. Further, the planning of human settlements near aquatic ecosystems should be considered alongside the sustainability of aquatic resources and ecosystem services in urban Africa, given the need to support life below water (UN SDG 14).

Keywords Aquatic ecosystems · Sawmill wastes · Abattoir wastewater · Stakeholders' environmental risk perception · Sustainable Development Goal 14

1 Introduction

Environmental, social and economic capacity is a key component of sustainable development. While industrialisation plays a critical role in strengthening economic capacity, it also propels urbanisation and improves urban settlements [12, 37]. Since the 1900s, during the upsurge of the Industrial Revolution and mass production to satisfy the demands of the teeming population globally, there has been a decline in the integrity of ecosystem, thus resulting in environmental crisis [33]. This environmental crisis is the consequence of deliberate human activities as well as of unintentional exposure to pollutants from anthropogenic activities [33]. In most African countries, inappropriate management of industrial practices is one of the contributing factors to the pollution of the aquatic environment, leading to an increase in environmental challenges such as the prevalence of waterborne diseases in human settlements [44].

Huge volumes of waste generated by various anthropogenic activities such as industrial, pharmaceutical, agricultural, domestic and municipal activities are being disposed into the immediate environment, including coastal lagoons and rivers [14, 40, 49]. The relative proximity of these aquatic ecosystems to human settlements has often resulted in adverse impacts on ecosystem services, thereby rendering them unbefitting, unsafe and unsuitable for primary and/or secondary usage [24]. The discharge of untreated wastewater into water bodies leads to further severe consequences on ecosystems owing to the inherent pollutants in them [18].

No doubt, the improper management of waste and wastewater from anthropogenic activities such as livestock processing in developing African countries contributes to the environmental pollution of cities and communities [12]. This pollution leads to the availability of surplus nutrients in surface waters, thus causing eutrophication and accumulation of enteric pathogens [25]. The poor state of public infrastructures and the unhygienic conditions of these processing activities therefore require appropriate public health interventions by regulatory agencies [9, 16]. Another anthropogenic activity is sawmilling, which provides processed wood and wood products for housing demands and for other industrial activities [41]. These sawmill activities result in the production of large quantities of wood wastes, which are often burnt or deposited near surface waters such as the Lagos lagoon (Fig. 1). This has potential adverse effects on the ambient air quality of human settlements around the area

Fig. 1 Okobaba area near the Lagos lagoon, Nigeria showing the nature of anthropogenic activity (*photo credit* Temitope O. Sogbanmu, December 2017)

[20]. Relatedly, biomass burning from formal and particularly informal settlements around this area potentiate the risk to human health. Furthermore, the dumping of wastewater and solid wastes from human settlements within and around such anthropogenic activities potentially increases the ecological and human health risks [20]. Examples include the dumping or disposal of human excreta into the Lagos lagoon, as well as the poor sanitary facilities in the slum settlements at the Makoko end of the Lagos lagoon, which are close to the sawmills. The dysfunctional—even sometimes absent—sewerage systems in most settlements, especially the informal settlements or slums, further worsens public health risks [12]. The detection of pharmaceuticals and personal care products in surface water and sediments of the Lagos lagoon [3] and Ogun River [39, 40] could be linked to the existence of human settlements near these aquatic ecosystems, as well as to the anthropogenic activities around them. In order to sustainably manage coastal ecosystem resources, particularly in the face of current global climate change, there is a need to involve stakeholders in such management programmes [51]. Implementing wastewater management programmes and adopting sustainable urban planning systems to improve the environmental condition of urban settlements located around water bodies are more effective solutions to making cities sustainable, safe and affordable [46], based on SDG 11 (sustainable cities and communities). The stakeholders include coastal aboriginal communities who consume an average of 15 times more seafood per capita than non-aboriginal people [11, 22]. However, the level of stakeholder awareness about the environmental and human health risks of their direct or indirect potentially polluting activities on coastal ecosystems is very low, particularly in developing nations [47]. In Nigeria, there is limited information on the level of awareness of Okobaba community stakeholders

on the environmental and human health risks of their activities on the lagoon [27]. Further, the implications of these anthropogenic activities and consequent human and environmental health impacts on the design and management of urban settlements requires consideration for evidence-informed sustainable development interventions.

Consequently, this study aims to investigate stakeholders' risk perception of anthropogenic activities at Okobaba Sawmill and Kara Cow Market in Lagos and Ogun states respectively. The results will provide information on the nexus among the planning of urban settlements (SDG 11—sustainable cities and communities), anthropogenic activities impacting aquatic ecosystems and environmental sustainability (SDG 14—[life below water]) and SDG 15 (life on land), as well as the strategies for regulatory interventions in these areas.

2   Literature Review

Lagos epitomises the future challenges that African cities face but on a scale that amplifies the stresses and strains of urbanisation [12]. With an estimated population of 24 million [21], to which 3000 new migrants are added each day, Lagos represents both the promise of economic opportunity and the burden on the ecosystem services sustaining the urban ecology. Lagos is one of the three megacities in Africa, as confirmed in the United Nations World Urban Prospects Revision Report of 2014 [4]. The Okobaba slum in Lagos is one of the settlements bordering the Lagos lagoon and is located around in Ebute-Metta, Lagos, Nigeria. The lagoon has been the subject of various studies [13, 17, 48, 50] elucidating its state of pollution from anthropogenic releases. Small- to large-scale industries, residential houses and stormwater are diffuse sources of pollutants released into the lagoon. Examples of these diffuse sources include industrial effluent [7], domestic wastes and sewage [32], wood burning and associated atmospheric emissions, solid wastes [36], heated water discharges from thermal plants [38], and run-off from the over-lying Third Mainland Bridge [14], as well as toxic chemicals from boats and jetties.

At the Okobaba waterfront, a dominant and unique activity in the area is sawmilling, which has a direct impact on lagoon water quality as well as on aquatic animals and humans [26]. Typical wastes generated from sawmills include sawdust, woodchip, bark, planer shavings, pole shavings, solvents, paints, and wood coatings. These wastes are disposed directly into the lagoon [8]. Solid wastes such as high- and low-density polythene, empty cans of food/pesticide sprays, glass bottles, used needles and syringes (hospital wastes), used car tyres, worn clothes and a host of others have equally contributed to the high level of pollution found in the lagoon [7]. Wood preservation chemicals, for instance, may include polycyclic aromatic hydrocarbons (PAHs), pentachlorophenol pesticides, and compounds of chromium, copper and arsenic [29]. The effect of this is contaminated water mixing with the underground water that is the source of drinking water in the community, thus threatening the health of residents, especially the children. Air pollution in urban areas such as Okobaba, Lagos is indeed a major issue. Air quality has been continuously degraded because of burning, which is the primary means of sawdust waste management in the area [1, 20]. Air pollutants associated with this burning include CO, NO_x, CO_2, SO_2, and H_2S, as well as volatile organic compounds (VOCs) [30].

A major challenge confronting environmental and public health authorities in developing countries like Nigeria is the continual pollution of surface water bodies by small- to large-scale livestock processing activities [35]. Abattoirs or slaughterhouses play a crucial role in livestock production, as they are established to receive and process livestock into consumable produce that is market-ready [34]. Typically, in Nigeria, most abattoirs are situated by riverbanks, where there is easy accessibility to water [34, 42]. Kara Cow Market, situated in Ojodu Berger, Ogun State, is one of many abattoirs in Nigeria; it is located along the Ogun River. It is a large processing facility that provides livestock produce to residents in Lagos and Ogun states, a fact that makes it not only important to the economy but also a major cause of environmental concern. As with other abattoirs such as the Bodija Abattior in Ibadan, Oyo State, untreated wastewater laden with pollutants and solid wastes from abattoir or slaughterhouse activities are often discharged into surface waters such as Ogun River in Ogun State, Nigeria [2, 10, 44]. These wastewaters have been shown to have deleterious effects, including histological and biochemical alterations, in the tissues of *Clarias gariepinus* [40] and *Poecilia reticulata* [49]. The continuous discharge of abattoir wastewater into Ogun River has the potential to cause eutrophication and to put a burden on the health condition of the river's ecosystem, thus negatively affecting biodiversity in the area. Furthermore, contaminants such as heavy metals could leak into the waterbed that provides water supply to residential communities around the market [2, 40].

3   Methods

3.1   Study Areas

The Okobaba Sawmill is located in Ebute-Metta Local Council Development Area (LCDA) of Lagos State, Nigeria. The sawmill sources its wood from different locations in southwest Nigeria. The wood is transported by rafting through the Lagos Lagoon to the sawmill.. Large quantities of sawdust are generated as waste products of the sawmill activities. The sawdust is consistently burnt (Fig. 1) to reduce the pile [20].

Kara Cow Market in Ogun State, Nigeria is located on the Ojodu/Isheri axis of Ogun River, at coordinates 6°38'48.0" N 3°22'46.5" E. Ogun River is an important river in the southwest of Nigeria that has its source close to Shaki in Oyo State, at coordinates 8°41" N 3°28" E. The river flows through Ogun State and discharges into the Ikorodu axis of the Lagos Lagoon at coordinates 6°38" N 3°22" E [40].

3.2   Questionnaire Administration to Stakeholders at Okobaba Sawmill (Lagos) and Kara Cow Market (Ogun)

A 5-point scale questionnaire was developed and administered. The questionnaire (Appendix 1) was developed to assess stakeholders' knowledge of the potential effects and risks of the sawmilling and cow market activities on their health and the aquatic environment [6, 52]. Aspects of the environment that were air quality, noise, land, water, lagoon ecosystem and waste disposal. The questionnaire was developed based on standards from both the British Medical Council questionnaire on occupational hazards and the Institute for Work and Health vulnerability questionnaire [5]. In the Okobaba community, 200 stakeholders received copies of the questionnaire, among whom were sawmillers, fishermen, residents, timber dealers, and traders.

The study population at Kara Cow Market, Ogun State consisted of herdsmen, butchers, fishermen, administrators, and traders. Fifty copies of the structured, pretested questionnaire (Appendix 2) were administered to respondents to obtain data on relevant socio-demographic variables of the stakeholders, the level of knowledge and awareness of pollution in their environment, as well as their practices and attitude towards their environment [19, 43]. The following formula [23, 53] was used to calculate the sample size:

$$\text{Sample size(n)} = \frac{Z^2 * (P) * (1 - P)}{C^2}$$

n = sample size
Z = Z statistic for a level of confidence
P = expected prevalence or proportion (in proportion of one; if 20%, P = 0.2)
C = precision (in proportion of one; if 5%, C = 0.05).

3.3   Data Analysis

The analysis of the response data from the questionnaire administration was carried out using MS Excel 2010, which data are presented in frequency distribution tables with percentages [15].

4 Results and Discussions

4.1 Sociodemographics and Questionnaire Responses of Stakeholders at Okobaba Sawmills, Lagos, Nigeria

The sociodemographics of the respondents showed a higher number of male respondents (62%) compared to females (38%). The age range with the highest number of respondents was between 36 and 45 years, with average age of respondents being 33 years (Table 1).

Responses to the questionnaire showed 56.5% of the respondents noted that sawdust contributes to wastes generated always, of which 82.5% is burnt and the run-off released into the lagoon. Other wastes include sewage (42%), wood chips (36%) and wood preservatives (26%). Ninety percent (90%) of respondents agree that air quality is the environmental component that is most affected by sawmilling activities. Moreover, 51 and 60% of the respondents noted that water and aquatic animals are always impacted by sawmill activities (Table 1).

Living conditions in urban slums and informal settlements such as Okobaba are particularly dire [37]. Limited access to education, good health care as well as water and sanitation results in endemic and high levels of poverty, inequality and deprivation [12]. In the case of Okobaba, where the predominant occupations of slum dwellers (timber dealing, fishing, saw milling) impact on the nearby lagoon, it is critical to consider the feedback from resident stakeholders on the sustainability of their practices. Responses from the questionnaire showed that sawdust contributes 84% of the waste, of which 90% (collated from % of respondents who ticked rarely to always) of the generated sawdust is burnt. Studies have shown that sawdust burning leads to the release of PAHs, which result in high particulate matter emission [31]. PAHs contained in charred sawdust also run off into the lagoon via winds and rainfall, thus exposing resident aquatic organisms to physiological changes as a result of this exposure [20]. Ninety percent (90%) of respondents perceived that air quality was the most impacted factor as a result of sawdust burning at the site. Consequently, the predominant health conditions often observed among residents were headaches (53%), fatigue (8%) and dermal conditions (7%).

4.2 Sociodemographics and Questionnaires Responses of Stakeholders at Kara Cow Market, Ogun State, Nigeria

The highest percentage of the respondents, i.e. 36%, fell within the age bracket of 30–45 years. Sixty percent (60%) of the respondents were males, 24% of whom were butchers and 18% of whom were herdsmen (Table 2). Sixty-two (62%) of the respondents temporarily reside in the market. The source of drinking water for 80% of the study group is sachet water while only 2% claim to drink water from the

Table 1 Sociodemographics and questionnaire responses of stakeholders at Okobaba Sawmills, Lagos, Nigeria

Variables		Number of respondents	Frequency (%)		
Gender	Male	124	62		
	Female	76	38		
Age range (in years)	15–25	44	22		
	26–35	56	28		
	36–45	64	32		
	46–55	24	12		
	56–65	5	2.5		
	66 and above	7	3.5		
Nature of business	Sawmilling	68	34		
	Fishing	18	9		
	Trader	36	18		
	Timber dealer	53	26.5		
	Civil servant	4	2		
	Others (student)	21	10.5		
Question group/description		% responses			
		Never	Rarely	Sometimes	Often	Always
Types of waste generated	Sewage	20	12	7.5	18.5	42
	Saw dust	16	8	12	7.5	56.5
	Wood chips	44	2.5	5.5	12	36
	Wood preservatives/chemicals	54	13	4	26	3
Sawmills waste disposal methods	Gutters	76	4	0	0	20
	Dump sites	48	0	32	20	0
	Lagos Waste Management Authority	44	8	20	17.5	10.5
	Cart pushers	56	17	12	8	7
	Burning	8	0	9.5	0	82.5
	Direct disposal into the lagoon	48	8.5	19.5	24	0
Predominant health issues for sawmillers	Headache	16	3	28	53	0
	Dizziness	56	36	7.5	0.5	0
	Fatigue	40	28	24	8	0
	Nausea/vomiting	52	40	8	0	0
	Abdominal pain	61	25.5	11	2.5	0
	Coughing with phlegm	75	20	5	0	0

(continued)

Table 1 (continued)

Question group/description		% responses				
		Never	Rarely	Sometimes	Often	Always
	Breathing difficulties	72	25.5	2.5	0	0
	Skin rashes	49	16	28	7	0
	Poor vision	95.5	1.5	0.5	2.5	0

n = 200 (number of administered questionnaires and received responses)

river. Thirty-four percent (34%) use water from the river for domestic purposes, including for washing the slaughter slabs and the slaughtered livestock. Although 56% of the respondents dispose their waste by discarding at dumpsites and utilising the services of contracted waste managers, 52% still prefer to always dump their waste directly into the river. Seventy-eight percent (78%) acknowledged that they deliberately dispose their animal waste into the river. The questionnaire survey also showed that 82% of the people present in the market acknowledged that the river is polluted (Table 2), while 68% acknowledge that the effluent is not treated. The breakdown for the 82% of respondents who acknowledged that the river is polluted is as follows: (38% = *the river is always polluted*, 18% = *it is usually polluted*, 20% = *it is occasionally polluted*, and 6% = *it is rarely polluted*).

The outcome of the study showed the poor sanitation behaviour of stakeholders in the market, as it was observed that, although a large number of the respondents (78%) acknowledged that the market is polluted, most of them deliberately dispose their waste into the river. These results support the conclusion of a study conducted in the Mekong Delta, Vietnam on perceptions of water sanitation and health, wherein it was found that people have a basic knowledge of proper hygiene and sanitation behaviour [28]. It was also observed that people in the market use the surface water for domestic activities such as general washing and cleaning, as well as for washing some of the animals after slaughter and as a source of drinking water for the animals. These practices reflect lack of understanding of the environmental and health consequences of waste management practices in the market or simply the lack of alternative options. Poor waste disposal practices in South Lunzu Township, an urban area in Malawi, have also been seen to have a negative impact on water quality, as copies of the questionnaire administered to residents reflect the adverse effects of the poor practices [45].

5 Conclusion

Findings from this study suggest that, although people are aware of the state of their immediate environment, they may not be much concerned about it because they may not understand the full implications of the environmental and human health risks of their activities. The adverse impact on aquatic resources is further exacerbated by the

Table 2 Sociodemographics and questionnaire responses of stakeholders at Kara Cow Market, Ogun State, Nigeria

Sociodemographics		Responses (%)		
Age (in years)	Under 15	4		
	16–29	30		
	30–45	36		
	>46	30		
Gender	Male	60		
	Female	40		
Occupation	Butcher	24		
	Fisherman	6		
	Trader	24		
	Cattle rearer	18		
	Others	28		
Residence in Kara	Yes	62		
	No	38		
Categories/responses	Never (%)	Rarely (%)	Sometimes (%)	Usually (%)	Always (%)

Categories/responses	Never (%)	Rarely (%)	Sometimes (%)	Usually (%)	Always (%)
Waste management					
Frequency of animal wastes disposal into the river	32	4	8	2	54
Frequency of animal feed remains disposal into the river	42	10	22	4	22
Frequency of human faeces disposal into the river	36	6	16	22	20
Frequency of solid wastes (bones, plastics and nylons) disposal into the river	38	4	26	6	26
Frequency of treatment of wastes generated from the slaughter slabs	54	16	14	6	10
Anthropogenic activities					
Frequency of perception of offensive odour in the market	12	4	16	24	44
Frequency of water hyacinths appearance in the river	20	10	14	16	40

(continued)

Table 2 (continued)

Categories/responses	Never (%)	Rarely (%)	Sometimes (%)	Usually (%)	Always (%)
Frequency of animal disease outbreak in the market	40	26	24	6	4
Frequency of burning of animals at the slaughter slabs	18	8	24	28	22
Frequency of fish catch from the river	22	6	32	22	18
Frequency of administration of pharmaceuticals to animals in the market	10	2	16	20	52
Perception of pollution of the Ogun River	18	6	20	18	38
Waste management	Burning	Dumpsites	Cart pushers	Waste managers	Disposal into the river
Waste disposal methods	18%	56%	14%	56%	52%

N = 50

poor planning and citing of settlements such as 'urban slums' and sawmills at Makoko and Okobaba near the Lagos Lagoon as well as at Kara Cow Market near Ogun River. This is particularly worthy of consideration for the development of intervention and management strategies within the contexts of SDG 11 on ensuring sustainable cities and communities and SDG 14 on supporting the sustainability of life below water (aquatic organisms). The study demonstrates the roles of stakeholders such as sawmillers, market leaders, fishermen, livestock traders, local government officials and community residents in assuring the sustainability of aquatic ecosystems in urban Africa. Very crucial in the effort to reverse the situation are environmental education, advocacy and risk communication, as well as demand for, and implementation of, evidence-based policies for the management of these ecosystems. The achievement of sustainable development requires inclusive stakeholders engagement and effective management of resources.

6 Recommendations

It is therefore recommended to adopt targeted solid waste management approaches, especially by recycling sawmill wastes into economically useful products such as pellets, carbon sequestering biofuels, and wood blocks. Abattoir wastewater should be pre-treated before being discharged into Ogun River while the sludge and solid wastes, especially livestock wastes, can be processed for the generation of biogas and manure. Further, there should be stakeholder education and involvement on the importance of sustainable use of the environment as well as targeted environmental interventions. Such efforts would fall within the framework of the UN SDGs 11 (resilient and sustainable cities and settlements), 14 (life below water) and 15 (life on land).

Acknowledgements The authors wish to acknowledge the Society for Environmental Toxicology and Pollution Mitigation (SETPOM) for research funding given to O. A. Ogunkoya.

Appendices

Appendix 1: Environmental Risk Assessment Questionnaire for Stakeholders at Okobaba Sawmills, Lagos

PUBLIC HEALTH RISK ASSESSMENT OF SAWMILLING ACTIVITY ON OKOBABA RESIDENTS

The purpose of this questionnaire is to assess stakeholders' perception of the effects of sawmilling on the health and environment of Okobaba residents. This questionnaire serves as part of a research conducted by Ogunkoya, O.A and Sogbanmu, T.O of the Department of Zoology, University of Lagos, Akoka, Lagos, Nigeria.

Participation in this questionnaire is voluntary and you may withdraw at any time. Your responses will be treated as highly confidential and results of this study will be used for academic/ research purposes only.

Kindly tick boxes accordingly.

Demographic information
i.   Gender : Male⬭ Female⬭
ii.   Age range   15– 25⬭ 26-35⬭ 36– 45⬭ 46- 55⬭ 56-65⬭66- Above⬭
iii.   What is the nature of your business? Sawmilling⬭   Fishing⬭   Trader⬭
Timber dealer ⬭   Civil servants⬭ Others (please specify) _____

Water sources
1.   How often do you use drink water from the following sources?

	Never	Rarely	Sometimes	Often	Always
Tap water					
Sachet water					
Well water					
Rain water					
Lagoon water					

2.   How often do you use domestic water (used for bathing, washing, cooking, cleaning e.t.c) from the following sources?

	Never	Rarely	Sometimes	Often	Always
Tap water					
Sachet water					
Well water					
Rain water					
Lagoon water					

3.   How often do you treat your drinking water using the following methods?

	Never	Rarely	Sometimes	Often	Always
Filtration					
Boiling					
Coagulation (use of alum)					
Sedimentation(settling of particles at the bottom)					
Disinfection(e.g dettol)					
No treatment					

4.   How often do you treat your domestic water (used for bathing, washing, cooking, cleaning e.t.c)?

	Never	Rarely	Sometimes	Often	Always
Filtration					
Boiling					
Coagulation (use of alum)					
Sedimentation(settling of particles at the bottom)					
Disinfection(e.g dettol)					
No treatment					

5.   Are you satisfied with the quality of water you use (drinking/ domestic)?
        Very dissatisfied ⬜   Satisfied ⬜ Neutral ⬜   Satisfied ⬜ Very satisfied ⬜

Waste management

6.   How often do you generate the following waste?

	Never	Rarely	Sometimes	Often	Always
Solid waste (paper,nylon,cans,plastics)					
Sewage(faecal and kitchen waste)					
Saw dust					
Wood chips					
Wood preservatives/ chemicals					

7.   How often do you dispose **sewage** (faecal and kitchen waste) using the following methods?

	Never	Rarely	Sometimes	Often	Always
Gutters					
Water closet					
Pit latrine					
Dump sites					
LAWMA					
Cart pushers					
Burning					
Directly into the lagoon					

8.   How often do you dispose **solid waste** (paper, nylon, can etc) using the following methods?

	Never	Rarely	Sometimes	Often	Always
Gutters					
Water closet					
Pit latrine					
Dump sites					
LAWMA					
Cart pushers					
Burning					
Directly into the lagoon					

9.   How often do you dispose **saw millingwaste** using the following methods?

	Never	Rarely	Sometimes	Often	Always
Gutters					
Water closet					
Pit latrine					
Dump sites					
LAWMA					
Cart pushers					
Burning					
Directly into the lagoon					

10.   How satisfied are you with your mode of waste disposal?
        Very dissatisfied ☐   Satisfied☐   Neutral ☐   Satisfied☐   Very satisfied☐

Effect on health
11.   How many times do you fall ill yearly   Never ☐ Rarely ☐ Sometimes ☐ Often ☐
Always ☐

12.   How often do you engage in the following activities?

	Never	Rarely	Sometimes	Often	Always
Cigarette smoking					
Tobacco smoking					
Alcohol drinking					
Use of hard drugs					
Skin bleaching					
Use of candles					
Body exercise					

13.   How often do you experience these symptoms:

	Never	Rarely	Sometimes	Often	Always
Headache					
Dizziness					
Fatigue					
Nausea/Vomiting					
Abdominal pain					
Diarrhea					
Coughing with phlegm					
Breathing difficulties					
Tremors and seizures					
Memory loss					
Stiff/Weak muscle					
Skin rashes					
Eye irritations					
Hearing difficulties					

14. How many people (including yourself) in your community do you have any of the following conditions?

	None	Few	Some	Many	Very many
Liver disease					
Kidney disease					
Nervous disorder					
Bronchitis					
Depression					
Dermatitis					
Diabetes					
High blood pressure					
Stroke					
Cancers					

15. How often do you visit a health center or hospital when you fall ill

Never ☐   Rarely☐   Sometimes☐   Often ☐ Always☐

Effect of Anthropogenic activities

16. To what extent do you feel **sewage disposal** (faecal and kitchen waste) in your community affects the environment?

	Not much	Little	Somewhat	Much	Very much
Water					
Air					
Noise					
Land					
Aquatic animals					
Land animals					

17. To what extent do you feel **solid waste** (paper, nylon, cans, plastics etc)disposal in your community affects the environment?

	Not much	Little	Somewhat	Much	Very much
Water					
Air					
Noise					
Land					
Aquatic animals					
Land animals					

18. To what extent do you feel **saw milling** activities affect your environment?

	Not much	Little	Somewhat	Much	Very much
Water					
Air					
Noise					
Land					
Aquatic animals					
Land animals					

19. To what extent do you feel **burning** activities in your community affects the environment?

	Not much	Little	Somewhat	Much	Very much
Water					
Air					
Noise					
Land					
Aquatic animals					
Land animals					

20. How satisfied are you with the current state of the environment in Okobaba community?

Very satisfied☐   Satisfied☐   Neutral☐   Unsatisfied ☐   Very unsatisfied ☐

Appendix 2: Environmental Risk Assessment Questionnaire for Stakeholders at Kara Cow Market, Ogun State, Nigeria

QUESTIONNNAIRE

Please answer the following questions by ticking the appropriate box.

1. Age: under 15yrs □   16-29yrs □ 30-45yrs □   46yrs and above □.
2. Gender: Male □   Female □.
3. Occupation: a. Butcher □ b. Fisherman □c. Trader □ d. Cattle rearer □. Others □
4. Do you reside in the Kara community? Yes □/ No □
5. How often do you drink water from the following sources?

	Always	Usually	Sometimes	Rarely	Never
Tap water					
Well water					
Borehole water					
Sachet water					
Rain water					
River					

6. How often do you use the following sources for domestic purposes (bathing, washing, cooking, cleaning as so on)?

	Always	Usually	Sometimes	Rarely	Never
Tap water					
Well water					
Borehole water					
Rain water					
River					

7. How often do you treat your water source(s)?

	Always	Usually	Sometimes	Rarely	Never
Boiling					
Filtration					
Coagulation (use of alum)					
Disinfectant (Water Guard, Dettol etc.)					

8. How often are animals allowed to drink the water from the river?

    Always □   Usually □   Sometimes □   Rarely □   Never □

9. Are you satisfied with the quality of water you use (drinking/ domestic)?

    Very satisfied □   Satisfied □   Indifferent □   Dissatisfied □ Very dissatisfied □

10. How often do you use the following waste disposal methods?

	Always	Usually	Sometimes	Rarely	Never
PSP waste trucks					
Burning					
Dump site					
Cart pushers					
Direct disposal into the river					

11. How often do you dispose the following waste into the river?

	Always	Usually	Sometimes	Rarely	Never
Animal waste (blood, feces, entrails etc.)					
Animal feed remains					
Human wastes (feces)					
Solid waste (bones, plastics and nylons)					

12. How often do you treat the waste generated from the slaughter slabs?

     Always □   Usually □   Sometimes □   Rarely □   Never □

13. How satisfied are you with your method of waste disposal?

     Very satisfied □   Satisfied □   Indifferent □   Dissatisfied □ Very dissatisfied □

14. How often do you get sick in a year?

     Always □   Usually □   Sometimes □   Rarely □   Never □

15. How often do you have the following illnesses?

	Always	Usually	Sometimes	Rarely	Never
Cholera					
Typhoid					
Dysentery					
Skin rashes					
Breathing difficulties					

16. How many people (including yourself) in your community have any of the following conditions?

	Very many	Many	Some	Few	None
Cancer					
Stroke					
High blood pressure					
Bronchitis					
Liver disease					
Kidney disease					
Skin problems					
	Very many	Many	Some	Few	None
Nervous disorder					
Depression					

17. How often do you visit the health center or hospital for an illness?

     Always □   Usually □   Sometimes □   Rarely □   Never □

18. How often do you perceive an offensive odour in this community?

     Always □   Usually □   Sometimes □   Rarely □   Never □

19. How often do hyacinths appear in the river?

     Always □   Usually □   Sometimes □   Rarely □   Never □

20. How often are animals burnt at the slaughter slabs?

     Always □   Usually □   Sometimes □   Rarely □   Never □

21. How often do animal disease break out in the market?

Always ☐   Usually ☐   Sometimes ☐   Rarely ☐   Never ☐
22. How often do animals receive medications in the market?

Always ☐   Usually ☐   Sometimes ☐   Rarely ☐   Never ☐
23. How often do you catch fish in the river?

Always ☐   Usually ☐   Sometimes ☐   Rarely ☐   Never ☐
24. Based on question 17, mention at least five (5) medications/drugs given to the animals.
....
....
....
....
....
....

25. Based on question 18, mention at least five (5) fish and/or aquatic animals caught in the river.
....
....
....
....
....
....

References

1. Adelagun ROA, Berezi EP, Akintunde OA (2012) Air pollution in a sawmill industry: the Okobaba (Ebute-Meta, Lagos) experience. J Sustain Dev Environ Prot 2(2):29–36
2. Adelegan JA (2002) Environmental policy and slaughterhouse waste in Nigeria. In: Proceedings of the 28th WEDC conference, Kolkata (Calcutta) India, vol 1, pp 3–6
3. Adeogun AO, Ibor OR, Chukwuka AV, Regoli F, Arukwe A (2019) The intersex phenomenon in *Sarotherodon melanotheron* from Lagos Lagoon (Nigeria): occurrence and severity in relation to contaminants burden in sediment. Environ Pollut 244:747–756
4. Adio-Moses D (2016) Smart city strategy and sustainable development goals for building construction framework in Lagos. In: International conference on infrastructure development in Africa, Johannesburg, South Africa, pp 1–12
5. Agbana BE, Joshua AO, Daikwo MA, Metiboba OL (2016) Knowledge of occupational hazards among sawmill workers in Kwara state, Nigeria. Niger Postgrad Med J 23:25–32
6. Agu AP, Umeokonkwo CD, Nnadu RC, Odusanya OO (2016) Health problems among sawmill workers in Abakaliki and workplace risk assessment. J Commun Med Prim Health Care 28:1–10
7. Ajao EA, Oyewo EO, Uyimad JP (1996) A review of the pollution of coastal waters in Nigeria. Nigerian Institute of Oceanography Technical Paper, no 107, 20 pp
8. Akhator P, Obanor A, Unege A (2017) Nigerian wood waste: a potential resource for economic development. J Appl Sci Environ Manag 21(2):246–251
9. Akinro AO, Ologunagba IB, Yahaya O (2009) Environmental implications of unhygienic operation of a city abattoir in Akure, Western Nigeria. J Eng Appl Sci 4(9):61–63
10. Alani R, Alo B, Ukoakonam F (2014) Preliminary investigation of the state of pollution of Ogun River at Kara Abattoir, near Berger, Lagos. Int J Environ Sci Toxicol Res 2(2):11–23
11. Alava JJ, Cheung WWL, Ross PS, Sumaila UR (2017) Climate change-contaminant interactions in marine food webs: towards a conceptual framework. Glob Change Biol 23:3984–4001
12. Aliyu AA, Amadu L (2017) Urbanization, cities and health: the challenges to Nigeria—a review. Ann Afr Med 16(4):149–158

13. Amaeze NH, Egonmwan RI, Jolaoso AF, Otitoloju AA (2012) Coastal environmental pollution and fish species diversity in Lagos Lagoon, Nigeria. Int J Environ Prot 2(11):8–16
14. Amaeze NH, Adeyemi RO, Adebesin AO (2015) Oxidative stress, heats shock protein and histopathological effects in the gills of African catfish, *Clarias gariepinus* induced by bridge runoffs. Environ Monit Assess 187(4):172
15. Awodele O, Popoola TD, Ogbudu BS, Akinyede A, Coker HAB, Akintonwa A (2014) Occupational hazards and safety measures amongst paint factory workers in Lagos, Nigeria. Saf Health Work 5:106–111
16. Ayoade F, Olayioye EO (2016) Microbiological assessment of housekeeping practices and environmental impact of selected abattoirs in Lagos and Ogun states of Nigeria. J Appl Biosci 99:9363–9372
17. Bawa-Allah KA, Saliu JK, Otitoloju AA (2018) Heavy metal pollution monitoring in vulnerable ecosystems: a case study of the Lagos Lagoon, Nigeria. Bull Environ Contam Toxicol 100(5):609–613
18. Bay S, Jones BH, Schiff K, Washburn L (2003) Water quality impacts of stormwater discharges to Santa Monica Bay. Mar Environ Res 56:205–223
19. Bradburn NM, Sudman S, Wansink B (2004) Asking questions: the definitive guide to questionnaire design. Wiley, New York
20. Buraimoh OM, Ilori MO, Amund OO, Michel FC Jr, Grewal SK (2015) Assessment of bacterial degradation of lignocellulosic residues (sawdust) in a tropical estuarine microcosm using improvised floating raft equipment. Int Biodeterior Biodegrad 104:186–193
21. Chikere C (2017) Nigeria's growing need for beach plastic audit. https://www.thenigerianvoice.com/news/258031/nigerias-growing-need-for-beach-plastic-audit.html
22. Cisneros-Motemayor AM, Pauly D, Weatherdon LV, Ota Y (2016) A global estimate of seafood consumption by coastal indigenous peoples. PLoS ONE 11:e0166681
23. Daniel WW (1999) Biostatistics: a foundation for analysis in the health sciences, 7th edn. Wiley, New York
24. Daso AP, Osibanjo O, Gbadebo AM (2011) The impact of industries on surface water quality of River Ona and River Alaro in Oluyole Industrial Estate, Ibadan, Nigeria. Afr J Biotechnol 10(4):696–702
25. Ekanem KV, Chukwuma GO, Ubah JI (2016) Determination of the physico-chemical characteristics of effluent discharged from Karu Abattoir. Int J Sci Technol 5(2):43–50
26. Elijah FB, Elegbede I (2015) Environmental sustainability impact of the Okobaba sawmill industry on some biogeochemistry characteristics of the Lagos Lagoon. Poult Fish Wildl Sci 3:131
27. Faremi OE (2018) Environmental risks knowledge, field evaluations and biological responses of macrobenthos and fish species at Okobaba area of the Lagos Lagoon, Nigeria. MSc thesis, University of Lagos, Nigeria, 75 pp
28. Herbst S, Benedikter S, Koester U, Phan N, Berger C, Rechenberg A, Kistemanmn T (2009) Perception of water, sanitation and health: a case study from the Mekong Delta, Vietnam. Water Sci Technol 60(3):699–707
29. Huff J (2001) Sawmill chemicals and carcinogenesis. Environ Health Perspect 109(3):209–212
30. Kar T, Keles S (2016) Environmental impacts of biomass combustion for heating and electricity generation. J Eng Res Appl Sci 5(2):458–465
31. Kim-Oanh NT, Nghiem-Le H, Phyu YL (2002) Emission of polycyclic aromatic hydrocarbons, toxicity and mutagenicity from domestic cooking using sawdust briquettes, wood and kerosene. Environ Sci Technol 36(5):833–839
32. Longe EO, Ogundipe AO (2010) Assessment of wastewater discharge impact from a sewage treatment plant on lagoon water, Lagos, Nigeria. Res J Appl Sci Eng Technol 2:274–282
33. Miller GW (2017) The international reach of toxicology. Toxicol Sci 157(2):274–275
34. Neboh HA, Ilusanya OA, Ezekoye CC, Orji FA (2013) Assessment of Ijebu-Igbo abattoir effluent and its impact on the ecology of the receiving soil and river. IOSR J Environ Sci Toxicol Food Technol 7(5):61–67

35. Nefarnda WD, Yaji A, Kubkomawa HI (2006) Impact of abattoir waste on aquatic life: a case study of Yola Abattoir. Glob J Pure Appl Sci 12:31–33
36. Nubi OA, Ajao EA, Nubi AT (2008) Pollution assessment of the impact of coastal activities on Lagos Lagoon, Nigeria. Sci World J 3:83–88
37. Nunez Collado JR, Wang HH, Tsai TY (2019) Urban informality in the Paris climate agreement: content analysis of the nationally determined contributions of highly urbanized developing countries. Sustainability 11:5228
38. Nwankwo DI, Chukwu LO, Onyema IC (2009) The hydrochemistry and biota of a thermal coolant water stressed tropical lagoon. Life Sci J 6:86–94
39. Oketola AA, Fagbemigun TK (2013) Determination of nonylphenol, octylphenol and bisphenol-A in water and sediments of two major rivers in Lagos, Nigeria. J Environ Prot 4:38–45
40. Olaniran EI, Sogbanmu TO, Saliu JK (2019) Biomonitoring, physico-chemical, and biomarker evaluations of abattoir effluent discharges into the Ogun River from Kara Market, Ogun State, Nigeria using *Clarias gariepinus*. Environ Monit Assess 191(1):44
41. Olawuni PO, Okunola OH (2014) Socioeconomic impacts of sawmill industry on residents. A case study of Ile-Ife, Nigeria. J Econ Dev Stud 2(3):167–176
42. Olowoporoku OA (2016) Assessing environmental sanitation practices in slaughterhouses in Osogbo, Nigeria: taking the good with the bad. MAYFEB J Environ Sci 1:44–54
43. Oppenheim AN (2000) Questionnaire design, interviewing and attitude measurement. Bloomsbury Publishing, London
44. Osibanjo O, Adie GU (2007) Impact of effluent from Bodija abattoir on the physico-chemical parameters of Oshunkaye Stream in Ibadan City, Nigeria. Afr J Biotechnol 6(15):1806–1811
45. Palamuleni LG (2002) Effects of sanitation facilities, domestic solid waste disposal and hygiene practices on water quality in Malawi's urban poor areas: a case study of South Lunzu Township in the city of Blantrye. Phys Chem Earth, Parts A/B/C 27(11–22):845–850
46. Parkinson J, Tayler K (2003) Decentralized wastewater management in peri-urban areas in low-income countries. Environ Urban 15(1):75–90
47. Shen H, Huang Y, Wang R, Zhu D, Li W, Shen G, Wang B, Zhang Y, Chen Y, Lu Y, Chen H, Li T, Sun K, Li B, Liu W, Liu J, Tao S (2013) Global atmospheric emissions of polycyclic aromatic hydrocarbons from 1960 to 2008 and future predictions. Environ Sci Technol 47:6415–6424
48. Sogbanmu TO, Nagy E, Phillips DH, Arlt VM, Otitoloju AA, Bury NR (2016) Lagos lagoon sediment organic extracts and polycyclic aromatic hydrocarbons induce embryotoxic, teratogenic and genotoxic effects in *Danio rerio* (zebrafish) embryos. Environ Sci Pollut Res 23:14489–14501
49. Sogbanmu TO, Sosanwo AA, Ugwumba AAA (2019) Histological, microbiological, physico-chemical and heavy metals evaluation of effluent from Kara Cow Market, Ogun State in Guppy Fish (*Poecilia reticulata*). Zoologist 17:54–61
50. Sogbanmu TO, Osibona AO, Otitoloju AA (2019) Specific polycyclic aromatic hydrocarbons identified as ecological risk factors in the Lagos lagoon, Nigeria. Environ Pollut 255:113295
51. Tompkins EL, Few R, Brown K (2008) Scenario-based stakeholder engagement: incorporating stakeholders' preferences into coastal planning for climate change. J Environ Manag 88(4):1580–1592
52. Ugheoke AJ, Wahab KW, Erhabor GE (2009) Prevalence of respiratory symptoms among sawmill workers in Benin City, Nigeria. Int J Trop Med 4:1–3
53. Young S, Goodwin wEJ, Sedgwick O, Gudjonsson GH (2013) The effectiveness of police custody assessments in identifying suspects with intellectual disabilities and attention-deficit hyperactivity disorder. BMC Med 11:248

Meeting the Sustainable Development Goals: Considerations for Household and Indoor Air Pollution in Nigeria and Ghana

Irene Appeaning Addo and Oluwafemi Ayodeji Olajide

Abstract Pollution being one of the symbolic features of unsustainable living in today's world, the UN's Sustainable Development Goals recognise the need to address air pollution for healthy cities and humans. Indoor air pollution in Africa is recognised as one of the leading causes of pulmonary diseases and death, given the high incidence of biomass fuel use in cooking. Yet, several African governments have not proffered effective solutions to the problem of indoor air pollution. In cases where there have been some forms of intervention, they have not been very successful and sustainable. Therefore, this chapter broadly reviews environmental air pollution and specifically indoor air pollution in two West African countries, Nigeria and Ghana. The findings indicate that there is a high incidence of indoor air pollution in sub-Saharan Africa due to widespread home-based enterprises and usage of solid fuels for cooking in poorly ventilated space. Other findings indicate that ambient air pollution has dominated policy documents while indoor air pollution is hardly recognised at the institutional level, with very little effort made in inventory, monitoring, enforcement and abatement. The chapter concludes that there is need for more research on the quality of indoor air. African countries need to establish agencies on indoor air pollution management if the continent is to achieve the Sustainable Development Goals related to air quality by 2030.

Keywords Indoor air pollution · Sustainable Development Goals · Sub-Saharan Africa · Nigeria · Ghana

I. A. Addo (✉)
Institute of African Studies, University of Ghana, Accra, Ghana
e-mail: iappeaningaddo@ug.edu.gh

O. A. Olajide
Department of Urban and Regional Planning, University of Lagos, Lagos, Nigeria

147

1 Introduction

As part of the effects of rapid development, urban areas in developing countries are increasingly becoming densely populated and overcrowded while displaying features of environmental pollution. Pollution, one of the symbolic features of unsustainable living in today's world, features prominently in the United Nations' (UN) Sustainable Development Goals (SDGs). This is so because its associated risks pose a serious threat to the realisation of the SDGs. Pollution has been described as a silent killer that has been damaging the natural environment and public health, causing about 7 million premature deaths annually. About 54% of these deaths are associated with indoor air pollution [65, 76]. It is projected that the overall deaths associated with air pollution may rise to 7.5 million by 2040, if nothing is done to stem the tide [65]. Therefore, air pollution and measures to address the challenges are essential components of the global sustainability discourse.

Following the United States Environmental Protection Agency [71] and the World Health Organisation [74], air pollution may be defined as contamination of the indoor or outdoor environment by any substances that modify the natural characteristics of the atmosphere with the capability to cause discomfort, harm or health-related risks to living organisms [72]. The challenge of air pollution, particularly indoor air pollution, is worsened by widespread usage of solid fuels for household cooking and home-based enterprises, a major component of the informal sector.

Home-based enterprises are economic activities operating within a dwelling unit and its premises rather than in a commercial or industrial building [39, 63, 66]. Home-based enterprises are a major source of livelihood for a significant proportion of developing countries' populations [18, 61, 66]. However, they come with both socioeconomic benefits and costs [39, 40], which expose households to poor environmental quality [49, 67].

Some of the Sustainable Development Goals explicitly address the need to enhance environmental quality and health. In this regard, access to clean air is suggested as one of the entry points to improving environmental quality and human health, as well as the attainment of several Sustainable Development Goals, particularly goals 3, 7, 11 and 13 [70], as discussed in the subsequent section.

Through a globally shared agenda, each country is expected to develop local and national policies and programmes aimed at translating the goals and indicators into reality. Several governments around the world, including African countries, have undertaken programmes and developed policies targeted at addressing air pollution. Yet the issue of air pollution, particularly indoor air pollution, which is largely associated with home-based activities, seems to be on the increase in many African countries. Against this background, drawing on the case studies of Ghana and Nigeria, this chapter examines government initiatives aimed at addressing the challenges of indoor air pollution. The discussion is informed by contextual analysis of secondary data, including articles, reports, census data and policy documents. After the introduction, the chapter offers a review of relevant literature regarding the connection between air pollution and the SDGs, with specific focus on global trends in indoor air pollution and the current state of indoor air pollution in Africa. The section also problematises indoor air pollution as a global environmental challenge and a threat to achievement of the SDGs. The section concludes with a review of African initiatives meant to address the challenges of indoor air pollution. The next section, without claiming to be exhaustive, focuses on the case studies of Nigeria and Ghana regarding the trends in indoor air pollution, as well as some of the major regulatory frameworks designed to address the challenge. The final section then discusses the findings and their impact on sub-Saharan Africa's bid to address air pollution in the SDGs; it also offers some policy recommendations for indoor air pollution mitigation.

2 Indoor Air Pollution

2.1 Positioning Indoor Air Pollution and the Sustainable Development Goals

The 2030 Global Agenda for Sustainable Development, as encapsulated in 17 goals and 169 targets, provides a global and unifying framework for a productive and prosperous planet for all across generations. Thus, at the core of the SDGs and their various targets are the need to eradicate poverty and create a healthy environment that supports individual and planetary well-being. Pollution, because of its associated health and death risks, as well as its link to global warming and climate change, has been identified both as one of the major challenges and solutions to the attainment of a healthy planet that supports individual well-being [20, 70, 75].

Air pollution is generally classified as outdoor or indoor. While the major source of outdoor pollution is attributed to the transport sector, indoor pollution is largely caused by household energy consumption. Inefficient household fuel combustion is a major source of air pollution. Globally, evidence suggests that approximately 40% of households rely on solid fuels for cooking and heating. The situation is acute in developing countries, particularly in sub-Saharan African countries, where the majority live in extreme poverty, rely on solid fuels for household cooking and engage in home-based enterprises, which is a major component of the informal sector [57]. Despite numerous intervention programmes that have been put in place, ambient air pollution in sub-Saharan Africa is estimated to be about 10–20 times above World Health Organisation (WHO) standards and above levels currently found in industrialised regions with high-income countries [35]. This may be explained by the rapidly increasing urbanisation occurring in Africa and the unvented burning of biomass indoors as well as household access to, and choice of energy technology, fuel-stove combinations and other energy-related behaviours [19].

The situation is exacerbated by the high incidence of home-based enterprises, which are a major source of livelihood for significant segments of developing countries' population [18, 61, 66]. While there are many commercial activities that fall within this description, food preparation accounts for a significant proportion of

Table 1 Sustainable development goals with corresponding targets and indicators related to air pollution

SDGs goals	Corresponding SDGs targets and indicators related to air pollution
Goal 3: good health and well-being	3.9—reduction in deaths and illnesses from pollution 3.9.1—air pollution-related mortality
Goal 7: affordable and clean energy	Target 7.1.2: access to clean energy in homes
Goal 11: sustainable cities and communities	11.6.2—air quality in cities 11.7: reduction of the environmental impact of cities by improving air quality
Goal 12: sustainable consumption and production	12.4—achieve the environmentally sound management of chemicals and all wastes, and significantly reduce their release to air, water and soil in order to minimise their adverse impacts on human health and the environment…
Goal 13: climate action	13.2.1—establishment of an integrated policy to adapt to the adverse impacts of climate change, and foster climate resilience and low greenhouse gas emissions development.

Sources United Nations [70]; The sustainable Development Goals Report 2018

home-based activities in developing countries. While home-based enterprises are a major source of income and employment for a significant part of the urban population [39, 40], the activities also generate environmental pollution, which exposes households and residents in the immediate environment to indoor air pollution [49, 67].

Several SDGs explicitly cite the need to enhance environmental quality and health (Table 1). To be sure, air quality is recognised as one of the entry points to achieving this goal, in spite of the associated challenges that pose a threat to human health and global prosperity.

2.2 Indoor Air Pollution as a Global Environmental Challenge

Previously more attention was paid to air pollution from industries than to indoor air pollution [64]. However, the situation changed around the 1970s, when indoor air pollution became a public health concern due to increasing use of biomass fuel for heating and cooking. Around the 1970s energy-conserving buildings began to emerge in North America and Europe, leading to some levels of improvement in energy efficiency and reduction in exchanges between outdoor fresh air and indoor air [78]. However, globally, air pollution remains a leading contributor to non-communicable respiratory diseases and a significant cause of morbidity and mortality [14].

Studies have shown the correlation between indoor air quality and health. Indoor air pollution may cause acute respiratory infections, chronic obstructive pulmonary disease (e.g. bronchitis), eye problems and cancer of the lungs. It also increases outdoor air pollution, especially in densely populated urban areas, in addition to leading to adverse effects on birth outcomes and causing chronic nutritional deficiencies [22, 32]. According to the WHO, more than half of the global urban population are exposed to air pollution levels that are 2.5 times higher than the recommended safety levels. The greatest health impact of air pollution is felt among the poorest and most vulnerable populations of the world, who live in the developing world, where approximately 76% of the global indoor exposure to Particulate Matter (PM) occurs [22].

Indoor air pollution exposure is a function of the complex interplay between human behaviour and household fuel consumption patterns, appliances, and housing design [8]. Reliance on traditional fuels such as coal, wood or kerosene as the primary source of domestic energy, as well as use of open fires for cooking and heating, exposes the household to high concentrations of particulate matter (PM) and gases, most times in significant excess of the WHO's air quality guidelines for 24-hour and annual mean concentrations [7, 48]. Inadequate ventilation and low exchange rates of indoor with outdoor air may significantly influence concentrations of pollutants indoors [2, 12, 22, 37]. For example, the number of windows in the kitchen, as well as the amount of firewood, determined levels of PM10 (particulate matter less than 10 mm in diameter) concentrations [12]. Indoor air quality is directly affected by the rate at which outdoor air enters a building and ventilation can be used to maintain low concentrations of indoor-generated pollutants. Thus, improving ventilation in rooms and introducing clean energy may reduce indoor air pollution.

2.3   Current State of Indoor Air Pollution in Africa

Indoor air pollution is becoming a critical issue for the majority of the population in sub-Saharan Africa, with about 90% of people's time spent indoors. High incidence of indoor air pollution in sub-Saharan Africa has been attributed to many factors, including use of solid fuels in poorly ventilated space and the widespread presence of home-based enterprises.

High incidence of indoor air pollution, especially among low-income households is attributed to the use of biomass as fuel for cooking. Almost half of the world's population and nearly 81% of Sub-Saharan Africa (SSA) depend on woody biomass energy for cooking and economic activities [9]. According to Amegah and Jaakkola [4], the global population using mainly solid fuel for cooking has increased over the last three decades, at around 2.7–2.8 billion, while between 1980 and 2010, the population exposed to household air pollution increased from 333 million to 646 million in sub-Saharan Africa and from 162 million to 190 million in the eastern Mediterranean and around 1 billion people in southeast Asia.

Nearly two billion kilograms of biomass is burnt daily in developing countries [6]. A study shows that in South Africa, despite having access to electricity, about 20% of homes continued to use fossil fuels as an alternate energy source [73]. Multiple fuel use by households to reduce the substantial costs associated with modern fuels (gas and electricity) is a common practice in Africa [9].

With decreasing formal-sector employment opportunities for many Africans, the importance of informal income-generating activities is being recognised, especially in poor communities [27, 30]. More and more urban low-income households are using the home or shelter as a place of work, from where they operate small-scale businesses [18, 28, 40, 77]. Ghana and Nigeria are not left out of the global phenomenon despite the prohibitive regulations on home-based enterprises [17, 56]. The most prevalent home-based enterprise is food preparation, sale of cooked food and food processing. Air pollution is one of the effects of home-based enterprises on the wider biophysical environment.

2.4   Addressing Indoor Air Pollution: African Initiatives and Meeting the SDGs

Longhurst et al. [41] have extensively reviewed and assessed the causes and impact of air pollution and its management through the lens of human needs, as collated under the United Nations Sustainable Development Goals (SDGs). They suggested that the issue of air pollution management is not clearly identified in the SDGs framework, although air pollution features in some of the targets: health (SDG 3), cities (SDG 11), energy (SDG 7), sustainable consumption and production (SDG 12) and climate action (SDG 13). In each case, air pollution is merely one of the targets associated with each goal and can be a barrier towards meeting the goals. These goals may be achieved by implementing the WHO indoor air quality guidelines on household fuel combustion; promoting and disseminating improved cookstoves; expanding liquefied petroleum gas production facilities and distribution networks; harnessing renewable energy potential; promoting biogas production at both household and community level; ensuring improved ventilation of homes through education and enforcement of building standards; and exploiting opportunities in the health and other sectors for changing health-damaging cooking behaviour [4]. However, Ezzati [19] asserts that the prohibitive cost associated with the transition to cleaner fuels for many low-income nations and households, is not a realistic option in the next two to three decades. Therefore, the author suggests the need for research and development on alternative technologies for accessible and clean energy sources and interventions that lower exposure to emissions, for example changes in housing design. It has been observed that affordability of modern energy fuels, accessibility of biomass fuel, house designs (apartments and shared units) as well as location (urban or rural) can either promote or constrain the adoption of modern energy services throughout the developing world, often resulting in multiple fuel use, careful fuel spending and fuel switching [32].

In sub-Saharan Africa, the use of solid fuels is still widespread, as it appears that intervention efforts are not keeping pace with population growth. The health impact of indoor air pollution is being worsened by climate change effects, leading to reduced biodiversity. Weather changes such as wind patterns, precipitation timing and intensity and increase of temperature have an effect on the severity and frequency of air pollution [59]. Baffour Awuah [5] asserts that planning can be used to design or redesign cities and promote efficient use of land resources and properties in ways that will address air pollution in sub-Saharan Africa.

3 Case Studies

3.1 Trends in Indoor Air Pollution and Policy Frameworks in Nigeria

Nigeria occupies a strategic position in the achievement of the SDGs in Africa, since it is the most populated country on the continent and it is rich in natural and human resources, as well as being the leading economy and one of the commercial hubs in Africa. Despite the country's impressive profile, the majority of the population live in multidimensional poverty as manifested in unemployment, inadequate infrastructure, as well as environmental degradation and unhealthy living conditions. Closely related to the discourse of poverty is the widespread informal economic activities. Like many other African countries, Nigeria is not immune to the incidence of informal economic activities, of which home-based business is a significant component. The informal sector is an essential economic sector in Nigeria, supporting the majority of the population for whom it serves as a means of livelihood.

Although Nigeria is rich in resources, particularly petroleum and gas resources, the country has a poor energy infrastructure, resulting in the majority of the population suffering from what scholars describe as energy poverty [45, 62], a condition of inadequate access to cleaner energy sources for household energy consumption [33]. Eventhough at household levels energy is required for various purposes, including heating, powering household appliances, cooking and lighting [55], cooking accounts for most of the household energy consumption in Nigeria [58]. For example, Gujba et al. [29] estimate that cooking accounts for approximately 80% of the total household energy consumption, with about three quarters of households relying on solid fuel, e.g. fuelwood and coal, for cooking [1]. A trend analysis of the sources of household energy for cooking, based on the Demographic and Health Survey [52, 53] and Malaria Indicator Survey between 2003 and 2015 [50, 51], indicates that the

Table 2 Sources of households' main cooking fuels

Cooking fuel	DHS 2003	DHS 2008	MIS 2010	DHS 2013	MIS 2015
Electricity	0.3	0.3	0.2	0.4	0.6
Kerosene	26.9	19.5	16.6	25.5	23.8
Straw/shrubs/grass	68.8	1.0	0.5	1.8	1.1
LPG/natural gas/biogas	–	1.1	0.7	2.3	4.7
Coal, lignite	–	0.4	0.1	0.3	0.1
Agricultural crop	–	0.2	0.0	0.0	0.6
Charcoal	–	2.6	2.2	3.2	3.3
Animal dung	0.5	0.0	0.0	0.1	0
Wood	–	74.1	79.4	63.7	65.0
No food cooked in house	–	0.7	0.1	2.6	0.6
Others	3.3	0.1	0.2	0	0
Total	100	100	100	100	100

Source Malaria Indicator Survey (2010, 2015); Demographic and Health Survey (2003, 2008, 2013)
Note Percentages may not add up to 100 due to missing cases

situation has not significantly improved, as the majority of the population still rely on solid fuels such as wood for cooking (Table 2).

Indeed, evidence seems to suggest that the situation is getting worse, with more people either stagnating or descending the energy ladder [62]. Although, while there is a widespread usage of solid fuels for cooking in Nigeria, the incidence is not evenly distributed across regions. The use of solid fuels is generally higher in the North than the South. The regional distribution shows that solid fuel usage in the north eastern Nigeria is about 94%, the central part of northern Nigeria is about 74% while the western part of northern Nigeria is about 92%. On the other hand, solid fuel usage in south eastern Nigeria is 66%, the south western Nigeria is also about 37% ([62]: 134).

A significant proportion of the Nigerian population is exposed to indoor air pollution because of the use of solid fuels for cooking and heating [31]. Several studies have shown that indoor air pollution is one of the major contributors to multidimensional poverty, particularly unhealthy conditions, for which home-based activities are major contributors [54, 60]. Indoor air pollution is a widespread phenomenon across Nigeria and it is associated with a high incidence of solid fuel use for open fires and poorly maintained kerosene stoves that release particulate matter and toxic fumes [1, 44]. A survey of the national disease burden shows that about 80,000 deaths are attributed to solid fuel usage [1]. Indoor air pollution imposes a significant burden on individual and public health, with 3.8% of the national burden of disease being attributable to usage of solid fuel for household cooking. Morbidity associated with solid fuel usage in Nigeria is the third highest globally, with 2.6 million disability [1]. Despite that indoor air pollution is one of the key environmental issues confronting the Nigerian population, particularly the poor and low-income communities, there is a lack of national data or specific inventory mechanisms and policy frameworks for dealing with the phenomenon. Most of the regulatory policy frameworks, as discussed later, have largely focused on ambient air quality.

The need to address environmental degradation, of which air quality is a major component, has been a major concern in Nigeria. The Nigerian Federal Government is a signatory to many international declarations and agreements on environmental issues, including "climate change, biodiversity, desertification, forestry, oil and gas, hazardous waste, marine and wildlife and pollution" [38]. Beyond the international declarations, the Nigerian government has developed various institutional regulatory and policy frameworks at the federal, state and local levels for dealing with the issues of environmental pollution in general and air pollution in particular. For example, in 1988, at the national level, the government established the Federal Environmental Protection Agency (FEPA) to regulate environmental quality, including air quality. On air quality, FEPA was statutorily responsible for formulating and enforcing air quality standards. In line with the international air quality guidelines, FEPA developed permissible limits of major ambient air pollutants and environmental standards. In 2007, owing to the weak institutional enforcement capacity of FEPA, the National Environmental Standards and Regulations Enforcement Agency (NESREA) was established to replace FEPA [38]. NESREA is statutorily responsible for restoring, preserving and improving environmental quality for the achievement of sustainable development. To this end, the agency is responsible for enforcing national environmental regulations, guidelines, policies and standards. In addition, NESREA is responsible for enforcing compliance with the provisions of the various international declarations and agreements on environmental issues of which Nigeria is a signatory. In relation to air quality, Section 20(1) of the NESREA Act states: "The Agency may make regulations setting specifications and standards to protect and enhance the quality of Nigeria's air resources, so as to promote the public health or welfare and the natural development and productive capacity of the nations' human, animal, marine or plant life."

In executing its mandate, NESREA has developed several environmental regulations. One of such regulations is the National Environmental (Air Quality Control) Regulations, 2014, which seeks to: "provide for improved control of the nation's air quality, enhance the protection of human health and other resources affected by air quality deterioration, the right to clean air and manage according to the principles of sustainable development." Similarly, within the national framework, each state of the federation is mandated to safeguard and improve environmental quality. Therefore, each state has at least a ministry or agency or parastatal that deals with environmental issues in the areas of water, air, land, forest and wildlife, as stipulated in Section 20 of the Nigerian Constitution. For example, the Lagos State Environmental Protection Agency (LASEPA), as an environmental regulator for the state, is responsible for protecting and improving environmental quality that is consistent with the social and economic needs of the state, with the overall aim of improving residents' health, well-being and quality of life. The overall vision of LASEPA is consistent with the

SDGs, as it seeks to formulate efficient environmental management strategies that support sustainable development.

Despite an array of institutional and policy frameworks, which are meant to provide for improved air quality, evidence shows that air quality falls below the permissible level for human health and well-being [42, 72]. The current statistics show that Nigeria is one of the nations with the highest exposure to air pollution, particularly particulate matter, with an exposure level that is higher than the global standards considered safe for human health [31]. Drawing on a recent document published by the Word Bank, Health Effects Institute [31] estimated that Nigeria experiences a mean annual exposure of $PM_{2.5}$ at 38 micrograms/cubic meter ($\mu g/m^3$) of air compared to the global safe standard of 20 μg/cubic meter ($\mu g/m^3$).

While there is an array of international conventions and national policy frameworks to monitor ambient air (outdoor air) quality and emission inventory in Nigeria, there is no specific policy on indoor air quality. However, while there seems not to be a coordinated national policy framework to monitor and take inventory of indoor air pollution, there is an increasing involvement of NGOs, research institutes and the private sector, sometimes in collaboration with the government, in promoting clean energy usage at the household level. For example, the Nigeria Alliance for Clean Cookstoves embarked on an ambitious project in 2011 to install 10 million stoves across the nation over a period of ten years. The involvement of government in this project is largely to create consumer awareness [1].

The National Energy Policy 2003 contains a number of provisions on the use of fuelwood, biomass, coal and natural gas (see Table 3), which can help reduce the incidence of indoor air pollution. For example, the policy, among other provisions, specifically advocates the use of alternatives to fuelwood, de-emphasises fuelwood in the nation's energy mix and provide incentives to households to convert to gas.

Table 3 The National Energy Policy 2003

Area of focus	Provisions
Fuelwood	Promote use of alternatives to fuelwood Promote improved efficiency in use of fuelwood De-emphasise fuelwood in nation's energy mix
Biomass	Harness non-fuelwood biomass such as coal Promote biomass as an alternative, especially in rural areas Reduce health hazards from combustion of biomass
Coal	Utilise coal as a viable alternative to fuelwood Provide incentives for large-scale production of coal stoves at affordable prices Organise awareness programmes for smokeless coal briquettes as fuelwood alternative
Natural gas	Expand utilisation of natural gas as domestic fuel Reduce gas flaring Provide incentives to domestic consumers to use or convert to gas

Source National Energy Policy by Energy Commission of Nigeria (April 2003); National Bureau of Statistics, Index mundi (cited in [1])

However, progress has been slow in this regard, as not much has been achieved in terms of clean energy for domestic cooking. An evaluation of the household sources of cooking fuel shows that about three-quarters of households still rely on wood [1]. The current situation is not in any way different from what existed in the past [31].

3.2 Trends in Indoor Air Pollution and Policy Frameworks in Ghana

Ghana occupies a strategic position in the achievement of the SDGs in Africa. According to UN-Ghana [68], Ghana is the first country in sub-Saharan Africa to meet the Millennium Development Goal One, i.e. reducing poverty by half. Yet, although the poverty incidence in the urban areas is declining, there remain some disparities and inequalities [21]. The relationship between poverty, growth of the informal sector and energy consumption cannot be overlooked. Studies show that energy prices, income, urbanisation and economic structure are significant demand drivers of the different energy types in Ghana [47]. A review of energy access in Ghana shows that biomass in the form of woodfuel remains the most prominent fuel in Ghana for cooking and heating and that even though liquefied petroleum gas (LPG) consumption in urban areas is relatively high, access to it is still low in the rural areas. (See [36] for detailed discussion.) Mensah and Adu [46] found that 89.2% of households in Ghana continued to use biomass fuel for cooking compared to 10.8% of the population using liquefied petroleum gas. These findings are consistent with Adusah-Poku and Takeuchi [3], who found that firewood and charcoal remain the predominant biomass fuel for household cooking activities with over 50% of the population using firewood as the main household cooking energy in 2005. This trend remained almost unchanged even in 2013. Similarly, about 30% of the population in Ghana continued to use charcoal as a household cooking energy. The use of electricity for household cooking was almost nonexistent and kerosene use was equally insignificant according to the findings of Adusah-Poku and Takeuchi [3].

Further analysis from the Ghana Living Standards Surveys (GLSS) five and six (Table 4) shows that the distribution of household cooking fuels at the rural and urban levels in Ghana indicates different dynamics. In 2008, about 80.2% of rural households as compared to 18.5% urban households used wood for cooking while charcoal was used more (52.6%) in the urban areas than in the rural areas (13.8%). The trend was similar in 2014, with a minor decrease in biomass fuel use. Generally, there is greater dependence on wood fuel for cooking in the rural areas.

Previously, Zhou et al. [79] studied four densely populated neighbourhoods in Accra and observed that biomass remains the predominant household fuel for small-scale commercial enterprises, e.g. commercial street food. They also found that household and community biomass fuel use was an important predictor of household particulate matter (PM) pollution. Earlier, Yankson [77] had observed that air

Table 4 Sources of cooking fuel and trends

Source of cooking fuel	2008 (GLSS V)			2014 (GLSS VI)		
	Urban	Rural	Ghana	Urban	Rural	Ghana
None, no cooking	7.2	2.2	4.4	5.4	2.2	3.9
Wood	18.5	80.2	53.5	14.3	74.8	41.3
Charcoal	52.6	13.8	30.6	43.6	16.5	31.5
Gas	20.2	1.5	9.5	35.8	5.5	22.3
Electricity	0.5	0.1	0.3	0.5	0.1	0.3
Kerosene	1.2	0.2	0.6	0.2	0.1	0.2
CROP residue	–	2.0	1.1	0.1	0.7	0.4
Other	–	–	–	0.1	–	–
Total	100	100	100	100	100	100

Source Ghana Statistical Service [24, 25]

pollution was the major health challenge associated with home-based enterprises, besides waste generation, noise and conflict in Accra.

Particulate matter concentrations and their effects vary among neighbourhoods in Ghana. For example, there is a high correlation between the levels of household PM pollution exposure and the socioeconomic status (SES) of urban communities [3, 34]. There is also a rising incidence of non-communicable diseases associated with air pollution in urban low-income communities in Ghana and Nigeria [11, 15, 16, 43]. In 2014 indoor air pollution caused an estimated 14,000 deaths every year and 3000 deaths in children below 5 years in Ghana [69]. According to the Ghana Health Service [23], respiratory tract infection caused by air pollution is the second highest cause of outpatient morbidity after malaria.

Despite several programmes that have been undertaken over the years to address energy poverty in Ghana, there is still a high dependence on biomass for cooking [3]. Unfortunately, women and children usually face the most exposure to household air pollution [26]. According to the Energy Commission of Ghana, a trend analysis of total biomass consumption at the household level, between 2000 and 2015, indicates that even though the situation is improving, there seems to be a gradual increase in the consumption of charcoal as a biomass fuel, For example, charcoal usage as a household fuel increased from 600 thousand tonnes of oil equivalent (ktoe) in 2000 to about 1000 ktoe in 2015. However, firewood as a household fuel decreased from about 2700 ktoe in 2000 to about 1500 ktoe in 2015 [3].

Despite indoor air pollution being one of the key environmental issues confronting Ghana, there are no comprehensive national data or specific inventory mechanisms and policy frameworks for measuring indoor air quality. Meanwhile, the government of Ghana is a signatory to many international declarations and legal frameworks aiming to reduce air pollution and global warming. The regulatory policy frameworks have largely focused on ambient air quality. This policy emanates from the Environmental Protection Agency (EPA) Act 490, which is meant to regulate

Table 5 Ambient air quality standards and pollutants limits + (hours)

Pollutants	Limits + (hours)	WHO guideline value ($\mu g/m^3$)	Ghana EPA value ($\mu g/m^3$)
Carbon monoxide (CO)	1	–	20
	24		35
Sulphur dioxide (SO$_2$)	1		10
	24	20	–
Nitrogen dioxide (NO$_2$)	1	200	–
	24	–	50
Volatile organic compounds (VOCs)	1		–
	24	–	15

Sources Compiled from Bright [10] and WHO (2018)

the environment and to ensure the implementation of government's environmental policy. The National Environmental Action Plan, under the National Environment Policy, extensively addresses ambient air pollution, conservation and sustainable development. The EPA is only now taking steps to include household air pollution in its existing air quality monitoring platform [13]. The ambient air quality guidelines for EPA as compared to the WHO guidelines, are shown in Table 5.

Historically, the Stockholm Conference of 1972 marked the start of Ghana's concerted and conscious efforts at environmental management. Ghana's first environmental policy was enacted in 1995, based on environmental conservation and sustainable development. The national environmental policy seeks to address environmental concerns resulting from rapid population growth, economic expansion, persisting poverty, poor governance and institutional weaknesses and failures. It also specifies the nature and causes of environmental problems while providing an appropriate legal framework. Although the policy recognises pollution as one of its operational policy areas, there is no specific mention of air pollution. However, waste management is clearly one of its areas of focus. Some of the past interventions that have been made include the following;

◆The introduction of new improved clean stove.
◆The promotion of use of LPG and subsidised LPG cylinders for rural areas.
◆The promotion of non-grid/grid electrification.
◆The promotion of rural electrification.
◆The promotion of cleaner cooking fuels and clean cook stoves.
◆The introduction of new improved clean stove.
◆The promotion of use of LPG and subsidised LPG cylinders for rural areas.

Other steps aimed at reducing indoor biomass burning or reducing its emissions through the Ghana Shared Growth and Development Agenda including expanding access to the national electric grid and developing oil and gas resources, which are steps projected to reduce reliance on wood fuels and charcoal. Another programme is the Efficient Lighting Initiative during which six million incandescent bulbs were replaced by CFLs, via government funding, thereby saving 200–240 MW in capacity. There is also the mandatory "Ghana Electrical Appliance Labelling and Standards Programme" (GEALSP) for CFLs and room air-conditioning as well as the introduction of a net metering code [69].

4 Conclusion

In both Nigeria and Ghana, indoor air pollution and the associated health risks are widespread phenomena, though with some country-specific variations. The major source of indoor air pollution is the household use of biomass fuels, which are mostly burnt in open fires or traditional stoves that release large proportions of the smoke indoors. The two case studies indicate that indoor air pollution is a major environmental challenge and a major contributor to poverty and unhealthy conditions.

Despite the associated health implications of indoor air pollution, not much attention is given to it at institutional levels. Most efforts, in terms of inventory, monitoring, enforcement and abatement of air pollution, are focused on ambient air, which is often taken as the national air quality standard. As such, there is insufficient knowledge on the quality of air that the population breathe inside their homes. While there is an array of institutional frameworks for regulating environmental quality at national and regional levels, there is evidence of weak enforcement of the various regulations. Moreover, evidence also shows that no agency or regulatory framework is directly responsible for formulating standards or regulating indoor air quality, even as efforts continue to be focused on ambient air quality.

Without knowledge about the current state of things, governments and other stakeholders cannot make effective plans towards achieving the SDGs related to air quality. Therefore, there is need to create an inventory of indoor air pollutants while identifying current levels of pollution and the acceptable limits of each pollutant. Doing so will ensure that policy options and strategies can be effectively tailored towards addressing pollution sources and benchmarking progress on the SDGs. One of the major hindrances to these is the lack of national policy frameworks. In this regard, one clear entry point is to develop a national policy framework that is directly responsible for formulating standards and guidelines as well as specifying monitoring and enforcement procedures.

References

1. Accenture Development Partnerships (2011) Global alliance for clean cookstoves: Nigeria market assessment sector mapping. Retrieved from: https://www.cleancookingalliance.org/bin ary-data/RESOURCE/file/000/000/168-1.pdf
2. Adamkiewicz G, Spengler JD, Harley AE, Stoddard A, Yang M, Alvarez-Reeves M, Sorensen G (2014) Environmental conditions in low-income urban housing: clustering and associations with self-reported health. Am J Public Health 104(9):1650–1656

3. Adusah-Poku F, Takeuchi K (2019) Energy poverty in Ghana: any progress so far? Renew Sustain Energy Rev 112:853–864
4. Amegah AK, Jaakkola JJ (2016) Household air pollution and the sustainable development goals. Bull World Health Organ 94(3):215
5. Baffour Awuah KG (2018) The role of urban planning in sub Saharan Africa urban pollution management. In: Charlesworth SM, Booth CA (eds) Urban pollution: science and management. Wiley, Hoboken, NJ, pp 385–395
6. Balakrishnan K, Sankar S, Ghosh S, Thangavel G, Mukhopadhyay K, Ramaswamy P, Johnson P, Thanasekaraan V (2014) Household air pollution related to solid cook fuel use: the exposure and health situation in developing countries. In: Pluschke P, Schleibinger H (eds), Indoor air pollution: the handbook of environmental chemistry, vol 64. Springer, Berlin, Heidelberg, pp 125–144. https://doi.org/10.1007/698_2014_260
7. Balakrishnan K, Ghosh S, Thangavel G, Sambandam S, Mukhopadhyay K, Puttaswamy N, Natesan D (2018) Exposures to fine particulate matter (PM2.5) and birthweight in a rural-urban, mother-child cohort in Tamil Nadu, India. Environ Res 161:524–531
8. Barnes B, Mathee A, Thomas E, Bruce N (2009) Household energy, indoor air pollution and child respiratory health in South Africa. J Energy South Afr 20(1):4–13
9. Bildirici M, Özaksoy F (2016) Woody biomass energy consumption and economic growth in Sub-Saharan Africa. Procedia Econ Finance 38:287–293
10. Bright S (2016) Diurnal rhythms of ambient air pollution due to vehicular traffic in Accra. A Master's thesis submitted to the University of Ghana, Legon in partial fulfillment of the requirement for the award of Master of Public Health degree
11. Chasant M (2019) A brief review of air pollution in Ghana, recent research findings and how cities around the world are tackling poor air quality. https://www.atcmask.com/blogs/blog/air-pollution-in-ghana. Accessed 30 Aug 2019
12. Clark NA, Allen RW, Hystad P, Wallace L, Dell SD, Foty R, Wheeler AJ (2010) Exploring variation and predictors of residential fine particulate matter infiltration. Int J Environ Res Public Health 7(8):3211–3224
13. Clean Cooking Alliance (2017) Ghana EPA incorporating household air pollution into air quality monitoring platform. https://www.cleancookingalliance.org/about/news/08-26-2017-ghana-epa-incorporating-household-air-pollution-into-air-quality-monitoring-platform.html. Accessed 30 Aug 2019
14. Climate and Clean Air Coalition (CCAC Secretariat) (2017) Reducing short-lived climate pollutants from domestic cooking, heating and lighting. https://ccacoalition.org/en/initiatives/household-energy. Accessed 24 Mar 2020
15. De Graft Aikins A, Addo J, Ofei F, Bosu WK, Agyemang C (2012) Ghana's burden of chronic non-communicable diseases: future directions in research, practice and policy. Ghana Med J 46(2):1–3
16. Ekpenyong CE, Ettebong EO, Akpan EE, Samson TK, Daniel NE (2012) Urban city transportation mode and respiratory health effect of air pollution: a cross-sectional study among transit and non-transit workers in Nigeria. BMJ Open 2(5):e001253
17. Esson J, Gough KV, Simon D, Amankwaa EF, Ninot O, Yankson PW (2016) Livelihoods in motion: linking transport, mobility and income-generating activities. J Transp Geogr 55:182–188
18. Ezeadichie N (2012) Home-based enterprises in urban space: obligation for strategic planning? Berkeley Plan J 25(1):44–63
19. Ezzati M (2005) Indoor air pollution and health in developing countries. Lancet 366(9480):104–106
20. Farmer A (2017) Tackling pollution is essential for meeting SDG poverty objectives. Perspective 27:1–8. United Nations Environment
21. Fosu AK (2015) Growth, inequality and poverty in Sub-Saharan Africa: recent progress in a global context. Oxf Dev Stud 43(1):44–59
22. Fullerton DG, Bruce N, Gordon SB (2008) Indoor air pollution from biomass fuel smoke is a major health concern in the developing world. Trans R Soc Trop Med Hygiene 102(9):843–851. Ghana Environmental Protection Agency, 2012

23. Ghana Health Service (2015) The health sector in Ghana: facts and figures, 2015
24. Ghana Statistical Service (2008) Ghana Living Standards Survey Round 5 (GLSS 5)
25. Ghana Statistical Service (2014). Ghana Living Standards Survey Round 6 (GLSS 6)
26. Gordon SB, Bruce NG, Grigg J, Hibberd PL, Kurmi OP, Lam KBH, Bar-Zeev N (2014) Respiratory risks from household air pollution in low and middle income countries. Lancet Respir Med 2(10):823–860
27. Gough KV, Tipple AG, Napier M (2003) Making a living in African cities: the role of home-based enterprises in Accra and Pretoria. Int Plan Stud 8(4):253–277
28. Gough K (2010) Continuity and adaptability of home-based enterprises: a longitudinal study from Accra, Ghana. Int Dev Plan Rev 32(1):45–70
29. Gujba H, Mulugetta Y, Azapagic A (2015) The household cooking sector in Nigeria: environmental and economic sustainability assessment. Resources 4(2):412–433
30. Güneralp B, Zhou Y, Ürge-Vorsatz D, Gupta M, Yu S, Patel PL, Seto KC (2017) Global scenarios of urban density and its impacts on building energy use through 2050. Proc Natl Acad Sci 114(34):8945–8950
31. Health Effects Institute (2018) State of the Global Air 2018: a special report on global exposure to air pollution and its disease burden. Health Effects Institute, Boston, MA
32. Heltberg R (2005) Factors determining household fuel choice in Guatemala. Environ Dev Econ 10(3):337–361
33. IEA (2017) Energy access outlook 2017: from poverty to prosperity. World Energy Outlook Special Report. Available from: http://www.iea.org/energyaccess
34. Jack DW, Asante KP, Wylie BJ, Chillrud SN, Whyatt RM, Quinn AK, Kaali S (2015) Ghana randomized air pollution and health study (GRAPHS): study protocol for a randomized controlled trial. Trials 16(1):420
35. Katoto PD, Byamungu L, Brand AS, Mokaya J, Strijdom H, Goswami N, Nemery B (2019) Ambient air pollution and health in Sub-Saharan Africa: current evidence, perspectives and a call to action. Environ Res 173:174–188. https://doi.org/10.1016/j.envres.2019.03.029
36. Kemausuor F, Obeng GY, Brew-Hammond A, Duker A (2011) A review of trends, policies and plans for increasing energy access in Ghana. Renew Sustain Energy Rev 15(9):5143–5154
37. Kraev TA, Adamkiewicz G, Hammond SK, Spengler JD (2009) Indoor concentrations of nicotine in low-income, multi-unit housing: associations with smoking behaviours and housing characteristics. Tobacco Control 18(6):438–444
38. Ladan M (2016) Appraisal of recent trends in environmental regulation of industrial pollution, energy sector and air quality control in Nigeria: 2011–2015. http://dx.doi.org/10.2139/ssrn.298 4266
39. Lawanson T, Olanrewaju D (2012) The home as workplace: Investigating home based enterprises in low-income settlements of the Lagos metropolis. Ethiop J Environ Stud Manag 4(5):397407
40. Lawanson T (2012) Poverty, home based enterprises and urban livelihoods in the Lagos metropolis. J Sustain Dev Afr 14:158–171
41. Longhurst J, Barnes J, Chatterton T, De Vito L, Everard M, Hayes EN, Prestwood E, Williams B (2018) Analysing air pollution and its management through the lens of the UN sustainable development goals: a review and assessment. WIT Trans Ecol Environ 230:3–14
42. Magaji JY, Hassan SM (2015) An assessment of air quality in and around Gwagwalada Abattoir, Gwagwalada, Abuja, Fct. J Environ Earth Sci 5(1):87–92
43. Maiyaki MB, Garbati MA (2014) The burden of non-communicable diseases in Nigeria; in the context of globalization. Ann Afr Med 13(1):1–10
44. Mbanya VN, Sridhar MKC (2017) PM_{10} emissions from cooking fuels in nigerian households and their impact on women and children. Health 9:1721–1733
45. Megbowon E, Mukarumbwa P, Ojo S, Olalekan OS (2018) Household cooking energy situation in Nigeria: insight from Nigeria malaria indicator survey 2015
46. Mensah JT, Adu G (2015) An empirical analysis of household energy choice in Ghana. Renew Sustain Energy Rev 51:1402–1411

47. Mensah JT, Marbuah G, Amoah A (2016) Energy demand in Ghana: a disaggregated analysis. Renew Sustain Energy Rev 53:924–935
48. Muller C, Yan H (2018) Household fuel use in developing countries: review of theory and evidence. Energy Econ 70:429–439
49. Nappier M, Balance A, Macozomo D (2000) Predicting the impact of home-based enterprises on health and the biophysical environment: observations from two South African settlements. Paper presented at the CARDO Conference on Housing, Work and Development: The Role of Home-Based Enterprises, University of Newcastle Upon Tyne, United Kingdom
50. National Malaria Elimination Programme (NMEP), National Population Commission (NPopC), National Bureau of Statistics (NBS), and ICF International (2016) Nigeria malaria indicator survey 2015. NMEP, NPopC, and ICF International, Abuja, Nigeria, and Rockville, Maryland, USA
51. National Population Commission (NPC) [Nigeria], National Malaria Control Programme (NMCP) [Nigeria], and ICF International (2012) Nigeria malaria indicator survey 2010. NPC, NMCP, and ICF International, Abuja, Nigeria
52. National Population Commission (NPC) [Nigeria] and ORC Macro (2004) Nigeria demographic and health survey 2003. National Population Commission and ORC Macro, Calverton, Maryland
53. National Population Commission (NPC) [Nigeria] and ICF International (2014) Nigeria demographic and health survey 2013. NPC and ICF International, Abuja, Nigeria, and Rockville, Maryland, USA
54. Oguntoke O, Opeolu BO, Babatunde N (2010) Indoor air pollution and health risks among rural dwellers in Odeda Area, South-Western Nigeria. Ethiop J Environ Stud Manag 3(2):39–46
55. Ogwumike FO, Ozughalu U (2012) Energy consumption, poverty and environmental linkages in Nigeria: a case of traditional and modern fuels for cooking. In: Adenikinju A, Iwayemi A, Iledare W (eds) Green energy and energy security: options for Africa. Atlantis Books, Ibadan
56. Onyebueke VU (2001) Denied reality, retarded perception or inaction? Official responses to the incidence of home-based enterprises (HBES) and its housing corollary in Nigerian cities. Cities 18(6):419–423
57. Osei-Boateng C, Ampratwum E (2011) The informal sector in Ghana. Friedrich-Ebert-Stiftung, Ghana Office, Accra
58. Oyedepo SO (2012) Energy and sustainable development in Nigeria: the way forward. Energy Sustain Soc 2(1):1–17
59. Pawankar R (2019) Climate change, air pollution, and biodiversity in Asia Pacific: impact on allergic diseases. Asia Pac Allergy 9(2):e11
60. Rim-Rukeh A (2015) An assessment of indoor air quality in selected households in squatter settlements in Warri, Nigeria. Adv Life Sci 5(1):1–11
61. Roy A (2005) Urban informality: toward an epistemology of planning. J Am Plan Assoc 71(2):147–158
62. Sa'ad S, Bugaje IM (2016) Biomass consumption in Nigeria: trends and policy issues. J Agric Sustain 9(2):127–157
63. Strassmann WP (1987) Home-based enterprises in cities of developing countries. Econ Dev Cult Change 36(1):121–144
64. Sundell J (2004) On the history of indoor air quality and health. Indoor Air 14(Suppl 7):51–58. https://doi.org/10.1111/j.1600-0668.2004.00273.x
65. The Lancet (2016) Air pollution-crossing borders. Lancet 388(10040):103. https://doi.org/10.1016/S0140-6736(16)31019-4
66. Tipple G (2005) The place of home-based enterprises in the informal sector: evidence from Cochabamba, New Delhi, Surabaya and Pretoria. Urban Stud 42(4):611–632
67. Tipple G, Coulson J, Kellet P (2002) The effects of home-based enterprises on the residential environment in developing countries. In: Romaya S, Rakodi C (eds) Building sustainable urban settlements: approaches and case studies in the developing world. ITDG Publications, London
68. UN-Ghana (2017). Goal 1: End poverty in all its forms everywhere. http://gh.one.un.org/content/unct/ghana/en/home/global-agenda-in-ghana/sustainable-development-goals/SDG-1-no-poverty.html Accessed 30 August 2019

69. UNEP (2015) Air quality policies in Ghana. https://www.unenvironment.org/resources/policy-brief/air-quality-policies-ghana. Accessed 30 Aug 2019
70. United Nations (2018) The sustainable development goals report 2018
71. United States Environmental Protection Agency (1994) Characterization of municipal solid waste in the United States: 1994 update
72. Urhie E, Odebiyi J, Popoola R (2017) Economic growth, air pollution standards enforcement and employment generation nexus in the Nigerian context. Int J Innov Res Dev 6(5):19–27
73. Vanker A, Barnett W, Nduru PM, Gie RP, Sly PD, Zar HJ (2015) Home environment and indoor air pollution exposure in an African birth cohort study. Sci Total Environ 536:362–367
74. WHO (World Health Organization) (2012) http://www.who.int/gho/phe/en/. Accessed 30 Aug 2019
75. WHO (World Health Organisation) (2019) Air pollution and the sustainable development goals
76. WHO (World Health Organization) (2017) 7 million premature deaths annually linked to air pollution. Retrieved from: http://www.who.int/mediacentre/news/releases/2014/air-pollution/en/
77. Yankson PW (2000) Houses and residential neighbourhoods as work places in urban areas: the case of selected low-income residential areas in greater Accra metropolitan area (GAMA), Ghana. Singap J Trop Geogr 21(2):200–214
78. Zhang J, Smith KR (2003) Indoor air pollution: a global health concern. Br Med Bull 68(1):209–225
79. Zhou Z, Dionisio KL, Arku RE, Quaye A, Hughes AF, Vallarino J, Spengler JD, Hill A, Agyei-Mensah S, Ezzati M (2011) Household and community poverty, biomass use, and air pollution in Accra, Ghana. Proc Natl Acad Sci USA 108:11028–11033. https://doi.org/10.1073/pnas.1019183108

A Study of Housing, Good Health and Well-Being in Kampala, Uganda

Emmanuel Musoke Mutyaba

Abstract This chapter presents adequate housing as a necessity for health and well-being. To the literature it adds a theoretical argument aimed at convincing African governments and peoples that adequate housing is a human right that needs to be respected in the bid to promote the health and well-being of people as demanded by the United Nations' Sustainable Development Goal 3. It is argued that human beings do not desire adequate housing for luxury or aesthetic reasons but as a natural imperative to which governments must respond. Given that a home should be a place where people feel safe and relaxed while feeling a sense of belonging and self-esteem and acquiring moral values therein, it is also noted here that only an adequate house can constitute a home. The chapter used a case study research design and a qualitative approach. It was mainly a desk study research. Further information was collected from field research. It was concluded that type of dwelling influences one's physical and emotional state as well as productivity. It is recommended that the right to adequate housing as not merely a dwelling place should be taken seriously for realization of SDG 3, which relates to people's health and well-being.

Keywords Dwelling · Housing · Home · Well-being · Human right

1 Introduction

Given that the United Nations' Sustainable Development Goal 3 emphasises human well-being and good health, this chapter presents literature that links health and well-being to adequate housing. The right to housing is indeed a human right, being inherent in our very nature and hence essential for our well-being. A house is a building that is constructed purposely to be the dwelling place of a person or a family [1]. A good house must be ventilated, have a door or doors, have a window or windows and firm walls and be located in a hygienic neighborhood. Such is housing

E. M. Mutyaba (✉)
Uganda Martyrs University, Kampala, Uganda
e-mail: emutyaba@umu.ac.ug

© The Author(s), under exclusive license to Springer Nature Singapore Pte Ltd. 2021
T. G. Nubi et al. (eds.), *Housing and SDGs in Urban Africa*, Advances in 21st Century
Human Settlements, https://doi.org/10.1007/978-981-33-4424-2_9

that transmits the benefits of a home to which human beings are inclined by nature. It is the kind of housing that can meet the targets of Sustainable Goal 3.

Sustainable goal 3.1 aims at reducing maternal mortality, which requires adequate housing. Many cases of maternal mortality occur due to mothers leaving hospitals after delivery and returning to homes in populated slums, which frequently lack hygienic toilet facilities, proper sewage disposal and waste management systems. Under such circumstances, such women often end up with infections such as the influenza A virus (H1N1) and bacterial infections such as streptococcus pneumonia, escherichia coli and clostridium perfringens. Moreover, according to [2] poor toilet facilities lead to genital infections.

Sustainable goal 3.2 seeks to end all preventable deaths of children under five years of age to at least 12 per 1000 live births and under five mortality to at least as low as 25 per 1000. According to World Health Organization findings, "Globally 2.5 million children died in the first month of life in [3], approximately 7000 newborn deaths occur every day, with about one third dying on the day of birth and close to three quarters dying within the first week of life" [3]. Needless to say that there is a link between poor sanitation, lack of clean water and hygiene (which characterize poor housing) and infant mortality. Sarina [4] notes that one out of every five newborn deaths in the developing world is due to lack of safe water, sanitation and hygiene. This points to the urgency of adequate housing that guarantees clean water, sanitation and hygiene.

Sustainable goal 3.3 is focussed on communicable diseases such as AIDS, tuberculosis, hepatitis, waterborne diseases, as well as malaria and other neglected tropical diseases. According to [5], "Approximately 940,000 Ugandans are living with HIV/AIDS. HIV prevalence is 10% in urban areas and 6% in rural areas. The HIV prevalence in Kampala is 8.5%." HIV is prevalent in urban areas because of the presence of slums and homeless people. King [6] observes that

> Housing is unique as a social determinant of health shaping our daily lives – while also a manifestation of broader, antecedent, structural processes of inequality and marginalization that are fundamental drivers of HIV vulnerability and poor HIV health outcomes....The rates of HIV incidence is as much as 16 times greater among persons experiencing homelessness. There is a significant association between housing instability and HIV sex and drug risk behaviors among persons with the same socio-demographic, clinical, substance use, mental health, and service use characteristics.

As a factor predisposing people to HIV infection, drug abuse is common among homeless people and causes them to lose control over their sexual desire. The homeless tend to be restless and frustrated, as they see themselves as social outcasts. As such, sustainable goal 3.5 seeks to address this situation via the provision of adequate housing for all. To be sure, adequate housing is a human right that needs to be respected in the bid to promote human health and well-being as demanded by SDG3.

This is a qualitative research involving a case study. The researcher relied mainly on secondary data sources such as books, newspapers and journals. Some primary data were also collected from three slum areas of Kampala: Kisenyi, Makerere Kivulu, and Bwayise, where 24 members from ten families were randomly selected for purposeful conversations and interactions. Data were also collected from six people purposefully chosen from six well-to-do families from Buziga, which is one of the rich zones of Kampala with adequate housing. The data collection process utilized conversations, open-ended questions and observation to get primary data.

There is rampant violation of the right to housing in Uganda today despite the existence of national legal and policy frameworks regarding housing such as the Rent Restriction Act (which prohibits unnecessary increment in rent charges), the Land Act Cap 227, Land Acquisition Act Cap 226, and the National Land Policy, etc. There is a high number of homeless people in Uganda, particularly in the capital city Kampala. According to a 2019 report, over 173,000 people around Kampala were displaced due to natural disaster and internal conflicts [7] and approximately 66% of Kampala residents live in inadequate housing [8]. People tend to think that housing is merely a dwelling place. Lack of proper knowledge of the right to housing has led the government of Uganda to respond to the increased demand for biofuel by the West and to accede to the World Bank's demand to boost agricultural production by 70% in the bid to feed the world's estimated 2 billion people by 2050 by encouraging landlords to sell off their land to local and foreign agricultural investors [9]. When these agro investors buy off land from landlords, they chase tenants out of the land and demolish their houses without proper compensation. Moreover, there is no government plan to house such people. Most of these former tenants flee to big cities, particularly Kampala, leading to increment in house rent due to high demand. Some landlords in Ugandan urban areas demand rent payments in dollars [10], a practice that has worsened the housing problem for many citizens. The situation has led to a state of conflict between investors, landlords, tenants, judicial institutions and the commissioners for land registration. It has also increased the number of slums on the city peripheries.

2   Literature Review

Brath [11] argues that the right to housing does not only have an ethical basis in the principle of justice and ideals of common wealth but it is also based on a highly pragmatic perspective, since housing plays a central role in shaping people's lives. Brath further noted that a commitment to the right to housing should be the foundation not only for the housing policy but also for a new social agenda. Brath insists that a host of new social relationships, personal health, safety, employment opportunities, decent education, security of tenure, economic security and so forth would emerge if the right to housing were granted to people. When a person changes from living in a shelter or from a poor house to a good house in terms of quality, their self-perception changes. They begin to feel that their status has been raised, so they begin to think big, e.g., by wanting a high-quality job, earning more money, offering their children quality education and quality medical care, etc. [11]. Indeed, decent housing plays a central role in fixing one's place in society and in the local community. Adequate or decent housing is a necessity for good living both physically and psychologically.

Physically, it is essential for good health and in psychological terms it gives one a sense of permanence, belonging and security. To feel at home is to feel welcome, since a home is a place where one can be himself or herself [12: 29]. Housing creates a sense of privacy while lack of privacy causes stress.

Perkins et al. [13] claim that a home plays a fundamental role in all kinds of narratives about human life. Moreover, regarding those benefits of a house as a home, earlier research insists on the comfort offered by the different places in the house: kitchen, bedrooms, drawers, cupboards, etc. [14: 136]. Consequently, those providing housing for homeless people should bear these aspects in mind.

Poor or inadequate slum housing, which usually lacks space and ventilation as well as a proper waste collection and management system, is known to cause bad health conditions for dwellers and their neighbors. For instance [15] linked the 1871 cholera outbreak in Chicago to the inadequate housing conditions of that time. According to the author, authorities responded by enacting tenement house laws in large cities such as Chicago and New York. It has been observed that failure to provide decent housing leads to vandalism and other criminalities, ill-health and underachievement by family members, a situation that ultimately affects society at large [16: 35–36]. According to [17] inadequate housing conditions contribute to family dysfunction, limit a family's capacity to receive guests and denies children the necessary space to do their homework, thus leading to failure in their studies.

Lerner [18] found that, compared to housed children, children who live in poor shelters miss classes quite often, show signs of anxiety and depression and demonstrate behavioural disturbances such as throwing tantrums and displaying aggressive behaviour. This means that poor housing significantly affects well-being.

In 1996 the Second United Nations Conference on Human Settlements, called Habitat II, confirmed the commitment of participating governments to guarantee the right to housing as a human right and more specifically to provide effective protection from forced evictions that are contrary to the law, thereby taking human rights into consideration [19: V]. The United Nations Human Rights Commission of 1993 also regards forced eviction as a gross violation of human rights. HABITAT International Coalition also promotes explicit recognition of the right to housing as a basic human right and describes forced eviction as a flagrant violation of this right [19: V].

3   Housing in Uganda: A Media Perspective

To promote the right to housing, the Ugandan Parliament enacted what is called the Rent Restriction Act, which seeks to consolidate the law on control of rents of dwelling houses and business premises. This Act seeks to ensure that the cost of a dwelling house, in the estimation of a competent board, is reasonable in relation to prevailing wage rates at the time of the erection of the dwelling house. The Rent Act states that

If the competent board evaluates the price of a dwelling house or premise as not reasonable, it sets the prices that it considers reasonable in all circumstances of the case. The competent board which in fixing the rent shall take into account the capital value of the site, as assessed by the commissioner of lands and surveys, and the capital cost of the dwelling house or premises but in no case shall the gross rent exceed 10% of the capital cost of the building plus 5% of the capital value of the site, except that subject to any right of appeal and to Sect. 9(2), the standard rent of a dwelling house or premises shall not be varied after it has been fixed by a competent board whether or not the rent was fixed before or after the commencement of this Act. No owner or lessee of a dwelling house or premises shall let or sublet that dwelling house or premises at a rent which exceeds the standard rent…..any person, whether the owner of the property or not, who in consideration of the letting or subletting of a dwelling house or premises to a person asks for, solicits or receives any sum of money other than rent or anything of value whether the asking, soliciting or receiving is made before or after the grant of a tenancy commits an offence and is liable on conviction to a fine not exceeding ten thousand shillings or to imprisonment not exceeding six months or to both such fine and imprisonment; except that a person acting bona fide as an agent for either party to an intended tenancy agreement shall be entitled to a reasonable commission for his or her services….Any person who makes it a term of any agreement to let or sublet any dwelling house or premises that more than six months rent should be paid in advance shall be deemed to commit an offence under subsection is liable on conviction to the same penalties as a person guilty of an offence under that subsection and is liable on conviction to the same penalties as a person guilty of an offence under that subsection [20].

In 2018 Uganda enacted the Landlord and Tenant Bill, which regulates the relationship between landlords and their tenants. The Bill seeks to accommodate the interests of tenants and landlords, aiming to check the intimidation of tenants by their landlords. This Bill requires that all rent shall be paid and recorded in Ugandan shillings. This was contrary to the initial proposal that parties can agree to transact rent payments and recordings in any other currency in the agreement. National newspapers such as *The Independent* (June 27, 2019) and *Daily Monitor* (June 26, 2019) noted that the most controversial provision is that all rent shall be settled and recorded in shillings, contrary to the initial proposal that parties can agree to transact in any other currency in the agreement. The Bill has several amendments related to the duties and rights of landlords and tenants in rented residential houses and commercial premises. The Bill also specifies that tenancy disputes shall be settled in local council courts and that no landlord should evict a tenant before seeking a court order to do so or else they will incur a penalty of five million Uganda shillings or a jail term of one year or both upon conviction.

However, despite the Landlord and Tenant Bill, illegal tenant evictions are still rampant in Uganda. Many victims of eviction are in Kampala and are neither compensated nor resettled; for instance, it was reported in *The Observer* (May 30, 2018) that Pastor Samuel Kakande evicted the people of Kaamaliba from their land to make way for his company Aqua World Uganda Limited. Many landlords complain about the Landlord and Tenant Bill of 2018, lamenting that it mostly favors tenants and puts landlords at a disadvantage, particularly where landlords have to seek consent from tenants to increase rent fees or in cases of eviction (Landlord and Tenants Bill of 2018. 1ff).

While real estate agents rejected the proposal to restrict rent payment in the local currency on the grounds that the shilling is often unstable, Kampala city traders support the Bill in the hope that it will help to resolve disputes between house owners and tenants in the city. Kampala City Traders Association (KACITA) argued that landlords in Kampala had been fleecing tenants in numerous ways. Therefore, the traders supported the outlawing of rent payment in foreign currency, arguing that it puts pressure on the local currency and reduces its value while overburdening tenants [21].

It is precisely due to those landlords who have refused to comply with the Landlord and Tenant Bill, together with the government's laxity in enforcing the bill, that the problem of housing affordability in Uganda's urban areas has worsened, particularly in Kampala. The attitude of such landlords and of the government reflects lack of adequate knowledge of the right to housing, since one cannot respect and fully implement what one does not understand well. According to *The Daily Monitor* (April 30, 2018), affordable housing continues to be a problem for many Ugandans even when there has been a substantial increase in the number of estate developers. *The Daily Monitor* (March 18, 2015) observed that the World Bank report on the Fifth Uganda Economic Update warned that Kampala might become a megaslum in ten years' time if housing is not planned well. The report also stated that failure to unlock the potential of cities might result in a deceleration of growth and the emergence of dysfunctional slum cities where people live in appalling conditions.

Furthermore, housing state minister Chris Baryomunsi was quoted in *The Observer* (April 10, 2018) as stating that government was aware of a report indicating that at least 30 percent of the cost of acquiring a house was due to the expense incurred by real-estate dealers in improving access to utilities such as roads, water, electricity and Internet supply. According to minister Talemwa:

> We would like to work with real estate dealers in improving access to these utilities, as a means of lowering the cost of acquiring a home....We are also looking at tax waivers for those in real estate, while calling for a mix in improved building technology. Plans are going on to see how public servants can be helped to obtain housing.... Consequently, the current housing deficit is estimated at 2.1 million housing units and is expected to reach 3 million by 2030. (*The Observer*, April 10, 2018)

The minister reported that a principal housing officer in the ministry of housing, Dave Khayangayanga, had also promised that the ministry would work with the poor homeless to ensure that they were involved in building their own houses. According to Khayangayanga, "Looking at population projections, more people will live in the city by 2025 due to high rate of urbanisation." Khayangayanga reported that it was regrettable that when in 2006 government set up houses for the poor homeless in Masese, the targeted beneficiaries rented out their new homes and returned to the slums. This example indicates the need for a theoretical argument to convince not only the government to respect the right to housing but also to convince people about the benefits of having a decent permanent house as one's home, not just having a four-walled structure and a roof, hence this study.

In *The Observer* (May 6, 2018) Kamoga Jonathan wrote:

> The Ministry of Lands, Housing and Urban Development has called upon both local and foreign investors to interest themselves in the country's largely ignored housing sector,

which has been underperforming every year. Last year (2017), the National Housing and Construction Company (NHCC), which is a government body mandated with developing the housing sector, had a shortfall of 140,000 housing units after constructing only 60,000 units out of the expected 200,000 per year. Authorities blame the housing shortfall on scanty government funding of the sector among other issues like land wrangles and compensation disputes in areas where development plans have been allocated to take place like Kasokoso zone in the out skirts of Kampala. (*The Observer*)

According to the first general report of AGFE to the executive director of UN-Habitat, in 2002, approximately 1500 people living on the Naguru and Nakawa estates in Kampala were threatened with eviction by Kampala City Council, which intended to use the land for construction of retail and middle-income housing [22: 39].

4 An Argument for the Right to Adequate Housing

Although the literature advocates housing for everyone, it is important to state that it is not a matter of giving houses to all but that such houses have to be adequate. An adequate house is more than four walls and a roof but a safe and secure house located in a decent, unpolluted neighborhood and having considerable space, hygienic toilets, kitchen and clean water sources. The conception of adequate housing in this work is based on the United Nations 'Human Settlement Programme of 2005 no. 7', which states that adequate housing must involve legal security tenure that guarantees protection against forced eviction and harassment (p. 21). It must be affordable in the sense that the cost of attaining a house must not compromise the attainment of other basic needs like food, clothing, medication, education and others. It must be habitable, with adequate space protecting inhabitants from climatic threats such as heavy heat, cold, flooding, earthquakes, etc. It has to have a dumping place and must be far from polluting sources, in addition to availability of infrastructure, safe drinking water, sanitation, food storage and refuse disposal facilities. It must also take into account the cultural values of occupants.

However, especially in Kampala, adequate housing has been available to only the rich. Lalaine [23] cites the Uganda Bureau of Statistics (UBOS) as stating that "the nationwide house prices rose strongly by 6.2% during the latest quarter, Q3 2019 (5.5% inflation-adjusted). In Kampala, Central and Makindye, residential property prices surged 18.5% during the year to Q3 2019 (16.3% inflation-adjusted)." The greater number of Ugandans in rural and urban areas live in poor housing conditions, as residents are either poorly housed or homeless. The majority of Kampala city dwellers live in slums, with unprecedented proliferation of slums and informal settlements and chronic lack of adequate housing characterizing Kampala suburbs and other Ugandan towns. This is due to both poverty and the Ugandan government's neglect of the right to adequate housing for citizens. Affordability to adequate housing remains a problem in Uganda due to poor salaries and wages as well as lack of jobs for many citizens. It is worrisome that despite Uganda being a signatory to the International Human Rights Covenant, the nation has a poor attitude towards the economic and social cultural rights article 11, which stresses the right to adequate housing. Indeed, having a secure place to live in is one of the elements that protect human dignity.

Based on Broth's ideas about how level of housing affects people's self-esteem, the researcher studied the three Kampala slums of Makerere Kivvulu, Bwayise, and Kisenyi, as well as one rich neighborhood, Buziga. The researcher selected three 10-year-olds from each of the three slums—two girls and a boy. The researcher also randomly selected three 10-year-olds from Buziga—two girls and a boy. The children were in Form Five and were Catholics according to their teachers. I took all of them to a well-attended mass at Lubaga Cathedral in Kampala City. I realized that those children from the slum areas immediately felt out of place as soon as we entered the filled cathedral. They went at once and stood in the back corner of the cathedral, sort of hiding from society because they felt unworthy of the place. They evinced a sense of being lost, as they told me after the mass. However, those from Buziga did not show any sign of intimidation or unworthiness, as they even moved around in the cathedral after mass to satisfy their curiosity about the beauty of the Basilica. The Buziga children kept asking questions about the statues while those from the slum areas quietly followed them with hands folded across their chests.

After the service the children from the slums rushed out of the cathedral faster than any of us. When I asked them why they had behaved the way they did, they said it was because of the huge size of the place and the crowd apparently staring at them. I had expected the kids from the slum areas to be more curious about seeing things they did not have at home but the contrary was true. Continuing the research, I joked with six youths from Bwayise by telling them we would find time to visit the parliament building. They all laughed saying that as soon as we got there we would be arrested immediately as thieves. One of them said that he even feared to walk from the middle of Kampala City because he thought he might be mistaken for a hooligan and that those were places for rich people. Yet, based on the researcher's observations, the slum boys were well behaved dressed smartly and were not different in appearance from those who came from adequate housing. The slum kids saw themselves as inferior. I received a similar response from 12 young women who sing in a Pentecostal church choir in Kisenyi when I suggested that they organize a gospel concert in Kampala City Square to collect some money to construct a better church. They said nobody would come from those Kampala houses to attend to a concert organized by a Kisenyi slum church. Yet I thought they sang well. I therefore concluded that denying people a right to adequate housing tends to damage their self-esteem, giving them a low self-gauging that can hinder their productivity.

Adequate housing constitutes a home but inadequate housing is merely a dwelling. A home is the sphere where one lives at peace and feels a sense of esteem. Ginsberg is perhaps right to say that humanity constitutes *homonity*/the aspect of having a home(cited in [12: 29]. Indeed, a home constitutes the elements that we need to express our humanity, hence the assertion that the right to adequate housing is a fundamental human right—a natural entitlement—since we do not desire housing merely for aesthetic or luxurious reasons. Housing as a home is an essential natural phenomenon for the creation of personhood (quality or status of a being that has conscience, rationality and moral sense). For that matter, the right to housing is a right to personhood and this is the basis of all other rights.

A home reveals the dynamism of human interaction with a supposed objective reality that cannot merely be reduced to a physical location but a metaphysical reality that is invested with human significance; it is the source of human values and of the moral sense instilled in an individual. A home is the unique future for analysis and acquisition of human values, beliefs and desires. A home creates identity, bringing about a sense of belonging via a physical space (adequate housing) even though it goes beyond a mere physical place since upon it one's internal life and physical and non-physical activities depend. Housing as simply a dwelling without the reality of a home does not communicate values that guide the inner formation to personhood, hence its leading to a feeling of lostness (one feels that he or she is nowhere; he or she does not feel fit to be anywhere in this world), loss of self-esteem and lack of value, which, in turn, lead to hooliganism, prostitution and other forms of sexual immorality, drug abuse, violence, robbery, etc. Each individual is an instant of a portable home; to be without a home is to be diminished in personhood. As [24: 11] notes, the reality of lacking a home forces upon its victims a linguistic and psychological disenfranchisement that bleeds into the social and psychological fabric of life.

Adequate housing is a physical place but it is inseparable from the interior psychological counterpart of a discrete and enduring metaphysical place known as a home, which gives one a sense of freedom and comfort or ease, hence an expression like "feel at home." This study shows that children living in shelters and in inadequate housing that does not communicate the sense of a home tend to face emotional disturbances. Therefore, lack of adequate housing that serves as a home jeopardizes the development of systems of meaning that are basic in the formation of an individual's identity. Interacting with the children from the slum area, I perceived indicators of a feeling of being out of place as well as a sense of perturbation and discomfort that was sometimes expressed in form of physical violence and hypersensitivity, all of which did not manifest in the children from Buziga.

Indeed, people who live in inadequate housing tend to feel a gap in their lives that produces a feeling of discomfort and anxiety; they feel a sense of misfortune that sometimes leads to feelings of envy and self-hate [18: 452]. They often become social threats due to lack of socialization into the civic and moral behaviors that a home offers, hence the reason why slums are well known for all sorts of immorality and criminality compared to decent residential areas. Inadequate housing, like those in slum areas, forces people to lead a survivor's life, with little emphasis on moral and civil values. Therefore, the researcher takes the position of existential philosophers such as Martin Heidegger, Fredrick Nietzsche, Jean-Paul Sartre and Emmanuel Levinus. In *Sein und Zeit* [25: 147], Heidegger notes that a dwelling constitutes a human way of being in the world, as very aptly expressed in German with phrases such as *ichbin* ("I am," literally "I dwell"), *du bist* ("you," literally meaning "you dwell"). *Bin* is a version of an old German word *bauen*, which means dwell. Heidegger's position supports the argument that houses are not built for aesthetic or luxurious reasons only.

Heidegger relates the state of homelessness to a state of estrangement, whereas Nietzche describes it as a wondering or peripatetic state where finding a dwelling place is compared to returning home [26: 219]. For Sartre, home is the product of a contingent historical-political situation of needs that is brought about by external necessities and is fundamental to the identity of human nature itself, an inevitable feature that may be abstracted from socioeconomic relations [12: 49–50]. Heidegger in his work "Building, dwelling, thinking" [25: 147–160], mourns modern architecture's detachment of man from his natural environment which he refers to as man's authentic dwelling. He calls this modern architecture, technological dasein which give nary a thought to natural elements such as earth and sky for man's surroundings. Holst J (27: 53) says that although Heidegger was not directly referring to physical architecture, nevertheless his ideas very much influenced the modern perspective of an adequate housing. It is therefore clear that these existential philosophers affirm that it is human nature to crave being in a home. Indeed, all of the foregoing arguments can be used to support the human right to adequate housing that constitutes a home.

5   Conclusion and Recommendations

The SDG3 is about adequate housing that promotes a sense of belonging, calmness, freedom, autonomy and ease, etc., in an environment that instills moral values, beliefs and acquisition of civic qualities. Inadequate housing is responsible for high maternal mortality rates and infant deaths due to poor conditions, HIV infections and premature pregnancies due to sex abuse, incest, hooliganism, drug abuse and other criminal activities. It is also responsible for diseases such as cholera, tuberculosis, malaria, typhoid, etc.

Based on the foregoing, the researcher makes a number of recommendations. One, the Ugandan government needs to recognize that the right to adequate housing is a fundamental human right which violation jeopardizes other human rights. The Ugandan government should institutionalize and implement the right to adequate housing. Two, the Ugandan government, through its ministry of housing, should utilize affordable local materials and engineers for the construction of adequate housing. Three, with regard to the problem of rampant evictions, the Ugandan government should activate the existing legal and policy framework on housing in order to punish violators.

References

1. Branhart KR (1995) The World book dictionary, vol 1. A-K INC, Chicago
2. Rigouzzo A (2017) Maternal deaths due to infectious cause, results from the French confidential enquiry into maternal deaths. Elsevier Masson, Paris

3. World Health Organization (2019) New born: reducing mortality. https://www.who.int/news-room/fact-sheets/detail/newborns-reducing-mortality. Accessed 8 Nov 2019
4. Sarina P (2019) One in five new born deaths could be prevented. https://www.wateraid.org/us/one-in-five-newborn-deaths-could-be-prevented. Accessed 4 May 2020
5. Nussbaum L (2010) Kampala, Uganda: housing for people with HIV/AIDS. https://www.thebodypro.com/article/kampala-uganda-housing-people-living-hiv-aids. Accessed 23 Nov 2019
6. King C (2019) Housing instability: a barrier to urban aids responses. New York City, Barbican
7. Amnesty International (2019) Uganda, hundreds left homeless by forced eviction. https://reliefweb.int/…/uganda/uganda-hundreds-left-homeless-forced-evictions. Accessed 4 May 2020
8. Habitat (2019) Annual Preogress Report 2019. Published in 2020 by UN-Habitat, HS Number HS/003/20E, ISBN 978-92-1-132864-6
9. Food and Agriculture Organization (2009) How to feed the world in 2050. www.fao.org/filead min/templates/wsfs/. Accessed 4th May 2020
10. The Monitor Uganda (2019) Parliament press bill. https://www.monitor.co.ug/News/National/Tenants-pay-rent-Uganda-shillings-Parliament-passes-Bill-/688334-5172914-15i3bwaz/index.html. Accessed 18 June 2019
11. Brath GR, Stone ME, Hartman C (eds) (2006) A right to housing: foundation for a new social agenda. Temple, Philadelphia
12. Abberno J (1999) Ethics of homelessness: philosophical perspectives. Atlanta, Amsterdam
13. Perkins HT, Winstanley AN (2002) The study of home from a social scientific perspective: an annotated bibliography. New Zealand, Canterbury
14. Bachelard G (1964) The poetics of space. Beacon Press, Boston
15. Freidman LM (1968) Housing urban America. 2ndn. Inc, Chicago
16. Hynes HP, Brugge D, Watts J, Lally J (2000) Public health and the physical environment in Boston public housing: a community-based survey and action agenda. Plann Prac Res 15(1–2):31–49
17. Kazol J (2011) Rachael and her children: homeless families in America. Crown, New York
18. Lerner MR (2002) The handbook of applied developmental science, vol 1. Sage, New York
19. Azuela AE, Ortiz E (1998) Evictions and the right to housing experience from Canada, Chile, the Dominiccan Republic, South Africa and South Korea, International Development Research Centre, Ottawa-Canada
20. Uganda Legal Information Institute (1949) Rent restriction act https://ulii.org/ug/legislation/consolidated-act/231. Accessed 4 May 2020
21. Kampala City Traders Association (2018) KACITA-supports landlord and tenant law. https://www.parliament.go.ug/news/3146/kacita-supports-landlord-and-tenant-law. Accessed 3 Aug 2019
22. UN-HABITAT (2005) Forced eviction-towards solutions? First report of AGFE to the executive director of UN-HABITAT., Nairobi
23. Lalaine CD (2019) Uganda's housing market gradually improving. https://thepromotar.com/ugandas-housing-market-gradually-improving-2/
24. Amir L (2018) New frontiers in philosophical practice. Cambridge University Press, Cambridge
25. Heidegger M (1971) Building, dwelling, thinking. (trans: Hofstadter A). Harper & Row, New York
26. Sirtori V (1996) Le grandi opere della filosofia. Garzanti, Milano
27. Holst J (2014) Rethinking dwelling and building. Universidad San Jorge. https://www.research gate.net/publication/323187928. Accessed 1 January 2021

Relocation and Informal Settlements Upgrading in South Africa: The Case Study of Mangaung Township, Free State Province

John Ntema

Abstract Despite being a global phenomenon, the practice and implementation of the informal settlement upgrading policy as a possible alternative to conventional public housing policy, has a unique history in post-apartheid South Africa. Some case studies have revealed that shack dwellers have experienced tenure security whether they were going to reside on the same stands with upgrading, move to transition camps while upgrading was in process or whether they had the possibility of being relocated to a new housing development project. Relocation is often resisted in that it threatens existing livelihoods, continued schooling and community networks and, not least, the home. Against this background, this chapter has a threefold aim. First, it assesses the knowledge and policy gap created by a pro-in situ approach to informal settlement upgrading in South Africa. Second, it contributes new evidence from an analysis of households' perceptions of basic service infrastructure, amenities and governance as expressed through a survey, in-depth qualitative interviews and focus group discussions in a relocation site in Mangaung Township (Bloemfontein). Third, it demonstrates how informal settlement upgrading through relocation may undermine the principle of participatory project planning and design.

Keywords Relocation · Informal settlement upgrading · Security of tenure · Infrastructure

1 Introduction

In the context of policy discourse on slum or informal settlement upgrading, Ziblim [51] and Mistro and Hensher [31] argue that South Africa advocates two dominant approaches: "in situ" and "total redevelopment". Other research including policy equates "total redevelopment" with "relocation" or "relocation to green-fields development" [9, 13, 14, 32] and "orderly development" in uninhabited land [14]. For the purpose of the chapter, "relocation" is a preferred concept. Upgrading of informal

J. Ntema (✉)
Department of Human Settlements, University of Fort Hare, Alice, South Africa
e-mail: lntema@ufh.ac.za

settlements is comprehensively outlined in the Breaking New Grounds (BNG) policy and its accompanying Upgrading Informal Settlement Plan (UISP) [9], while is mentioned no more than five times and with no details in the Housing White Paper [8]. For upgrading purposes, the BNG policy gives preference to the in situ approach as first choice [9]. Consequently, the policy dynamics around implementation of in situ approaches are comprehensive while those on relocation are largely superficial. A significant amount of research and literature focuses on implementation of in situ upgrading compared to relocation sites. A pro-in situ focus in policy, literature and research seems to have created not only a knowledge gap but also an opportunity for studies such as this one.

Against this backdrop, the aim of the chapter is to show how most studies on upgrading and relocation seem to have failed to evaluate the performance of upgrading in relocation sites from a policy and governance point of view. Essentially, the chapter argues that informal settlement upgrading in a relocation site may promote partial and selective access to basic infrastructural development at the expense of the policy stance on "incremental upgrading" and the principle of "good governance". To achieve this aim, the chapter is structured as follows. First there is a brief literature overview on informal settlements upgrading in developing countries. This is followed by a contextualisation of the South African policy on informal settlements upgrading within selected Sustainable Development Goals (SDGs). Next follows a literature overview on informal settlements upgrading in South Africa. Then follows an analysis of perceptions of government officials and households in the Phase 6 area. Finally, there is a conclusion and policy recommendations section.

2   A Historical Overview of Informal Settlements Upgrading in the Developing Context

The prevalence of informal settlements is a global phenomenon. As such, it is ideal for countries to strive towards attaining a universal and conventional approach to it. In Africa and to some degree Latin America, governments' response to informal settlements has been twofold. First, there has been direct action that included either demolition and replacement or demolition without replacement. Second, there has been provision of mass housing through site and service schemes [29]. While the 1950s and early 1960s saw the widespread adoption of demolition and replacement of squatter settlements in different parts of Africa [1], the late 1960s and early 1970s reflected, in particular, Turner's influence on low-income housing policies [12]. The involvement of the World Bank in funding site and service schemes saw the beginning of the recognition of squatter settlements as a base for settlement upgrading [1, 44].

The literature on settlement upgrading strategies indicates that the widespread adoption of settlement upgrading programmes in different countries required a change in government's attitude and role in rolling out these programmes. For instance, sound intergovernmental relations between the national and the local sphere of government made it possible for the cities of Shangai and Mumbai to move swifter with the relocation of households and the construction of megaprojects [18]. Underpinned by the principle of "community participation", several activities could be ascribed to Brazil's successful implementation of some of its initial upgrading projects. First, meetings with the affected communities were held on a regular basis and all programmes were carried out with community agents who had been elected by the communities in question [7]. While the community remained the key stakeholder, also driving this mutual relationship was the continuous institutional presence of local government in areas that had been upgraded—something always seen as imperative. It would thus be appropriate for the chapter to argue that underpinning the success of the settlement upgrading schemes in Brazil could be the application of a constant state-citizen interaction. At the heart of these interactions, including the principle of a community-based approach (see also [3, 7, 19]), is the view that projects should not be seen as one-off interventions but rather as long-term engagements. Other countries with similar success in settlements upgrading include Zambia, Kenya, Sri Lanka, Pakistan, New Guinea, and Indonesia [1, 47].

Notwithstanding the successes discussed above, the implementation of policies and programmes on informal settlement upgrading in developing countries has not been without criticism. For instance, despite the widely held view that the success of any informal settlement upgrading project is dependent on the extent to which the affected community is allowed to participate in key decision-making processes, particularly during crucial project phases such as planning and implementation, Skinner [43] advances a twofold argument. First, he argues that community participation processes in upgrading are commonly viewed as a way to organise communities. Second, he notes that such processes seldom lead to local control within site-and-service projects and upgrading programmes.

3 Sustainable Development Goals, Informal Settlement Upgrading and Policy Discourse in South Africa: Is There Room for Relocation?

Despite being criticised for a lack of policy implementation due, among other things, to a lack of political will and capacity (see also [16, 17]), South Africa embraces informal settlements upgrading as an alternative to the conventional public housing model. Both the 1994 Housing White Paper [8] and the 2004 Breaking New Grounds (BNG) policy with its accompanying Upgrading of Informal Settlement Programme (UISP) [9:12] directly or indirectly advocate settlements upgrading. Although indirect and without any strategic details, a superficial reference seemed to have been made to settlement upgrading in the 1994 Housing White Paper. The following are some of the phrases used: "promoting integrated communities situated in areas allowing *progressive access* to secure tenure; portable water; adequate sanitary facilities; waste disposal"; and "promoting integrated communities in areas allowing *convenient* and *progressive access* to economic opportunities; educational and social amenities" [8: 19]. The 1994 Housing White Paper also advocates promotion of "continuous housing improvements through consolidation and upgrading" (p. 26). Contrary to the 1994 Housing White Paper, the 2004 BNG policy, as well as its accompanying UISP, seems to have prioritised the implementation of informal settlements upgrading as a devoted strategy or tool. At the core of the 2004 BNG and its accompanying UISP are phrases such as "progressive eradication of informal settlements through phased *in situ* upgrading in desired locations, coupled to the *relocation* of households where development is not possible or desirable" and the recognition of housing as a "key strategy for poverty alleviation; promoting social cohesion and combating crime" [9: 17]. The policy further advocates "incremental provision of services, social amenities and tenure." Thus the 1994 Housing White Paper is silent on both the "in situ" and "relocation" approaches but refers to "progressive access to...." Similarly, the 2004 BNG and its accompanying UISP advocate "incremental provision of ..." through the in situ or relocation approach. Yet, it may be appropriate to argue that South Africa has a bias for the policy of upgrading. Despite being given a "last resort" status in policy, "relocation", it may further be appropriate to argue, is to some extent considered as a possible strategy for the eradication of informal settlements. This is despite the fact that government has been criticised for its tendency to conveniently evoke the policy principle on "eradication" to encourage unpopular 'forced' relocation, as opposed to either in situ upgrading [17, 51] or what a researcher calls 'negotiated' relocation. Implementation of the 2004 BNG policy, with its emphasis on "eradication", is also being criticised for being no different from apartheid's Prevention of Illegal Squatting Act of 1951 and its emphasis on the "eviction" and "demolition" of informal settlements [17].

The policy shift from the Housing White Paper of 1994 to the BNG in 2004 did not happen in a vacuum. In the main, the influence of the United Nation's 2030 Agenda and its 17 Sustainable Development Goals (preceded by the Millennium Development Goals [MDGs]) is evident in policy formulation and subsequent reviews. Relevant to this chapter are the following Sustainable Development Goals: SDG 1 (end poverty in all its forms everywhere); SDG 3 (healthy lives and well-being for all at all ages); SDG 4 (inclusive and equitable quality education and lifelong learning opportunities for all); SDG 6 (availability and sustainable management of water and sanitation for all) and SDG 11 (making cities and human settlements inclusive, safe, resilient and sustainable). Other than responding to the unique housing needs in informal settlements, several policy formulation, shifts and reviews are being undertaken to align the South African agenda on national development with the UN 2030 Agenda, among other things.

The 2004 BNG policy on settlement upgrading refers to a phased in situ upgrading, in line with international best practice. Thus a decision was made by the National Department of Human Settlements to design and implement the National Upgrading Support Programme in partnership with the World Bank Institute and Cities Alliance [9:17]. As a member of the United Nations, South Africa committed herself to the United Nation's earlier Millennium Development Goal 7, Target 11 in 2000 [17]. Persuading developing countries to acknowledge informal settlements as an integral part of their urban fabric, rather than an 'eye sore', the UN through the earlier Millennium Development Goal 7, Target 11 advocates "significant improvement" in the lives of over 100 million informal or slum dwellers by 2030. In pursuit of possible alignment, it should probably not come as a surprise to have had 2014 as an initial target date to eradicate all informal settlements in South Africa [9, 31]. The latest National Development Plan (NDP) 2030 Vision is another effort towards a broad alignment of national priorities with the United Nation's 2030 Agenda and its Sustainable Development Goals. Acknowledgement of the African Union (AU), Agenda 2063, as well as the decision by the UN 2030 Agenda to consider it as an integral part, is another high-level reflection of indirect commitment by South Africa to the UN 2030 Agenda.

Despite policy shifts, reviews and an attempt to align her national agenda and priorities with the UN-2030 Agenda, South Africa continues to experience growing backlogs in the provision of some of key services, thus risking the possibility of failing to meet both her NDP-2030 Agenda and the UN-2030 Agenda. For instance, the number of households in informal settlements increased from 1.4 million in 1994 to over 2 million in 2017 [8, 10]. The growing backlog means that the country failed to meet the initial target of eradicating informal settlements by 2014. This figure also does not bode well with the vision and targets set by the NDP-2030, AU-Agenda 2063 and the UN-2030 Agenda, particularly the five SDGs on which the chapter and its empirical findings on upgraded Phase 6 in Mangaung Township are anchored. It would seem that South Africa is struggling (particularly from the perspective of SDG 7, Target 11) to ensure that by 2030 "no one would be left behind" [48]. Despite being in contrast to SDG 7, Target 11, by implication, the 2 million plus households without adequate shelter might mean that the same number of households is functionally living in poverty (contrary to SDG1), being without adequate access to health facilities and clinics (contrary to SDG 3), far from schools (contrary to SDG 4) and without adequate sanitation and portable water (contrary to SDG 6). The extent to which the growing backlog in infrastructural development could be attributed to a lack of policy implementation and governance, particularly in relocation sites, is the focus of the empirical analysis below.

4   South Africa and Informal Settlements Upgrading: Literature Overview

The focus now shifts to an analysis of relevant literature from a South African perspective. In previous discussions, policy analysis shows the extent to which the country has adopted a twofold approach: prioritisation of the in situ approach to upgrading, with relocation to be considered as a last resort [9]. Despite the policy recommendation that it should be considered as a last resort where the in situ approach is impractical, it would seem that relocation continues to dominate the South African housing landscape on upgrading. The impact of relocation on transforming the housing landscape and incremental access to basic service infrastructure and amenities in upgraded areas, shall be the focus of discussion and analysis in the next sections.

4.1 Behind the Unintentional Shift from the In Situ to the Relocation Approach

The history of informal settlement upgrading in post-apartheid South Africa seems to have followed an interesting trajectory. It was only in 2004, with the promulgation of the BNG policy and its accompanying UISP, that almost a decade (1994–2003) of policy gap on informal settlement upgrading was eventually cut short [15]. However, the literature shows that in the absence of a policy devoted specifically to informal settlement upgrading between 1994 and 2003, the country's programme was guided by the 1994 Housing White Paper [4, 32]. Following a mismatch between key policy principles enshrined in the 1994 Housing White Paper and the actual housing needs of informal settlers, it should not come as a surprise to have only 28% of all post-apartheid settlements upgrading projects being in situ, while the remainder are greenfield projects, during which residents in informal settlements were relocated [13, 14]. The estimated 72% of all post-apartheid settlement upgrading projects, which are in relocation sites, further shows the dominance of relocation as the most common approach to upgrading. The dominance of relocation sites could be ascribed to numerous factors. The following are some of the factors: lack of political will to agitate rate payers in affluent suburbs adjacent to informal settlements [14, 32], city officials consciously or unwittingly acting as defendants of orderly urban and housing development [16], absence of agreements governing cooperation, lack of communication and coordination at all three levels of government [18, 45, 49: 1], shortage of well-located, state-owned land [45], as well as unpopular and top-down relocations [18].

While some academics and researchers maintain that the peripheral location of upgrading projects, as usually perpetuated by relocation, is a flaw in the initial policy [16, 17, 20], others argue that the cost-benefit analysis related to the implementation of the policy—one that emphasises aspects such as size and the quality of the final product over locational considerations—played a role [17]. Essentially, the important point that emerges from the context above is that both the location and the locational context of a housing development project are crucial aspects to be considered. The literature shows that the widespread adoption of relocation as opposed to the in situ approach is not immune from criticism.

4.2 Relocation Sites and Experiences of Ordinary Community Members

Literature and research studies show that proximity to employment and employment opportunities, as well as proximity to schools and cheaper living costs, are some of the pull factors to informal settlements [16, 32, 51]. Yet, government's upgrading in relocation sites usually causes disruptions to the livelihoods of informal settlers. Disruptions manifest through a reduced pace of infrastructural development and a dampened sense of place attachment, including households' investment in housing [2, 11, 33, 40]. The implementation of relocation is characterised by some far-reaching implications on the socio-economic well-being of informal settlers. As noted in the previous section, widespread shortage of well-located, state-owned land usually leads to urban sprawl or the peripheral location of relocation projects [2, 14, 32, 33]. This is so despite location in relation to job opportunities, service infrastructure and social amenities being some of the key requirements in any low-income housing policy, including informal settlement upgrading. Subsequent to the peripheral location of most upgraded areas, a number of spatially related challenges emerge in most relocation sites. For instance, households in these upgraded areas expressed their frustration at government's failure to prioritise and fund the installation of bulk infrastructure and basic amenities, such as water provision, sanitation, electricity and proper roads [14, 36, 52]. Where basic service infrastructure is installed, particularly water, this is done by means of communal taps that are usually not within the prescribed RDP walking distance of 200 m [52]. There is concern about the absence of social amenities such as public schools, clinics and police stations and of employment opportunities [35, 42]. Demonstrating a lack of social amenities in relocation sites, [4] provides an example of relocation in Gauteng Province that left the children 14 km farther from their school. At some point, the children erected their own shack near the school and returned home only over weekends. The unsuitability of locations further negatively influences investment in housing [2, 33].

To counter criticisms levelled against both relocation as a strategy and its actual performance, a number of valuable points have been raised in the literature and policy. For housing projects to succeed, including those in relocation sites, [32] emphasises the need for high-level political mobilisation and buy-in. A prominent scholar in settlement upgrading and self-help housing, JFC Turner, argues that any housing programme may be capable of successfully delivering, provided it adopts a "bottom-up" approach or "dweller control" [46, 50]. Although minimal, there is evidence in the literature of some success in certain relocation sites. Some case studies revealed that shack dwellers experienced tenure security whether they were going to reside on the same stands with upgrading, move to transition camps while upgrading was in process or whether there was the possibility of their being relocated to a new housing development project. One dominant view expressed in policy [9] and the literature [32, 37, 40, 51] is that informal settlement upgrading has the potential to be used as a tool to alleviate poverty in upgraded areas. Even the World Bank has considered poverty alleviation to be a primary goal of any upgrading [1]. It has the potential to also promote, among other things, housing consolidation [25, 39], positive health outcomes [22, 51], security of tenure [2, 11], and place attachment [26].

5 Research Methodology

Methodologically, the chapter followed a case-study approach, focusing on an upgraded area called Phase 6 area in Mangaung Township (Bloemfontein). Bloemfontein is the capital city of the Free State Province (see Fig. 1). There are over 3000 households in the Phase 6 area. The mixed method was employed for data gathering. The choice of a mixed research method was informed by the need to add insight and understanding that might be overlooked when only a single method is used, thereby producing more comprehensive knowledge necessary to inform the policy and practice. A household survey of 200 heads of households was undertaken in 2016. One key criterion in selecting study participants was that such individuals should be adults and (one way or the other), owners of dwellings. A systematic random sampling method was employed in identifying participants. Making systematic random sampling possible was the fact that the researcher could obtain, from Mangaung Metropolitan Municipality, a map with site numbers for all households in Phase 6 area. The survey was complemented with two focus group discussions,

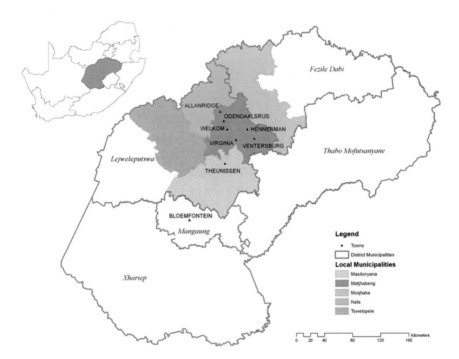

Fig. 1 Location of Mangaung Township (Bloemfontein) in the Free State Province. *Source* [21]

each group comprising nine participants and five in-depth interviews with heads of households. There were further in-depth interviews with five senior municipal officials responsible for informal settlement upgrading in Mangaung Metropolitan Municipality, as well as with a ward councillor in the Phase 6 area.

6 Mangaung Township and Informal Settlements Upgrading

The history of informal settlement upgrading in Mangaung Township (Bloemfontein) is well documented. Other than the upgrading in Phase 6 area, the research and literature on Mangaung upgrading projects stretches over more than two decades. The upgrading of Freedom Square in Mangaung Township is probably the most researched informal settlements upgrading project in South Africa. Research on the Freedom Square upgrading project includes two Ph.D. theses [5, 38] and several articles [6, 23–25, 27, 28, 30, 34, 39]. Making the study on the Phase 6 upgrading project interesting could be the fact that it adopted a 'relocation' approach instead of the 'in situ' approach that Freedom Square upgrading adopted in 1993. Against this brief background, the focus now shifts to an analysis of governance and policy issues in the provision of service infrastructure and social amenities, together with the extent to which such provision has caused changes in the socioeconomic and environmental profile of the upgraded Phase 6 area since 1999.

7 Households' Livelihoods and Poverty Alleviation

An analysis of livelihoods and the poverty level should largely be understood within the context of the current 45.8% unemployment rate among respondents in the Phase 6 area. Unique to the relocation process of the Phase 6 area in 1999 was a strong sense of community buy-in. In-depth interviews showed that through a 'negotiated relocation', community leadership, local politicians and government officials managed to avert any possible community resistance. Consequently, there was minimal disruption of households' livelihoods. For instance, 17% of respondents cited proximity to the workplace as the main reason for their relocation to the Phase 6 area. A further 41% of respondents cited security of tenure as the main reason for their relocation there. This is probably the reason for the decline in the perceived poverty level among respondents in the Phase 6 area. The number of respondents who perceived themselves to have lived in poverty when they first came to live in Phase 6 in 1999 declined from 68.2 to 29.4% in 2016. This is a significant decline in the perceived poverty level.

8 Social Cohesion, Place Attachment and Spatial Transformation

Both the survey findings and in-depth interviews show the significant role played by 'negotiated relocation' in addressing issues of social cohesion, place attachment and apartheid spatial planning. Upgrading in the Phase 6 area seems to have played a role in bridging a 60 km distance between former 'white' Bloemfontein and the R292 settlement, Botshabelo. Survey findings show that through in-migration, the upgraded Phase 6 area attracted 16% of respondents who originated from Botshabelo, thus making upgrading in the Phase 6 area a key driver of 'spatial infilling'. One interviewee noted: "I used to spend almost half of my salary commuting daily between Bosthabelo and Bloemfontein for work…. Since I relocated to Phase 6, I am now able to save for my family."

Such remarks may be one of the reasons for the strong sense of place attachment among respondents. The survey shows that 84.6% of respondents have no intention of leaving the Phase 6 area. This should mainly be understood within the context of the discriminatory apartheid laws that sought to displace and confine Black Africans to underdeveloped and un-serviced homelands with no tenure security [41]. One interviewee said:

> I go nowhere; we have suffered to be where we are today…. At times we had to literally walk more than 20 km from Ipopeng to Glass House (municipal offices in the city) fighting for this place to be approved. I cannot just leave my house and good neighbours after that particular struggle…. We may not have enough but is better than in Ipopeng informal settlement.

Perception on social cohesion among respondents has also improved since relocation. When asked about 'goodwill' in their neighborhood, the percentage rose from 55.2% for those who believe there was 'goodwill' at the time of settling in 1999 to 60.7% for those who believe there was 'goodwill' in 2016. Typical comments included: "Our male neighbours formed night patrolling groups to look after our area…. We also make occasional donations to buy cheap coffins and funeral groceries for families who are destitute and unable to afford burials." Thus, it comes as no surprise to have 29.7% of respondents citing "good neighborhood" as the 'Best' thing about living in the Phase 6 area.

9 Access to Service Infrastructure and Social Amenities

Despite the alleged promises by some prominent local politicians in 1999—and despite being key drivers of social cohesion—place attachment and spatial in-filling in Mangaung Township, an upgrading project in the Phase 6 area continue to experience backlogs in the provision of service infrastructure and social amenities. The Phase 6 area is still without a secondary school, a clinic (except what is alleged to be an unreliable mobile clinic that visits the area once every two weeks) or police station. At least 85.6% of the respondents are without running water and are forced to rely on communal taps, with a further 68% of respondents being without flushing toilets and having to rely on pit latrines. In a community without a police station, it would not be surprising to have 45.5% of respondents cite crime as the worst thing about living in the Phase 6 area, even as a further 86.1% of households have burglar proofing.

A very interesting observation about the study's findings is that despite the backlog, there are more households with access to complete state-funded housing than those with access to flushing toilets and on-site water supply. It may thus be argued that government investment in housing development in the Phase 6 area is being done at the expense of adequate provision of sanitation and water. A government official acknowledged that "the biggest weakness of our upgrading plan and strategy is its narrow focus on housing provision that excludes other basic services and amenities."

It is probably unthinkable that in an upgraded project area, more than half of over 3000 households could spend more than a decade without sanitation and on-site water supply, let alone being without a clinic, secondary school and police station. In the main, these findings are in contrast to policy advocacy for "progressive access to adequate housing and basic services and social amenities" [8] and "incremental provision of services and social amenities" [9]. Findings also contrast with the municipality's key strategic document on the Integrated Development Plan (IDP), including the National Spatial Development Perspective (NSDP).

10 Possible Impact of Political Rhetoric and Governance Dynamics in Implementation of the Upgrading Policy

In-depth interviews with all participants confirmed that a 'negotiated relocation' to the Phase 6 area occurred early in 1999, with most project beneficiaries citing March and April 1999 as the exact date. An interesting thing about the timing is the fact that it was two months before the national elections of June 1999. Making the observation above more significant could be the following two comments from the interviewees, the first of whom says: "Our leaders were supportive and were always with us on the ground.... We were all moved free of charge.... Government hired trucks to move us from Ipopeng informal settlement to this area, Phase 6." The second interviewee provided some form of despair: "Our comrades who were in the forefront of our struggle to get this area approved and upgraded have all disappeared.... We are now on our own.... Even our current councillor, most of us do not know him or her."

With the two comments above, and the current backlog in the provision of basic services and social amenities in Phase 6, one cannot help but ponder the usual political rhetoric that characterises political campaigns and mobilisation at the height of any political election season. With more than half of the residents still without promised sanitation and on-site water supply, it may be appropriate to argue that even the 1999 election campaigns were no exception, making relocation on the eve of the elections more politically convenient than a genuine response to the plight of informal settlers. This represents what Misselhorn [32: 15] described as a "political and developmental powder-keg."

Further responsibility for the current backlogs in basic service infrastructure and social amenities in the Phase 6 area could be alleged weak intergovernmental relations, existence of a silo mentality and a lack of capacity on the part of government. Explaining the possible impact of a lack of joint project planning between the Mangaung Metropolitan Municipality and various Provincial Departments on the implementation of the upgrading project, one government official mentioned:

> Contrary to the concept of Intergovernmental Relations (IGR), you find that other stakeholders, such as various Provincial Departments, do not consult one another but also with our municipality during the initial project phase when planning and designing is done.... Thus, even today we still have most upgraded communities travelling long distances to schools and clinics.

At the city level, officials cited the working-in-silos mentality and a lack of internal capacity as some of the contributing factors. It is alleged that key directorates for informal settlement upgrading, such as Human Settlements, Town Planning, as well as Engineering and Infrastructure, have a tendency to plan separately for the same upgrading project. This is probably why upgrading projects such as Phase 6 end up with housing without bulk infrastructure such as water and sanitation. Consequently, good governance in terms of project implementation in the Phase 6 area seems to have been negatively affected by a combination of weak intergovernmental relations, the working-in-silos mentality and lack of internal capacity.

11   Conclusion

An increase in the number of households residing in informal settlements in the country and those currently without sanitation, on-site water supply, clinic, police station and secondary school in the Phase 6 area, does not bode well for the vision and targets set by the UN-2030 Agenda, particularly the five SDGs on which the chapter is anchored. However, despite being a relocation site, there are key policy and governance lessons to be learnt from implementation of the upgrading project in the Phase 6 area. Notwithstanding existing backlogs in sanitation, water supply and social amenities, the seamless and resistance-free relocation of early 1999 in particular was remarkable. It presents Phase 6 upgrading as a model for 'negotiated relocation'. This may be a compelling reason for possible consideration by policymakers for a conceptual shift in policy from 'relocation' to 'negotiated relocation', so that it becomes mandatory and compulsory for project implementers to consistently negotiate, rather than simply impose, relocation.

The expressed sense of social cohesion among respondents may largely be ascribed to the notion of 'negotiated relocation'. For an upgraded area with over 3000 households to have spent more than a decade without a clinic, secondary school, police station and a significant backlog in sanitation and on-site water supply, there could at least be three possible conclusions. First, it might have been the result of a working-in-silos mentality among directorates in Mangaung Metropolitan Municipality. Second, it might have been due to a lack of intergovernmental relations among national departments (Education, Health, Police, Housing, Water and Sanitation), including the Mangaung Metropolitan Municipality. As things stand, there is exclusive investment in housing development by the National Department of Housing at the expense of basic services and social amenities in the upgraded Phase 6 area. Third, it might have reflected the extent to which the policy principle on 'incremental upgrading' may be used especially by politicians' in the post-relocation process to endlessly avoid accountability for some of the infrastructural development promises they might have made during their campaigns for office. Eventually, those in power resort to the political rhetoric of "things are in the pipeline", which may be interpreted as an attempt to equate such rhetoric with policy principles on "incremental upgrading" or "progressive access to service infrastructure." This is informed by the fact that "negotiated relocation" to the Phase 6 area happened on the eve of the 1999 national elections, making it appropriate to further suggest that politicians should refrain from making infrastructural development promises that are not well researched, during their political campaigns.

Study findings further show that security of tenure and proximity to workplace are among possible key drivers for a strong sense of place attachment, spatial in-filling and a perceived decline in poverty levels in the upgraded Phase 6 area. It is worth noting that both security of tenure and a strong sense of place attachment did not translate into tangible housing investment and consolidation in the upgraded Phase 6 area. To ensure that there is accountability, efficiency and improved working relations between politicians and government officials in particular, those responsible for monitoring and evaluating the state's funded upgrading projects may have to consider introduction of consequence management and sound intergovernmental relations across the three spheres of government, i.e., national, provincial, and municipality. In ensuring accountability and making upgrading projects to be responsive to the community needs, it may be appropriate for government to consider compulsory consumer education for beneficiaries and internal training and capacity building in project management for project implementers/government officials.

Acknowledgements I acknowledge the financial contribution of the National Research Foundation (South Africa) for the fieldwork for this chapter (Grant no: 89788).

References

1. Abbott J (2002) An analysis of informal settlement upgrading and critique of existing methodological approaches. Habitat Int 26(3):303–315

2. Adebayo P (2008) Still no room in the inn: post-apartheid housing policy and the challenges of integrating the poor in South African Cities. http://www.courses.arch.ntua.gr/fsr/134809/pos tapartheid-1.doc.pdf. Accessed 20 Septe 2014

3. Bassett EM (2001) Institutions and informal settlements: the planning implications of the community land trust experiment in Kenya. Unpublished Doctoral thesis, University of Wisconsin-Madison, Wisconsin

4. Baumann T (2003) Harnessing people's power: policy-makers' options for scaling up the delivery of housing subsidies via people's housing process. Research paper, Housing Finance Resources Programme, Johannesburg

5. Botes L (1999) Community participation in the upgrading of informal settlements: theoretical and practical guidelines. Research paper, University of the Free State, Bloemfontein

6. Botes L, Krige S, Wessels J et al (1991) Informal settlements in Bloemfontein: a case study of migration patterns, socio-economic profile, living conditions and future housing expectations. Research report, Urban Foundation, Bloemfontein

7. Daniel C (2001) Integrated programme for social inclusion in Santo Andre, Brazil. Research paper, Un-Habitat Thematic Committee, Istanbul

8. Department of Housing (1994) White paper on housing: a new housing policy and strategy for South Africa. Government Gazette No, Pretoria, p 16178

9. Department of Housing (2004) "Breaking new ground" a comprehensive plan for the development of sustainable human settlements. Department of Housing, Pretoria

10. Farber T (2017) Informal settlements are here to stay. Sunday Times, p 16

11. Govender T (2011) The health and sanitation status of specific low-cost housing communities as contrasted with those occupying backyard dwellings in the City of Cape Town, South Africa. Unpublished Doctoral thesis, University of Stellenbosch, Stellenbosch

12. Harris R (2003) A double irony: the originality and influence of John F.C. Turner. Habitat Int 27(2):245–269

13. Huchzermeyer M (2001) Concent and contradiction: scholarly responses to the capital subsidy model for informal settlement intervention in South Africa. Urban Forum 21(1):71–106

14. Huchzermeyer M (2004) From "contravention of laws" to "lack of rights": redefining the problem of informal settlements in South Africa. Habitat Int 28:333–347

15. Huchzermeyer M (2006) The new instrument for upgrading informal settlements in South Africa: contributions and constraints. In: Huchzermeyer M, Karam A (eds) Informal settlements: a perpetual challenge?. UCT Press, Cape Town, pp 41–61

16. Huchzermeyer M (2009) The struggle for in situ upgrading of informal settlements: a reflection on cases in Gauteng. Dev South Afr 26(1):59–73

17. Huchzermeyer M (2010) A legacy of control? The capital subsidy for housing and informal settlement intervention in South Africa. Int J Urban Reg Res 27(3):591–612

18. Jordhus-Lier D (2015) Community resistance to megaprojects: the case of the N2 Gateway project in Joe Slovo informal settlement, Cape Town. Habitat Int 45:169–176

19. Kessides C (1997) World Bank experience with the provision of infrastructure services for the urban poor: preliminary identification and review of the best practices. Research paper, World Bank, Washington

20. Lalloo K (1999) Arenas of contested citizenship: housing policy in South Africa. Habitat Int 23(1):35–47

21. Marais L (2013) Resources policy and mine closure in South Africa: the case of the free state goldfields. Res Policy 38(3):363–372

22. Marais L, Cloete J (2014) "Dying to get a house?" The health outcomes of the South African low-income housing programme. Habitat Int 43(1):48–60

23. Marais L, Krige S (1999) Post-apartheid housing policy and initiatives in South Africa: reflections on the Bloemfontein-Botshabelo-Thaba 'Nchu Region. Urban Forum 10(2):115–136

24. Marais L, Krige S (1997) The upgrading of freedom square informal settlement in Bloemfontein: lessons for future low-income housing. Urban Forum 8(2):176–193

25. Marais L, Ntema J (2013) The upgrading of an informal settlement in South Africa: two decades onwards. Habitat Int 39:85–95

26. Marais L, Ntema J, Cloete J, Lenka M et al (2018) Informal settlement upgrading, assets and poverty alleviation: evidence from longitudinal research in South Africa. Dev South Afr 35(1):105–125

27. Marais L, Ntema J, Cloete J, Venter A et al (2014) From informality to formality to informality: extralegal land transfers in an upgraded informal settlement of South Africa. Urbani Izziv 25:148–161

28. Marais L, Van Rensburg N, Botes L et al (2003) An empirical comparison of self-help housing and contractor-driven housing: evidence from Thabong (Welkom) and Mangaung (Bloemfontein). Urban Forum 14(4):347–365

29. Matovu G (2000) Upgrading urban low-income settlements in Africa: constraints, potentials and policy options. Research report, Municipal Development Programme, Johannesburg

30. Mehlomakulu T, Marais L (1999) Dweller perceptions of public and self-built houses: some evidence from Mangaung. J Family Ecol Consum Sci 27(2):92–102

31. Mistro RD, Hensher DA (2009) Upgrading informal settlements in South Africa: policy, rhetoric and what residents really value. Hous Stud 24(3):333–354

32. Misselhorn M (2008) Position paper on informal settlement upgrading. Research paper, Urban LandMark, Cape Town

33. Mkhize N (2003) An investigation of how construction skills transfer leads to sustainable employment and housing improvements in incremental housing projects. Unpublished Masters dissertation, University of KwaZulu-Natal, Durban

34. Mokoena M, Marais L (2008) An evaluation of post-apartheid housing delivery in Mangaung: a view from local government. In: Marais L, Visser G (eds) Spatialities of urban change: selected themes from Bloemfontein at the beginning of the 21st century. Sun Press, Stellenbosch, pp 97–114

35. Moolla R, Kotze N, Block L et al (2011) Housing satisfaction and quality of life in RDP houses in Braamfischerville, Soweto: a South African case study. Urbani Izziv 22(1):138–143

36. Narsai P, Taylor M, Jinabhai C, Stevens F et al (2013) Variations in housing satisfaction and health status in four lower socio-economic housing typologies in the eThekwini Municipality in KwaZulu-Natal. Dev South Afr 30(3):367–385

37. Ntema J (2017) Service infrastructure, housing consolidation and informal settlement upgrading: reflections from longitudinal research in free state goldfields, South Africa. Afr Insight 47(2):77–95

38. Ntema LJ (2011) Self-help housing in South Africa: paradigms, policy and practice. Unpublished Doctoral thesis, University of the Free State, Bloemfontein

39. Ntema J, Marais L (2013) Comparing low-income housing outcomes in self-help and contractor-driven projects: the case for longitudinal research. Urban Forum 24(3):389–405

40. Ntema J, Marais L, Cloete J, Lenka M et al (2017) Social disruption, mine closure and housing policy: evidence from the Free State Goldfields, South Africa. Nat Res Forum 41(1):30–41

41. Platzky L, Walker C (1985) The surplus people: forced removals in South Africa. Ravan Press, Johannesburg

42. Sebake N (2010) Does participating in the development of medium-density mixed housing projects make a difference in the residents' satisfaction with the quality of their environment?. Research paper, CSIR, Pretoria

43. Skinner RJ (1983) Community participation: its scope and organization. In: Skinner RJ, Rodell MJ (eds) People, poverty and shelter: problems of self-help housing in the Third World. Methuen, London, pp 125–150

44. Sliuzas R (2003) Opportunities for enhancing communication in settlement upgrading with geographic information technology-based support tools. Habitat Int 27:613–628

45. Tissington K (2011) Demolishing development at gabon informal settlement: public interest litigation beyond modderklip? South Afr J Hum Rights 27(1):192–205

46. Turner J (1976) Housing by people: towards autonomy in building environment. Marion Byers, London

47. Uji ZA (1998) The squatter house in Nigeria: a triumph of self-determination, self-help and decision making. J Environ Sci 1(2):104–114

48. United Nations (2015) The world population prospects: 2015 revision. United Nations, New York
49. Van der Walt S (2016) Ageing workforce in water sector a worry-AG. News 24:1
50. Ward PM (1982) Self-help housing: a critique. Mansell, London
51. Ziblim A (2013) The dynamics of informal settlements upgrading in South Africa: legislative, and policy context, problems, tensions and contradictions. Research report, Habitat for Humanity International, Slovakia
52. Zunguzane N, Smallwood J, Emuze F et al (2012) Perceptions of the quality of low-income houses in South Africa: defects and their causes. Acta Structillia 19(1):19–38

Green Bonds and Green Buildings: New Options for Achieving Sustainable Development in Nigeria

Oluwaseun James Oguntuase and Abimbola Windapo

Abstract Buildings are a critical component of our society, providing countless benefits to the populace, but also having significant negative impacts arising from utilisation of massive natural resources and energy throughout the building life cycle. This has necessitated the adoption of green-building practices in construction project procurement. However, such green-building practices are not yet well developed in Nigeria. With the current development in the Nigerian green bond market, it is important to explore the role that green buildings could play in achieving the SDGs in Nigeria. Therefore, this chapter examines the concept of green buildings and green bonds as an investment vehicle for achieving the Sustainable Development Goals, including Affordable and Sustainable Energy for Human Settlements in Nigeria. The chapter employs a systematic literature review approach in obtaining information that addresses the research objectives. The chapter concludes that employing green bonds as investment vehicles for green buildings in Nigeria will help in providing sustainable and affordable housing while also achieving SDG 3 (Good Health and Well-being), 6 (Clean Water and Sanitation), 7 (Affordable and Clean Energy), 12 (Responsible Consumption and Production) and 13 (Climate Action), among others.

Keywords Clean energy · Green bonds · Green buildings · Investment vehicles

1 Introduction

The Nigerian housing sector is bedeviled by a myriad of challenges that are compounded by population explosion, increased household formation and the rapid rate of urbanization. The housing deficit has been estimated to be up to 17 million [69]. And with population growth now expected to be in the range of 2.5%, about 1.5

O. J. Oguntuase (✉)
Lagos State University, Lagos, Nigeria
e-mail: oluwaseunoguntuase@gmail.com

A. Windapo
University of Cape Town, Cape Town, South Africa
e-mail: abimbola.windapo@uct.ac.za

million new homes would be required annually between 2012 and 2025 [74]. Yet, it is important to meet this housing demand only by constructing more sustainable and affordable residential buildings.

In that regard, green-building practices have now gained traction in several countries as a way to improve building environmental performance and to mitigate the adverse effects of building stock on the environment, society and economy. However, despite the glaring challenges of unsustainable building practices, mismanagement of buildings and the prevalence of a poor maintenance culture with no consideration for environmental impact in Nigeria [107], the country, like most African countries, is still lagging behind in adopting green-building practices [42, 43, 89]. Therefore, this chapter aims to present green building bonds as a veritable instrument to finance and promote green residential building developments in Nigeria, towards sustainably meeting the growing demands for new buildings to negate the substantial housing deficits while achieving the SDGs in the country.

In recent years, building-related environmental and health issues have become increasingly important because buildings are large-scale consumers of resources and big producers of waste. The significant adverse impacts come from massive natural resources and energy consumption as well as from greenhouse emissions released throughout building life cycles—from the design and production of construction materials to the operational stages, maintenance, demolition and materials recycling to reconstruction [50, 76, 124, 131].

The negative impacts of residential buildings on the environment are widely acknowledged in the literature. Cuéllar-Franca and Azapagic [39], as well as Kumar et al. [90], submitted that residential buildings, being major shareholders of entire land use and serving the highest number of consumers, contribute significantly to resources and energy consumption, as well as causing other environmental impacts such as generation of wastewater, emissions and solid waste. Other authors have documented the negative effects of residential buildings in several countries. For example, the impact of emission of greenhouse gases in Bangladesh, China, Colombia, Malaysia, Portugal, and Turkey have been studied by Alam and Ahmad [8], Chang et al. [27], Ortiz-Rodríguez et al. [111], Abd Rashid et al. [1], Bastos et al. [20], Atmaca and Atmaca [17] respectively. Similarly, poor air conditions have been examined in China by Liu and He [93], while acidification and eutrophication processes have been investigated in Central Europe, Malaysia, and South Korea by Estokova et al. [59], Abd Rashid et al. [1], Kim and Chae [88]. Studies have also been conducted on solid wastes in Egypt, Malaysia, Taiwan, and Turkey by Ali et al. [12], Abd Hamid et al. [2], Lai et al. [91], Altuncu and Kasapseckin [14] respectively.

Different Nigerian studies have also examined the situation. Ezema et al. [61] found that the embodied and operational emissions of a typical urban residential apartment building in the commercial city of Lagos were significant when compared with baseline scenarios in other countries. Ononiwu and Nwanya [109] calculated the embodied energy and carbon footprint of a one-bedroom and one-storey flat as $2878.32 \, MJ/m^2$ and $367.21 \, MJ/m^2$ respectively. Ede et al. [52] undertook a comparative environmental impact of concrete and steel—the basic building materials for the construction of residential houses in Nigeria—and concluded that timber structures are more eco-friendly than concrete ones. Ezema et al. [60] found that the embodied energy intensity of Lagos State Development and Property Corporation (LSDPC) estates of 7, 378 MJ/m^2 is comparable to that of Brazilian and Malaysian estates as determined in studies by Paulsen and Sposto [116], Mari [98]. However, the recurring embodied energy component in proportion to the total embodied energy is higher in Ezema et al. [60] study than in previous Nigerian studies. As such, there is need for more sustainable building practices in Nigeria.

Sustainable development has emerged as an important issue during the last two decades, leading to the United Nations' Millennium Development Goals (MDGs) in September 2000 and the Sustainable Development Goals (SDGs) in September 2015. The SDGs emerged in the context of a rapidly urbanizing world with huge housing deficits. Globally an estimated 1.6 billion people live in inadequate housing globally, according to the United Nations Human Settlements Programme (UN-Habitat [129], and about two-thirds of the global population is expected to live in cities by 2050, as projected by the United Nations Department of Economic and Social Affairs [128]. These statistics therefore place the provision of safe and affordable homes at the centre of the sustainable development agenda.

2 The Concept of Green Buildings

While there are varied definitions and rating systems for green buildings around the world, it is generally accepted that the concept of 'green building' is the outcome of planning, better site location, design, construction, operation, maintenance and removal of buildings. Green buildings have several principal considerations, including minimizing the use of resources (energy and water) while reducing the impact of the building on environmental and human health [28, 45, 83, 103, 118, 130, 136, 143].

According to Buys and Hurbissoon [25], the first truly green buildings dated from AD 1. These were stone dwellings of the Anasazi Indians that were built based on local knowledge of the sun and heating as well as of natural ventilation and how to capture water. The only materials used were stone, mud and wood. Anasazi buildings were completely free of toxins and thus healthy. However, despite their long years of existence, green buildings still account for only a tiny proportion of the total building stock in the developed and developing countries [66, 142], cited in [139].

Komolafe and Oyewole [89] described the green building as environmentally friendly as well as socially and economically viable from design to the end of its useful life. They listed other features of green buildings as including energy conservation; indoor air climate improvement; material conservation; water, rainwater and sewage collection for reutilization; waste management and recycling; consideration of green issues in site selection, design and landscaping; and owner and occupant education. These features make green buildings a priority sustainable development matter in the developed and developing nations [46].

Green buildings usually undergo a rigorous certification process by green-building rating systems such as LEED (Leadership in Energy and Environmental Design), BREEAM (Building Research Establishment Assessment Method), CASBEE (Comprehensive Assessment System for Building Environmental Efficiency), DGNB (Deutsche Gesellschaft für Nachhaltiges Bauen), CEEQUAL (Civil Engineering Environmental Quality Assessment and Awards Scheme), the International Finance Corporation Excellence in Design for Greater Efficiencies (EDGE), Code for Sustainable Homes and Building Control Authority and the Green Mark Scheme. These rating systems measure relative levels of compliance or performance against sustainability goals and requirements to create construction projects that are environmentally responsible and use resources efficiently throughout the project life-cycle [133]. The various categories and criteria are constantly revised to keep in line with sustainable trends in building development.

3   Green Building and the United Nations Sustainable Development Goals

Global sustainability goals have led to the development of the green building movement. Green building development is a concept of sustainable development that emphasizes adaptation to local conditions, times and issues Zhang et al. [141], and this extends to green residential buildings [123]. Green buildings promote environmental awareness of building practices and lay down the fundamental direction for the building industry to move towards environmental protection and to achieve the goal of sustainability [49], with green building metrics providing quantifiable metrics to measure contribution towards sustainable development [126].

Highlighting the pivotal roles of green building practices in the drive towards sustainable development are studies by Balaban and Puppim de Oliveira [18], Bombugala and Atputharajah [22], Buys and Hurbissoon [25], Chatterjee [28], Isopescu [80], Loumer [96]. As sourced from the literature, Table 1 shows the contributions of green buildings to efforts to realise the United Nations' Sustainable Development Goals (SDGs).

Table 1 shows that green buildings have been found to contribute to Good Health and Well-being, Clean Water and Sanitation, Affordable and Clean Energy, Decent Work and Economic Growth, Industry Innovation and Infrastructure, Sustainable cities and communities, Responsible Construction and Production, Climate Action, Life on land, and Partnerships to achieve the goal.

Table 1 Green building contributions to SDGs

SDG	References
Goal 3—good health and well-being	Allen et al. [13], Balaban and Puppim de Oliveira [18], Breysse et al. [24], Colton et al. [37, 36], Jacobs et al. [82], MacNaughton et al. [97], Singh et al. [125]
Goal 6—clean water and sanitation	Alawneh et al. [9], Cheng et al. [29], Eisenstein et al. [57], Gabay et al. [67], Zhang and Zhang [140]
Goal 7—affordable and clean energy	Ahn et al. [7], Alawneh et al. [9], Balaban and Puppim de Oliveira [18], Balaras et al. [19], Galante and Pasetti [68], Loius [95]
Goal 8—decent work and economic growth	Bersson et al. [21], Edwards [53], Eichholtz et al. [55, 56], Gibbs and O'Neill [70], Heerwagen [72], Issa et al. [81], Keeton [85], Loius [95], Mondor et al. [99], Ries et al. [119], Robichaud and Anantatmula [120], Zhang and Altan [138]
Goal 9—Industry, innovation and infrastructure	Alawneh et al. [9], Balaras et al. [19], Bersson et al. [21], Edwards [53], Eichholtz et al. [56], Heerwagen [72], Keeton [85], Li et al. [92], Mondor et al. [99]
Goal 11—Sustainable cities and communities	Alhorr et al. [11], Balaban and Puppim de Oliveira [18], Bersson et al. [21], Cidell [30], Di Foggia [48], Galante and Pasetti [68], Gouldson et al. [71], Li et al. [92] Mondor et al. [99]
Goal 12—responsible consumption and production	Alawneh et al. [9], Arslan [15], Durmus-Pedini and Ashuri [51], Hendrickson and Roseland [73], Kamar and Hamid [84], Loius [95], Park et al. [114]
Goal 13—climate action	Affolderbach et al. [5], Alhorr et al. [11], Di Foggia [48], Gouldson et al. [71], Hossain [75], Loius [95], Nguyen et al. [104], Park et al. [114], Vanakuru and Giduthuri [132]
Goal 15—Life on land	Alawneh et al. [9], Louis [95], Eisenstein et al. [57]
Goal 17—partnerships to achieve the goal	Heerwagen [72], Li et al. [92]

4 The State of Green Building Practices in Nigeria

In the Nigerian context, a growing body of works exists on the concept of green buildings. Otegbulu [112] decried the failure of Nigerian designers and contractors to incorporate green concepts, sustainability and environmental issues into new building designs and the retrofitting of old buildings. Ogunde et al. [105] also established that sampled buildings in the country were not sustainable and that the available natural resources were not being adequately utilized.

While Nduka and Ogunsanmi [101] reported that practitioners in the Nigerian building industry were familiar with green-building principles and the benefits, Olumide and Adjekophori [108], Oladokun et al. [106], Dahiru et al. [40] reported a low level of awareness of green-building practices among the country's building industry practitioners. Challenges to the adoption of green-building practices on construction projects were found to include the poor economic situation in the country [41], lack of an enabling environment in the form of policies or legislation that will encourage prospective clients to use green buildings [3, 40], lack of political will and support for green buildings at the top level of government, lack of good examples of green buildings and green technologies in the country [106], insufficient research on materials and indigenous building technologies, lack of requisite human resource capacity [3], unwillingness of tenants to pay for additional green-building features [113].

Oladokun et al. [106] therefore called for public awareness to enable the populace to appreciate the negative impact of buildings as currently constructed and of the potential environmental and commercial benefits of green buildings. For Komolafe and Oyewole [89], public sensitization on features and benefits of green buildings will improve homeowners' orientation and enhance the spread of green buildings in Nigeria.

Olumide and Adjekophori [108] identified careful orientation, low-energy lighting design, maximum use of natural energy, use of renewable energy and use of energy-efficient and eco-friendly equipment as green features to be adopted by practitioners in Nigeria's building industry. However, Dalibi et al. [42, 43] reported that the adaptable and affordable green features found to be favoured by end users are energy-efficient systems, building envelope, water-efficient systems, as well as indoor environmental quality and day-lighting systems.

Studies conducted in Nigeria have highlighted the need for green-building practices and standards that ensure the contribution of the construction sector to the sustainable development of the country (see, for example, [16, 41, 101, 102]. However, Nigeria is yet to produce any building assessment and rating tool(s) for assessing the 'greenness' levels of office, retail, multi-unit residential, public and education building projects. Shaba and Noir [122] noted that the Green Building Council of Nigeria (GBCN), established by the Nigerian Institute of Quantity Surveyors (NIQS), temporarily adopted the Green Star SA of Green Building Council of South Africa (GBCSA), pending the establishment of its own rating system.

The prevailing regulations in Nigeria are the National Building Code enacted in 1987, the Environmental Impact Assessment Act of 1992 (EIA Act) and the National Environmental Standards and Regulations Enforcement Agency Act of 2007 (NESREA Act). In 2016 the then Federal Ministry of Power, Works and Housing commissioned the Building Energy Efficiency Guidelines in collaboration with the Nigerian Energy Support Programme (NESP), to embed energy efficiency into the National Building Code. The core element of the Guidelines is the Nigerian Building Energy Efficiency Code (BEEC), which is the new legal framework for addressing levels of energy efficiency in the country's buildings. The Guidelines give practical advice to professional practitioners on how to design, construct and operate more energy-efficient buildings. The Guidelines also aim to educate the public on energy efficiency measures and provide clients with information that can help them to choose energy-efficient buildings [63].

Adegbile [4], Ogunde et al. [105] posited that the Leadership in Energy and Environmental Design (LEED) green building rating system is appropriate for new projects and renovations in Nigeria because it has a strong base, significant investments and proven advantages, in addition to allowing homeowners to determine the desired environmental performance. Nduka and Sotunbo [102] noted that most building industry professionals in Nigeria are familiar with green building rating systems and prefer LEED for possible adoption in Nigerian construction projects. According to Atanda and Olukoya [16], the LEED assessment criteria should be synthesized into the Nigerian National Building Code to improve its sustainability criteria towards ensuring that the construction sector contributes to sustainable development in the country.

5 The Emergence of Green Bonds

A bond is a fixed income debt instrument used to borrow funds for a defined period of more than one year. As a recent innovation in sustainable finance, the green bond label can be applied to any debt format, including private placement, securitization, covered bond, and Sukuk, as well as labeled green loans that comply with the Green Bond Principles (GBP) or the Green Loan Principles (GLP) (Filkoya et al. 2019). The main difference between conventional bond issues and green issues is in the use of the proceeds. As with standard bonds, green bond issues represent borrowed funds over a period and investors/creditors receive a coupon with a fixed or variable rate of return. Green bonds can be structured as asset-backed securities, linked to specific green infrastructure projects, but are most often issued for raising capital allocated to a portfolio of green projects [10] and committed to the financing of low-carbon, climate-friendly projects [65]. For the International Capital Market Association [79], a green bond is any debt instrument whose proceeds are specifically applied to finance or re-finance partially or entirely new and existing eligible green projects resulting in environmental benefits.

Early issuance of green bonds was primarily to finance projects that aim to cut carbon emissions. The usage spans renewable energy, energy efficiency, pollution prevention and control, biodiversity protection, environmentally sustainable land and resource management, bio-diversity, water and waste management and clean transport. Hyun et al. [77] stated that introduction of the Green Bond Principles (GBPs) in January 2014 by the International Capital Market Association (ICMA) had catalyzed green bond market development. The green bond market grew from US$15 billion in 2013 [100] to about US$343 billion at the end of 2017 [121], when new annual issuances worldwide reached US$156 billion [6]. The green bond market is projected to reach US$1 trillion annual issuances by the year 2020 [135].

Environmental Finance [58] put the global green bond issuance for green buildings at nearly US$60 billion.

The global investment opportunities for green building are reflected as follows: Table 2 shows that issuers of green bonds range from banks and municipalities to real-estate companies and universities. The issuance was for rehabilitation, real-estate development, energy improvement projects and for pursuing green-building certification, while the value of the bonds ranged widely from 700,000 to two billion dollars. As a socially responsible investment (SRI) instrument, labeled green bonds are preferred for sustainable (responsible) investment and long-term investment for individual and institutional investors looking for new financial instruments to achieve their investment targets [115, 117, 137]. Socially responsible investment (SRI), a form of investment that takes environmental and societal indicators into account, has not only emerged as a popular trend for socially responsible investors but has also made its way into mainstream investment because it addresses the financial risks caused by environmental and societal issues [134]. In 2018, the labeled green bond issuance grew steadily from 4% year-over-year to US$162.1 billion to reach a cumulative green bond issuance of US$521 billion since inception [64].

Figure 1 shows the various types of labeled green bonds in use. The first green-labelled bond—Green Use of Proceeds—was issued in 2007 by the Word Bank to connect financing from investors to climate change-related projects. Nevertheless, green bonds still account for a minuscule fraction of the fixed-income market, with a share of less than 1.6% of the global debt issuance in 2016 [54] and only 0.3% of the entire bond market in 2017 [121].

Compared to other continents, the African green bond market remains underdeveloped, with the USA, China, France, Germany, and the Netherlands having market shares of 20%, 18%, 8%, 5% and 4% respectively [64]. Table 2 shows a list of labeled green bond issuers in Africa, the year of issue, jurisdiction of issuance, sector criteria,

Fig. 1 Various types of labeled green bonds [78]

as well as value and tenor. Until 2017 most issuances on the African continent had been from development finance institutions such as the African Development Bank, the International Finance Corporation and the European Investment Bank, as well as some foreign development banks issuing in the South African rand and targeting the Japanese Uridashi market. Within the period there were only four local issuances: two in Morocco and two in South Africa, asides a private placement [6].

Issuers of labeled green bonds in Africa include the Republic of Seychelles (world's first blue sovereign bond), Federal Government of Nigeria, Nigeria's Access

Table 2 Selected green bonds for green buildings

Issuer	Issue month	Jurisdiction of issuance	Use of proceeds	Value	Reference
Commonwealth of Massachusetts	Several issuances since June 2013	United States of America	Municipal green bonds for public building energy efficiency improvements, habitat restoration and water quality improvements	US$12 billion	Climate Bonds Initiative [31]
Regency Centers Corporation	May 2014	United States of America	To fund eligible green-building projects, including the acquisition, construction, development or redevelopment of projects that will pursue LEED certification	US$250 million	Borchersen-Keto [23]
Tandem Health Partners	July 2014	Canada	To build two new hospitals in Campbell River and the Comox Valley of British	US$218 million	Ordonez et al. [110]
Massachusetts Institute of Technology (MIT)	September 2014	United States of America		US$370 million	Kidney [86]
Stockland Corporation Limited	October 2014	Australia	Green building projects in line with the Green Star rating system	US$380 million	Darcy et al. [44]

(continued)

Table 2 (continued)

Issuer	Issue month	Jurisdiction of issuance	Use of proceeds	Value	Reference
Fastighets AB Förvaltaren	October 2014	Sweden	For certified green building and redevelopment of Banana House at Sundbyberg Square in Stockholm	US$55 million	Kidney [87]
National Australia Bank	February 2018	Australia	Residential mortgage-backed securitization that includes green Australian residential properties tranche that meets the Climate Bonds Standard Certification criteria for Australian low-carbon residential buildings	US$234 million	Filkova et al. [64]
	August 2019	Australia	Improve the environmental performance of shopping centres owned by QIC Shopping Centre Fund (QSCF)	A$300 million	Fatin [62]
Columbia University	April 2016	United States of America	Finance the LEED-certified Jerome L. Greene Science Center on the Manhattanville campus	US$50 million	Columbia University [38]
Swire Properties	January 2018	Hong Kong	To upgrade the energy efficiency of one of the Hong Kong-listed company's properties	US$500 million	Swire Properties [127]

(continued)

Table 2 (continued)

Issuer	Issue month	Jurisdiction of issuance	Use of proceeds	Value	Reference
Mitsubishi Estate	June 2018	Japan	Construction of the Tokyo Tokiwabashi Tower-A, which is expected to receive 4 or 5 stars under the DBJ Green Building Certification	US$91 million	Climate Bonds Initiative [32]
Fraser Property Limited	September 2018	Singapore	Refinance existing loans relating to the development of Frasers Tower, which received the Singapore Building and Construction Authority ("BCA") Green Mark Platinum Award	US$876 million	Climate Bonds Initiative [33]
City of Minneapolis, Minnesota	October 2018	United States of America	Public LEED Gold facilities	US$96.22 million	Climate Bonds Initiative [34]
Dormitory Authority of the State of New York	September 2019	United States of America	Construction of LEED Silver or Gold certified buildings at Cornell University	US$121.4 million	Climate Bonds Initiative [35]
CPI Property Group	October 2019	Luxembourg	Acquire and refurbish certified green buildings, improving energy efficiency, and promoting sustainable farming.	US$831 million	CPI Property Group [26]

Bank PLC, North-South Power Company Limited, the South African cities of Johannesburg and Cape Town, the South African banks Nedbank Limited and Industrial Development Corporation (IDC), as well as Bank Windhoek of Namibia, Moroccan Agency for Solar Energy (Masen) and BMCE Bank of Morocco. The challenges faced by issuers and investors in Africa include lack of clear guidelines on green bond issuance, lack of awareness of green bond opportunities, lack of projects for investment and lack of government incentives to support market development [94] (Table 3).

Table 3 Selected labeled green bonds issuance in Africa

Issue month	Issuer	Jurisdiction of issuance	Sector criteria	Value	Tenor
April 2012	Industrial Development Corporation (IDC)	South Africa	Clean energy projects	ZAR 5.2billion (US$700 million)	5 years
June 2014	Johannesburg City	South Africa	Renewable energy, transportation	ZAR 1.46billion (US$140 million)	10 years
November 2016	Banque Centrale Populaire (BCP)	Morocco	Renewable energy	MAD 2billion (US$15.76 million)	7–10 years
November 2016	Moroccan Agency for Sustainable Energy S.A. (MASEN)	Morocco	Solar	MAD 1.15billion (US$115 million)	18 years
November 2016	BMCE Bank of Africa	Morocco	Eco-responsible projects	MAD 500million (US$50.5 million)	5 years
February 2017	ReNew Power	Mauritius	Solar, wind	US$475 million	5 years
July 2017	City of Cape Town	South Africa	Water infrastructure, Low-carbon transport	ZAR 1billion (US$76 million)	10 years
August 2017	Azure Power	Mauritius	Solar	US$500 million	5 years
April 2018	Growth Point Properties	South Africa	Environmental initiatives	ZAR1.1 billion (US$94 million)	5–10 years
September 2018	Casablanca Finance City (CFC)	Morocco	Real estate	MAD 335 million (US$35.8 million)	15 years

(continued)

Table 3 (continued)

Issue month	Issuer	Jurisdiction of issuance	Sector criteria	Value	Tenor
October 2018	Republic of Seychelles	Seychelles	Land use and marine resources	(US$15 million)	–
November 2018	Groupe Al Omrane	Morocco	Housing and essential services	MAD 500million	10 years
December 2018	Bank Windhoek	Namibia	Eco-friendly projects	NAD 66.6million (US$4.74 million)	–
April 2019	Nedbank Limited	South Africa	Solar, wind	ZAR 1.66billion (US$116 million)	Multiple tranches, up to 7 years
September 2019	Acorn Holdings	Kenya	Green and environmentally friendly accommodation for 5000 university students in Nairobi	KES 4.3 billion (US$41.45 million)	5 years

6 The Nigerian Green Bond Market

Certain guidelines must be followed by issuers of green bonds in Nigeria, hence the Federal Ministry of Environment's publication of the Nigerian Green Bond Guidelines (GBGs) in 2016. The guidelines, which are broadly aligned with the Green Bond Principles, are meant to guide the process for issuance of green bonds targeted at the Nigerian market to ensure that resources raised are channeled towards activities that complement the country's Nationally Determined Contributions (NDCs) to reduce carbon emissions.

In December 2017, Nigeria became the first country in Africa to issue a sovereign green bond. The ₦10.69billion 5-year Federal Government Sovereign Green Bond, offered at a coupon rate of 13.48%, was subsequently listed on the Nigerian Stock Exchange (NSE) by the Debt Management Office (DMO) on July 20, 2018. The Federal Government of Nigeria returned to the market in June 2019 to issue Series II Green Bonds worth ₦15 billion (US$42 m). The issuances are part of the Nigerian government's ₦150 billion green bond programme for renewable energy, sustainable forestry management and clean transport in Nigeria. According to the Debt Management Office [47], beyond helping the government to achieve the country's commitment to the Paris Agreement, the issuances are part of the Federal Government's Economic Recovery and Growth Plan (ERGP) and have a target of creating 1.5 million green jobs by 2020.

In February 2019, Access Bank PLC issued the very first Standard Certified Corporate Green Bond in Africa. The issuance, a five-year fixed rate senior unsecured green bond of up to ₦15 billion (US$41.8 m), was meant for funding a coastal flood defence system against sea-level rise as well as for building solar energy generation facilities in the country and boosting agriculture projects. The latest corporate move was the listing of the NSP-SPV Power Corp PLC series 1—₦8.5 billion (US$24 million) 15.16% 15-year Fixed Rate Senior Green Infrastructure Bond on the NSE by North-South Power Company Limited in May 2019. The net proceeds of the bond will be deployed in respect of Shiroro Hydroelectric Power Plant, as shown in Table 4.

7 Overview of Sustainable Projects Procured Using Green Building Concept

The following case studies show that green building bonds are available for housing developments that comply with green building standards and certifications.

7.1 Metrics-Driven Green Mortgages by Bancolombia

Medellín-based Colombia's largest commercial bank, Bancolombia, with total assets of Colombian pesos (COP) 220.1 trillion (about US$65 billion) at December 31, 2018, is ranked as one of the most sustainable banks in the world by the Dow Jones Sustainability Index. The bank provides a variable loan for green construction financing from 0.5% to up to 2% less than conventional market rates for projects certified with EDGE or LEED certification system. Certification services for EDGE are offered by the Colombian Chamber of Construction (CAMACOL), a leading construction trade group in Colombia. The more measurably green the project is, the better the financing rate the project enjoys.

Bancolombia offers green mortgages to qualifying home buyers who purchase qualified properties at a discounted rate of 65 basis points for the first seven years of the loan. To respond to demand, the bank supplements bond proceeds with its capital, with US$175 million already invested in green buildings. The first green bond of 350 billion Colombian pesos (US$117.1 million) was issued in 2016. The second local currency-denominated bond for 300 billion Colombian pesos (US$100 million) issued in 2018 was over-subscribed 2.8 times, with 72 new investors participating.

Bancolombia remains a model for how to move towards a future with better performing buildings through green bonds. The impressive result achieved by the bank has attracted another leading bank in Colombia, Banco Davivienda S.A., to issue green bonds for green buildings.

Bogotá-based bank Davivienda issued 433 billion Colombian pesos (US$150 million) green bond in April 2017. This largest green bond issuance by a private

Table 4 Green bonds in Nigeria

Issue month	Issuer	Value	Tenor	Use of proceed
December 2017	Federal Government of Nigeria	NGN10.69billion (US$29.7 million)	5 years	Energising education program—₦8554million Renewable energy micro utilities—₦146million Afforestation programme—₦1990million
March 2019	Access Bank PLC	NGN15billion (US$41.8 million)	5 years	Refinancing water flood defence project—₦12,845million Refinancing solar generation facilities—₦1,675million New financing of solar generation facilities—₦104 million Refinancing agriculture project—₦192million Cost of bond issuance—₦184million
May 2019	North-South Power Company Limited	₦8.5billion (US$24 million)	15 years	Refinance existing domestic and foreign currency debts in respect of Shiroro Hydroelectric Power Plant—₦4510 million Overhaul maintenance of the Power Plant Turbine IV to extend the power plant useful life—₦3050 million Funding the minimum reserve account to cover at least one coupon and principal payment obligation of the Bond—₦740million Cost of bond issuance—₦190million
June 2019	Federal Government of Nigeria	₦15 billion (US$42 million)	7 years	Renewable power generation—₦8264million Irrigation systems—₦2,818million Rail mass transit—₦1,597million Ecological restoration—₦896million Agro-forestry farmers' regeneration—₦600million Solar powered tricycle—₦500million Afforestation programme—₦324million

financial institution in Latin America was acquired entirely by the International Finance Corporation (IFC). The bonds have a term of 10 years (at a rate of 2.13% above income-based repayment, IBR). Davivienda uses the funds from the green bonds to finance the construction of three types of green buildings—housing, offices, and hotels—to reduce resource (energy and water) consumption. Buildings with recognized local or international environmental certifications are qualified for the bonds.

7.2 Vesteda's Sustainable Low-Income Homes

Vesteda is an entrepreneurial and service-oriented institutional investor with a clear focus on the Dutch middle-income residential real estate market. Vesteda manages the largest Dutch residential fund, an investment portfolio of more than 27,800 sustainable homes worth €7.0 billion euros at the end of 2018. The firm key investment regions are the Randstad urban conurbation and the Noord Brabant province.

Vesteda reported six times oversubscription of its €500 million green bond, which has a term of 8 years and a coupon of 1.5% issued in May 2019. The bond, which will be listed on Euronext Amsterdam, is the second benchmark size bond under its €2.5 billion EMTN programme and the first EUR green bond by a residential fund. In their statement, Vesteda advised that the proceeds of the green bond would be fully allocated to the eligible green assets (homes). Vesteda will report on the estimated energy savings and greenhouse gas emission avoidance of the assets under this green bond.

7.3 Swedish Real-Estate Services Company

Fastighets AB Förvaltaren, a real estate rental agency owned by the municipality of Sundbyberg in Stockholm, issued its first green bond worth SEK 400 m (US$55 m) with a five-year tenor in October 2014 [87]. Swedish financial group Skandinaviska Enskilda Banken AB was the book runner for the transaction rated AA-by Standards and Poor's (S&P).

Proceeds from the issuance are earmarked to eligible green-building projects and in part for the redevelopment of the Banana House at Sundbyberg Square, Stockholm County, Sweden. To qualify, municipal buildings, new or refurbished, must meet a specific green-building standard as reviewed by CICERO: Miljöbyggnad (a Swedish green building standard)—minimum silver; LEED—minimum gold; and BREEAM SE (operated by the Swedish Green Building Council under licence from BRE Global)—minimum very good [87].

7.4 Combating Unhealthy Housing as al Omrane's Social Vocation

The demand for Al Omrane Group's first issue of the social and green bond reached a level of 8.2 billion dirhams (US$2.23 billion), 8.2 times of 1 billion dirhams (US$272 million) for a maturity of 10 years and rate of 4.5%. The demand was mainly by institutional investors, including undertakings for collective investment in transferable securities (UCITS), credit institutions, insurance companies and pension funds.

In consolidating its citizen and environmental commitments, the public operator-developer allows refinancing two categories of projects resulting from a clearly defined evaluation and selection process. Five hundred million dirhams is allocated to the group development plan and another 500 million dirhams will be awarded to projects with positive economic, social and environmental impact.

Al Omrane plans to equip all of its real-estate assets with new construction technologies, thus enabling a significant reduction of its environmental footprint, including the energy efficiency of the group's new buildings and headquarters.

In furtherance of their social responsibility of fighting against unhealthy housing, the group plans to accelerate the deployment of relocation programmes and resettlement of the poor and disadvantaged groups. Al Omrane social ho has been able to improve the housing conditions of 5 million Moroccans as part of the urban upgrade, including 1.5 million slum dwellers. The group plans to accelerate relocation programmes and resettlement of poor and disadvantaged groups by building 22,000 homes. The aggregate budget for this strategic vision with a strong ecological dimension is 31 billion dirhams by 2022.

7.5 Casablanca Finance City's Attractive Green Estate

During the 2016 United Nations Climate Change Conference (COP22), hosts Morocco announced the country's plan to turn the Casablanca Finance City Authority (CFC) into a Pan-African Climate Finance Centre. Morocco's finance and business hub Casablanca Finance City was created in 2010 to become a leading international financial centre capable of attracting global capital flows, supporting economic actors and facilitating deals across the continent.

CFC, in September 2018, through a 15-year green bond, raised 335 million dirhams ($US35.8 m) to finance the extension of the CFC real-estate project. There is plan to obtain international LEED certification for the project, which includes the development of one tower and two buildings.

8   Conclusion

This chapter has examined the concept of green buildings and green bonds as an investment vehicle for achieving the Sustainable Development Goals. The chapter reviewed the literature on green-building practices and green bonds and provided evidence through case studies of how sustainable development can be realised through the interaction of green practices and bonds. It was noted that Nigeria's first green bond was issued only in 2017. It was also noted that, with its over 180 million people, the country will need the capital to meet the 17 million housing deficits in Nigeria, thereby opening significant market potential for green bonds.

The review established that the need to implement sustainable development in the construction industry has made sustainability an inherent characteristic of green-building practices. This has also prompted the creation of green-building standards, certifications and rating systems aimed at mitigating the impact of buildings on the natural environment through sustainable design.

The chapter therefore concludes that when combined, green buildings are the perfect asset for green bonds as investors seek socially responsible investment options for contributing to the achievement of SDGs in a warming planet. Financing green buildings with green bonds will help in the provision of sustainable and affordable housing in Nigeria while also achieving SDG 3 (Good Health and Well-being), 6 (Clean Water and Sanitation), 7 (Affordable and Clean Energy), 12 (Responsible Consumption and Production), and 13 (Climate Action), among others. Furthermore, green buildings are the perfect assets for green bonds as investors seek socially responsible investment options for contributing to the achievement of sustainable development.

9   Recommendations

With the Nigerian Green Bond Guidelines (GBGs) and investors' wide acceptance of green bond issuances in the country, and drawing from case studies, the authors submit that it is imperative to make green building bonds a priority in the public policy around urbanisation and achievement of the SDGs in Nigeria. The study acknowledges that realisation of the green residential building concept should be the first step towards achieving green building bonds since houses are a primary human need.

Making green buildings affordable is quite a challenge and economic incentives that can improve affordability are important drivers. The study therefore recommends that the Nigerian government provide incentives to attract investments such as green bonds for housing. The economic incentives could be in the form of subsidies, tax reduction, rebate systems and other concessions.

The review of existing studies showed that the concept of green buildings is rapidly gaining momentum in the construction industry following recognition of many negative environmental challenges posed by the sector and the potential social and economic benefits of green buildings. However, the development of green buildings is a complicated process consisting of interactions and feedbacks among several stakeholders and institutional arrangements. Moreover, green buildings require considerable funding. Future research is therefore needed to address current challenges and provide awareness of green-building practices and how to access funds for green building in Nigeria.

In all, this chapter provides a preliminary view on using green building bonds for implementation of the SDGs. As such, the chapter does not comprehensively address the complex tasks of programme design, budgeting, financing, service delivery, monitoring and evaluation. More comprehensive and robust evidence about life cycle costing, cost-benefit analysis, market size and value, and corporate reputation improvement is therefore urgently required. Such research will provide holistic knowledge about the economic returns from green-building investments in the Nigerian context, thus helping to attract socially responsible investments.

References

1. Abd Rashid AF, Idris J, Yusoff S (2017) Environmental impact analysis on residential building in Malaysia using life cycle assessment. Sustainability 9:329. https://doi.org/10.3390/su9 030329
2. Abd Hamid KB, Ishak MY, Samah MAA (2015) Analysis of municipal solid waste generation and composition at administrative building café in Universiti Putra Malaysia: a case study. Polish J Environ Stud 24(5):1969–1982. https://doi.org/10.15244/pjoes/39106
3. Adebowale AP, Adekunle AO, Omotehinse OJ, Ankeli IA, Dabara DI (2017) The need for green building rating systems development for Nigeria: the process, progress and prospect. Acad J Sci 7(2):35–44
4. Adegbile M (2012, June) Development of a green building rating system for Nigeria. Paper presented at sustainable futures: architecture and urbanism in the Global South, Kampala, Uganda
5. Affolderbach J, Schulz C, Braun B (2018) Green building as urban climate change strategy. In: Affolderbach J, Schulz C (eds) Green building transitions: Regional trajectories of innovation in Europe, Canada and Australia. The Urban Book Series. Springer, Cham, pp 3–14. https://doi.org/10.1007/978-3-319-77709-2_1
6. African Local Currency Bond Fund (2018) Understanding the African green bond market. African Local Currency Bond Fund, Port Louis, Mauritius
7. Ahn YH, Pearce AR, Wang Y, Wang G (2013) Drivers and barriers of sustainable design and construction: the perception of green building experience. Int J Sustain Build Technol Urban Devel 4(1):35–45. https://doi.org/10.1080/2093761X.2012.759887
8. Alam MS, Ahmad SI (2013) Analysis of life cycle environmental impact for residential building in Bangladesh. Int J Technol Enhancements Emerg Eng Res 2(1):1–4
9. Alawneh R, Ghazali FEM, Ali H, Asif M (2018) Assessing the contribution of water and energy efficiency in green buildings to achieve United Nations Sustainable Development Goals in Jordan. Build Environ 146:119–132. https://doi.org/10.1016/j.buildenv.2018.09.043
10. Aleksandrova-Zlatanska S, Kalcheva DZ (2019) Alternatives for financing of municipal investments—green bonds. Rev Econ Bus Stud 12(1):59–78. https://doi.org/10.1515/rebs-2019-0082

11. Alhorr Y, Eliskandarani E, Elsarrag E (2014) Approaches to reducing carbon dioxide emissions in the built environment: low carbon cities. Int J Sustain Built Environ 3:167–178. https://doi.org/10.1016/j.ijsbe.2014.11.003
12. Ali AAM, Negm AM, Bady MF, Ibrahim MGE (2015) Environmental life cycle assessment of a residential building in Egypt: a case study. Procedia Technol 19:349–356
13. Allen JP, MacNaughton P, Laurent JGC, Flanigan SS, Eitland ES, Spengler JD (2015) Green buildings and health. Curr Environ Health Rep 2(3):250–258. https://doi.org/10.1007/s40572-015-0063-y
14. Altuncu D, Kasapseckin MA (2011) Management and recycling of constructional solid waste in Turkey. Procedia Eng 21:1072–1077. https://doi.org/10.1016/j.proeng.2011.11.2113
15. Arslan F (2017) The role of green building in sustainable production: example of Inci Aku Industrial Battery Factory, Turkey. Araştırma Makalesi 19(1):119–145
16. Atanda JO, Olukoya OAP (2019) Green building standards: opportunities for Nigeria. J Clean Prod 277:366–377. https://doi.org/10.1016/j.jclepro.2019.04.189
17. Atmaca A, Atmaca N (2015) Life cycle energy (LCEA) and carbon dioxide emissions ($LCCO_2A$) assessment of two residential buildings in Gaziantep, Turkey. Energy Build 102(1):417–431. https://doi.org/10.1016/j.enbuild.2015.06.008
18. Balaban O, Puppim de Oliveira JA (2017) Sustainable buildings for healthier cities: assessing the co-benefits of green buildings in Japan. J Clean Prod 163(1):S68–S78. https://doi.org/10.1016/j.jclepro.2016.01.086
19. Balaras CA, Gaglia AG, Georgopoulou E, Mirasgedis S, Sarafidis Y, Lalas DP et al (2007) European residential buildings and empirical assessment of the Hellenic building stock, energy consumption, emissions and potential energy savings. Build Environ 42:1298–1314
20. Bastos J, Batterman SA, Freire F (2014) Life-cycle energy and greenhouse gas analysis of three building types in a residential area in Lisbon. Energy Build 69:344–353
21. Bersson TF, Mazzuchi T, Sarkani S (2012) A framework for application of system engineering process models to sustainable design of high-performance buildings. J Green Build 7(3):171–192
22. Bombugala BAWP, Atputharajah A (2010, December). Sustainable development through green building concept in Sri Lanka. In: Paper presented at international conference on sustainable built environment (ICSBE-2010), Kandy, Sri Lanka
23. Borchersen-Keto S (2014) Regency centers sell $250 million of 10-year 'green bonds. https://www.reit.com/news/articles/regency-centers-sells-250-million-10-year-green-bonds#:~:text=Regency%20Centers%20Sells%20%24250%20Million%20of%2010%2DYear%20'Green%20Bonds',5%2F16%2F2014&text=Shopping%20center%20REIT%20Regency%20Centers,investment%20in%20environmental%20sustainability%20projects. Accessed 5 Aug 2019
24. Breysse J, Jacobs DE, Weber W, Dixon S, Kawecki C, Aceti S et al (2011) Health outcomes and green renovation of affordable housing. Public Health Rep 126(1):64–75. https://doi.org/10.1177/00333549111260S110
25. Buys F, Hurbissoon R (2011) Green buildings: a Mauritian built environment stakeholders' perspective. Acta Structilia 18(1):81–101
26. CPI Property Group (2019) Green bond framework. https://www.cpipg.com/uploads/95c09ab2e3599de8decc51a6f588eda638ffa195.pdf. Accessed 5 Aug 2019
27. Chang Y, Ries RJ, Wang Y (2013) Life-cycle energy of residential buildings in China. Energy Policy 62:656–664
28. Chatterjee AK (2009) Sustainable construction and green buildings on the foundation of building ecology. Indian Concr J 83(5):27–30
29. Cheng C, Peng J, Ho M, Liao W, Chern S (2016) Evaluation of water efficiency in green building in Taiwan. Water 8:236. https://doi.org/10.3390/w8060236
30. Cidell J (2015) Performing leadership: municipal green building policies and the city as role model. Environ Plann C Gov Policy 33(3):566–579. https://doi.org/10.1068/c12181
31. Climate Bonds Initiative (2018) Can US municipals scale up green bond issuance? Likely, "yes". https://www.climatebonds.net/files/reports/us_muni_climate-aligned_bonds_11-07-2018.pdf. Accessed 5 Aug 2019

32. Climate Bonds Initiative (2018) Green bond fact sheet. https://www.climatebonds.net/files/files/2018-06%20JP%20Mitsubishi%20Estate.pdf. Accessed 5 Aug 2019
33. Climate Bonds Initiative (2018) Green bond fact sheet. https://www.climatebonds.net/files/files/2018-09%20SG%20Fraser%20Property%20Limited.pdf. Accessed 5 Aug 2019
34. Climate Bonds Initiative (2018) Green bond fact sheet. https://www.climatebonds.net/files/files/2018-10%20US%20City%20of%20Minneapolis%2C%20Minnesota.pdf. Accessed 5 Aug 2019
35. Climate Bonds Initiative (2018) Green Bond Fact Sheet. https://www.climatebonds.net/files/files/2019-09%20US%20Dormitory%20Authority%20Of%20The%20State%20Of%20New%20York%C2%A0.pdf. Accessed 5 Aug 2019
36. Colton MD, Laurent JGC, MacNaughton P, Kane J, Bennett-Fripp M, Spengler J et al (2015) Health benefits of green public housing: associations with asthma morbidity and building-related symptoms. Am J Publ Health 105(1):2482–2489. https://doi.org/10.2105/AJPH.2015.302793
37. Colton M, MacNaughton P, Vallarino J, Kane J, Bennett-Fripp M, Spengler J et al (2014) Indoor air quality in green vs conventional multifamily low-income housing. Environ Sci Technol 48(14):7833–7841. https://doi.org/10.1021/es501489u
38. Columbia University (2016) Columbia issues first green bonds. https://sustainable.columbia.edu/GreenBonds. Accessed 5 Aug 2019
39. Cuéllar-Franca RM, Azapagic A (2012) Environmental impacts of the UK residential sector: life cycle assessment of houses. Build Environ 54:86–99
40. Dahiru D, Dania AA, Adejoh A (2014) An investigation into the prospects of green building practice in Nigeria. J Sustain Dev 7(6):158–167
41. Dahiru D, Bala K, Abdul-Aziz AD (2013, October) Professionals' perception on the prospect of green building practice in Nigeria. In: Paper presented at SB13 Southern Africa: creating a resilient and regenerative built environment, Cape Town, South Africa
42. Dalibi SG, Feng JC, Shuangqin L, Sadiq A, Bello BS, Danja II (2017a) Hindrances to green building developments in Nigeria's built environment: The project professionals' perspectives. IOP Conf Ser Earth Environ Sci 63:012033. https://doi.org/10.1088/1755-1315/63/1/012033
43. Dalibi SG, Feng J-C, Sadiq A, Sani Bello B (2017b, June) Adopting green building concepts in housing estate development projects in Abuja F.C.T., Nigeria: exploring the potentialities of end-users' preferences. In: Paper presented at world sustainable built environment conference, Hong Kong
44. Darcy J, Kelly T, Webb J, Horan J (2015) Green bonds: emergence of the Australian and Asian markets. https://data.allens.com.au/pubs/pdf/baf/report-baf23oct15.pdf. Accessed 20 July 2019
45. Darko A, Chan AP (2016) Critical analysis of green building research trend in construction journals. Habitat Int 57:53–63
46. Darko A, Chan AP (2017) Review of barriers to green building adoption. Sustain Dev 25:167–179. https://doi.org/10.1002/sd.1651
47. Debt Management Office (2017) Debt Management Office and Federal Government of Nigeria (FGN) Bonds reports. https://www.dmo.gov.ng/fgn-bonds. Accessed 5 July 2019
48. Di Foggia G (2018) Energy efficiency measures in buildings for achieving sustainable development goals. Heliyon 4:e00953. https://doi.org/10.1016/j.heliyon.2018.e00953
49. Ding GKC (2008) Sustainable construction—the role of environmental assessment tools. J Environ Manage 86:451–464
50. Dixit MK (2017) Life cycle embodied energy analysis of residential buildings: a review of literature to investigate embodied energy parameters. Renew Sustain Energy Rev 79:390–413. https://doi.org/10.1016/j.rser.2017.05.051
51. Durmus-Pedini A, Ashuri B (2010) An overview of the benefits and risk factors of going green in existing building. Int J Facility Manag 1(1):1–15
52. Ede AN, Adebayo SO, Ugwu EI, Emenike CP (2014) Life cycle assessment of environmental impacts using concrete or timber to construct a duplex residential building. J Mech Civ Eng 11(2):62–72

53. Edwards B (2006) Benefits of green offices in the UK: analysis from examples built in the 1990s. Sustain Dev 14(3):190–204
54. Ehlers T, Packer F (2017) Green bond finance and certification. BIS Quar Rev 89–104
55. Eichholtz P, Kok N, Quigley JM (2013) The economics of green building. Rev Econ Stat 95(1):50–63. https://doi.org/10.1162/REST_a_00291
56. Eichholtz P, Kok N, Quigley JM (2016) Ecological responsiveness and corporate real estate. Bus Soc 55:330–360
57. Eisenstein W, Fuertes G, Kaam S, Seigel K, Arens E, Mozingo L (2017) Climate co-benefits of green building standards: water, waste and transportation. Build Res Inf 45(8):828–844. https://doi.org/10.1080/09613218.2016.1204519
58. Environmental Finance (2019) Opportunity awaits—understanding Asia's green building bond journey. https://www.environmental-finance.com/content/market-insight/opportunity-awaits-understanding-asias-green-building-bond-journey.html. Accessed 10 July 2019
59. Estokova A, Vilcekova S, Porhincak M (2017) Analyzing embodied energy, global warming and acidification potentials of materials in residential buildings. Procedia Eng 180:1675–1683. https://doi.org/10.1016/j.proeng.2017.04.330
60. Ezema IC, Olotuah AO, Fagbenle OI (2015) Estimating embodied energy in residential buildings in a Nigerian context. Int J Appl Eng Res 10(24):44140–44149
61. Ezema IC, Opoko AP, Oluwatayo AA (2016) De-carbonizing the Nigerian housing sector: the role of life cycle CO_2 assessment. Int J Appl Environ Sci 11(1):325–349
62. Fatin L (2019) QIC Shopping Centre GB takes Australian green issuance over AUD15bn (USD11.3bn) mark—full Aus/NZ market analysis coming in late August. https://www.climatebonds.net/2019/08/qic-shopping-centre-gb-takes-australian-green-issuance-over-aud15bn-usd113bn-mark-full-ausnz. Accessed 5 Aug 2019
63. Federal Ministry of Power, Works and Housing (2016) Building energy efficiency guideline for Nigeria. Federal Ministry of Power, Works and Housing, Abuja, Nigeria
64. Filkova M, Frandon-Martinez C, Giorg A (2019) Green bonds: The state of the market 2018. Climate Bonds Initiative, London
65. Flammer C (2019) Green bonds: effectiveness and implications for public policy. NBER Working Paper No. 25950
66. Fuerst F, Kontokosta C, McAllister P (2014) Determinants of green building adoption. Environ Plan 41(3):551–570. https://doi.org/10.1068/b120017p
67. Gabay H, Meir IA, Schwartz M, Werzberger E (2014) Cost-benefit analysis of green buildings: an Israeli office buildings case study. Energy Build 76:558–564
68. Galante A, Pasetti G (2012) A methodology for evaluating the potential energy savings of retrofitting residential building stocks. Sustain Cities Soc 4:12–21
69. Geissler S, Österreicher D, Macharm E (2018) Transition towards energy efficiency: developing the Nigerian building energy efficiency code. Sustainability 10:2620. https://doi.org/10.3390/su10082620
70. Gibbs D, O'Neill K (2014) Rethinking sociotechnical transitions and green entrepreneurship: the potential for transformative change in the green building sector. Environ Planning A 46(5):1088–1107. https://doi.org/10.1068/a46259
71. Gouldson A, Colenbrande S, Sudmant A, Papargyropoulou E, Kerr N, McAnulla F et al (2016) Cities and climate change mitigation: economic opportunities and governance challenges in Asia. Cities 54(2016):11–19. https://doi.org/10.1016/j.cities.2015.10.010
72. Heerwagen J (2000) Green buildings, organizational success and occupant productivity. Build Res Inf 28(5–6):353–367
73. Hendrickson DJ, Roseland M (2010) Green buildings, green consumption: do "green" residential developments reduce post-occupancy consumption levels? Simon Fraser University Centre for Sustainable Community Development, Vancouver
74. Hogarth JR, Haywood C, Whitley S (2015) Low carbon development in sub-Saharan Africa: 20 cross-sector transitions. Oversea Development Institute, London
75. Hossain MF (2017) Green science: independent building technology to mitigate energy, environment, and climate change. Renew Sustain Energy Rev 73:695–705. https://doi.org/10.1016/j.rser.2017.01.136

76. Huang T, Shi F, Tanikawa H, Fei J, Han J (2013) Materials demand and environmental impact of buildings construction and demolition in China based on dynamic material flow analysis. Resour Conserv Recycl 72:91–101. https://doi.org/10.1016/j.resconrec.2012.12.013

77. Hyun S, Park D, Tian S (2019) Differences between green bonds versus conventional bonds—an empirical exploration. In: Sachs J, Woo W, Yoshino N, Taghizadeh-Hesary F (eds) Handbook of green finance, sustainable development. Springer, Singapore, pp 91–134

78. International Capital Market Association (2016) Green bond principles 2016: voluntary process guidelines for issuing green bonds. http://www.icmagroup.org/Regulatory-Policy-andMarket-Practice/green-bonds/green-bond-principles/. Accessed 20 July 2019

79. International Capital Market Association (2017) The green bond principles 2017: voluntary process guidelines for issuing green bonds. Annual Report. International Capital Market Association, Switzerland. https://www.icmagroup.org/assets/documents/Regulatory/Green-Bonds/-GreenBondsBrochure-JUNE2017.pdf. Accessed 20 July 2019

80. Isopescu DN (2018) The impact of green building principles in the sustainable development of the built environment. IOP Conf Ser Mater Sci Eng 399:012026. https://doi.org/10.1088/1757-899X/399/1/012026

81. Issa MH, Rankin JH, Christian AJ (2010) Canadian practitioners' perception of research work investigating the cost premiums, long-term costs and health and productivity benefits of green buildings. Build Environ 45(7):1698–1711

82. Jacobs DE, Ahonen E, Dixon SL, Dorevitch S, Breysse J, Smith J et al (2015) Moving into green healthy housing. J Public Health Manag Pract 21(4):345–354. https://doi.org/10.1097/PHH.0000000000000047

83. Kamana CP, Escultura E (2011) Building green to attain sustainability. Int J Earth Sci Eng 4(4):725–729

84. Kamar KAM, Hamid ZA (2011) Sustainable construction and green building: The case of Malaysia. WIT Trans Ecol Environ 167:15–22. https://doi.org/10.2495/ST110021

85. Keeton JM (2010) The road to platinum: using the USGBC's LEED-EB® green building rating system to retrofit the US Environmental Protection Agency's Region 10 Park Place Office Building. J Green Build 5(2):55–75

86. Kidney S (2014a) MIT issues green property bonds to refinance green buildings, $370 m, 24 yr, 3.959% coupon, Aaae. We like! https://www.climatebonds.net/2014/09/mit-issues-green-property-bonds-refinance-green-buildings-370m-24-yr-3959-coupon-aaae-we. Accessed 20 July 2019

87. Kidney S (2014b) Sweden's Förvaltaren issues $55 m (SEK 400 m) green property bond, 5 yr, AA-. Another Swedish green city bond! https://www.climatebonds.net/2014/10/swe den%E2%80%99s-f%C3%B6rvaltaren-issues-55m-sek-400m-green-property-bond-5yr-aa-another-swedish-green. Accessed 20 July 2019

88. Kim TH, Chae CU (2016) Environmental impact analysis of acidification and eutrophication due to emissions from the production of concrete. Sustainability 8:578. https://doi.org/10.3390/su8060578

89. Komolafe MO, Oyewole MO (2015) Perception of estate surveyors and valuers on users' preference for green building in Lagos, Nigeria. In: Laryea S, Leiringer R (eds) Proceedings of the 6th West Africa Built Environment Research (WABER) Conference. WABER Conference, Accra, Ghana, pp 863–886

90. Kumar V, Hewage K, Sadiq R (2015) Life cycle assessment of residential buildings: a case study in Canada. Int J Energy Environ Eng 9(8):1017–1025

91. Lai Y, Yeh L, Chen P, Suig P, Lee Y (2016) Management and recycling of construction waste in Taiwan. Procedia Environ Sci 35:723–730. https://doi.org/10.1016/j.proenv.2016.07.077

92. Li X, Strezov V, Amati M (2013) A qualitative study of motivation and influences for academic green building developments in Australian universities. J Green Build 8(3):166–183

93. Liu Y, He X (2015) Embodied environmental impact assessments of urban residential buildings in China based on life cycle analyses. J Tsinghua Univ (Sci Technol) 55(1):74–79

94. London Stock Exchange Africa Advisory Group (2018, November) Developing the green bond market in Africa. https://www.lseg.com/documents/africa-greenfinancing-mwv10-pdf. Accessed 20 July 2019

95. Louis I (2017, September) Contributions of the building sector to the 2030 Sustainable Development Goals (SDGs). Paper presented at international green building conference, Marina Bay Sands, Singapore
96. Loumer SH (2015) An evaluation of green building components and their relationship with sustainable development objects. World J Manag Art 2(2):74–79
97. MacNaughton P, Satish U, Laurent JGC, Flanigan S, Vallarino J, Coull B et al (2017) The impact of working in a green certified building on cognitive function and health. Build Environ 114:178–186. https://doi.org/10.1016/j.buildenv.2016.11.041
98. Mari TS (2007) Embodied energy of building materials: a comparative analysis of terraced houses in Malaysia. In: Proceedings of the 41st annual conference of the Architectural Science Association (ANZAScA), Deakin University, Australia
99. Mondor C, Hockley S, Deal D (2013) The David Lawrence convention center: how green building design and operations can save money, drive local economic opportunity, and transform an industry. J Green Build 8(1):28–43
100. Morere L (2018, June) Green building bonds. Urban Chronicles 2.0 #01
101. Nduka DO, Ogunsanmi OE (2015) Construction professionals' perception on green building awareness and accruable benefits in construction projects in Nigeria. Covenant J Res Built Environ 3(2):30–52
102. Nduka DO, Sotunbo AS (2014) Stakeholders' perception on the awareness of green building rating systems and accruable benefits in construction projects in Nigeria. J Sustain Devel Africa 16(7):118–130
103. Nguyen H-T, Skitmore M, Gray M, Zhang X, Olanipekun AO (2017) Will green building development take off? An exploratory study of barriers to green building in Vietnam. Resour Conserv Recycl 127:8–20. https://doi.org/10.1016/j.resconrec.2017.08.012
104. Nguyen H-T, Gray M, Skitmore M (2016, January) Comparative study on green building supportive policies of pacific-rim countries most vulnerable to climate change. Paper presented at 22nd annual Pacific-Rim real estate society conference, Sunshine Coast, Queensland
105. Ogunde AO, Amos V, Tunji-Olayeni P, Akinbile B, Ogunde A (2018) Evaluation of application of eco-friendly systems in buildings in Nigeria. Int J Civ Eng Technol 9(6):568–576
106. Oladokun TT, Gbadegesin JT, Ogunba OA (2010) Perceptual analysis of the benefits and implementation difficulties of green building in Lagos Metropolis, Nigeria. In: Proceedings of international research conference on sustainability in built environment. Columbia, Sri Lanka: Commonwealth Association of Surveyors and Land Economist, pp 166–178
107. Olaniyi OO, Smith A, Liyanage C, Akintoye A (2014) Facilities management approach for achieving sustainability in commercial buildings in Nigeria. In: Paper presented at 13th EuroFM research symposium, Berlin, Germany
108. Olumide SO, Adjekophori B (2018) Adoption of green building practice in commercial properties in Lagos, Nigeria. Int J Eng Manag Res 8(6):182–191
109. Ononiwu NH, Nwanya S (2016) Embodied energy and carbon footprints in residential buildings. Int J Adv Eng Res Sci 3(8):49–54
110. Ordonez CD, Uzsoki D, Dorji ST (2015) Green bonds in public-private partnerships. International Institute for Sustainable Development, Winnipeg, Canada
111. Ortiz-Rodríguez Ó, Castells F, Sonnemann G (2012) Environmental impact of the construction and use of a house: assessment of building materials and electricity end-uses in a residential area of the province of Norte de Santander, Colombia. Ingeniería y Universidad Bogotá (Colombia) 16(1):147–161
112. Otegbulu A (2011) Economics of green design and environmental sustainability. J Sustain Devel 4:240–248
113. Oyewole MO, Komolafe MO (2018) Tenants' willingness to pay for green features in office properties. Nigerian J Environ Sci Technol 2(2):233–242
114. Park M, Tae S, Suk S, Ford G, Smith ME, Steffen R (2015) A study on the sustainable building technologies considering to performance of greenhouse gas emission reduction. Procedia Eng 118:1305–1308. https://doi.org/10.1016/j.proeng.2015.08.492

115. Park S (2018) investors as regulators: green bonds and the governance challenges of the sustainable finance revolution. University of Connecticut School of Business Research Paper No. 18–12
116. Paulsen JS, Sposto RM (2013) A life cycle energy analysis of social housing in Brazil: case study for the program "My House My Life". Energy Build 57:95–102
117. Pradiptarathi P (2017) Green bond: A socially responsible investment (SRI) instrument. Inst Cost Accountants India Res Bull 43(1):97–113
118. Qian QK, Chan EHW (2010) Government measures needed to promote building energy efficiency (BEE) in China. Facilities 28(11/12):564–589
119. Ries R, Bilec MM, Gokhan NM, Needy KL (2006) The economic benefits of green buildings: a comprehensive case study. Eng Econ 51:259–295
120. Robichaud LB, Anantatmula VS (2011) Greening project management practices for sustainable construction. J Manag Eng 27(1):48–57
121. Schneeweiß A (2019) Credibility and additionality of green bonds. Südwind e.V, Bonn, Germany
122. Shaba V, Noir E (2014) Local content report: green star SA for use in Nigeria. WSP Group Africa (Pty) Ltd. Bryanston, Johannesburg, South Africa
123. Shafiei MWM, Abadi H (2017) The impacts of green building index towards energy consumption in Malaysia. Aust J Basic Appl Sci 11(4):131–139
124. Sharma A, Saxena A, Sethi M, Varun V (2011) Life cycle assessment of buildings: a review. Renew Sustain Energy Rev 15(1):871–875. https://doi.org/10.1016/j.rser.2010.09.008
125. Singh A, Syal M, Grady SC, Korkmaz S (2010) Effects of green buildings on employee health and productivity. Am J Public Health 100(9):1665–1668. https://doi.org/10.2105/AJPH.2009.180687
126. Sinha A, Gupta R, Kutnar A (2013) Sustainable development and green buildings. Drvna Industrija 64(1):45–53
127. Swire Properties (2018) Green bond report 2018. https://www.swireproperties.com/-/media/files/swireproperties/green-bond/swire_green-bond_final.ashx. Accessed 5 Aug 2019
128. United Nations Department of Economic and Social Affairs (2015) World urbanization prospects—the 2014 revision. United Nations, New York. https://esa.un.org/unpd/wup/Publications/Files/WUP2014-Report.pdf. Accessed 20 July 2019
129. United Nations Human Settlements Programme (2016) World cities report. United Nations, Nairobi, Kenya
130. United States Green Building Council (2003) Draft LEED-EB reference guide material—EA prerequisite 1 existing building commissioning. U.S. Green Buildings Council, Washington, D.C.
131. Ürge-Vorsatz D, Petrichenko K, Staniec M, Eom J (2013) Energy use in buildings in a long-term perspective. Curr Opin Environ Sustain 5(2):141–151. https://doi.org/10.1016/j.cosust.2013.05.004
132. Vanakuru R, Giduthuri VK (2017) Practicing green building techniques in reducing greenhouse gases: an over view. Int J Eng Technol 9(3):2595–2597. https://doi.org/10.21817/ijet/2017/v9i3/1709030196
133. Vierra S (2019) Green building standards and certification systems. https://www.wbdg.org/resources/green-building-standards-and-certification-systems. Accessed 19 November 2019
134. Weber O, Saravade V (2019, January) Green bonds: current development and their future. Centre for International Governance Innovation (CIGI) Papers No. 210
135. Whiley A (2017, June) Climate bonds initiative. Climate Bonds Blog; Green Bond Policy Highlights from Q1–Q2 2017
136. World Green Building Council (2013) The business case for green building—a review of the costs and benefits for developers, investors and occupants. http://www.worldgbc.org/news-media/business-case-green-building-review-costsand-benefits-developers-investors-and-occupants. Accessed 15 July 2019
137. Zerbib OD (2019) The effect of pro-environmental preferences on bond prices: evidence from green bonds. J Bank Finance 98:39–60. https://doi.org/10.1016/j.jbankfin.2018.10.012

138. Zhang Y, Altan H (2011) A comparison of the occupant comfort in a conventional high-rise office block and a contemporary environmentally-concerned building. Build Environ 46:535–545
139. Zhang L, Wu J, Liu H (2018) Turning green into gold: a review on the economics of green buildings. J Clean Prod 172:2234–2245. https://doi.org/10.1016/j.jclepro.2017.11.188
140. Zhang L, Zhang W (2012) A case study of water conservation evaluates for green building. Adv Mater Res 374–377:62–65
141. Zhang D, Liu D, Xiao M, Chen L (2011) Research on the localization strategy of green building. In: Paper presented at international conference on Civil Engineering and Building Materials, Kunming, China
142. Zhou Y (2015) State power and environmental initiatives in China: analyzing China's green building program through an ecological modernization perspective. Geoforum 61:1–12
143. Zuo J, Zhao Z (2014) Green building research—current status and future agenda: a review. Renew Sustain Energy Rev 30:271–281

Homeownership in a Sub-Saharan Africa City: Exploring Self-help via Qualitative Insight to Achieve Sustainable Housing

Andrew Ebekozien

Abstract Globally, several studies have shown that Sub-Saharan Africa is the leading continent in terms of the percentage of urban population living in informal settlements. One of the possible reasons is that governments' policies have hindered accessibility and affordability of homes to the people; thus, resulting in "self-help" approaches on the people's part. This paper examines the lived experiences of Nigerian middle-income groups in Lagos, Nigeria, highlighting the multifaceted encumbrances they face in attempting to become homeowners, as well as the strategies they employ. Methodologically, the study is a phenomenology type of qualitative research. Forty participants interviewed were in three categories: homeowners, prospective homeowners, and government officials. Findings show that the desire for homeownership has created residential mobility among the middle-income earners, leading to the city's expansion. Moreover, land purchase and government residential building approval process were identified as the most cumbersome obstacle faced by intending homeowners in Lagos. Findings also show that organised self-help housing provision can achieve sustainable housing but should be supported by pragmatic policy and implemented by the government. The implication of this study among others will stir up policymakers and other stakeholders to tailor their drive towards achieving sustainable homeownership for all on or before the year 2030.

Keywords Homeownership · Middle-income · Self-help housing · Sustainable housing

1 Introduction

The urban population in Sub-Saharan Africa cities are growing speedily both in population and significant inflows of migrants from the rural areas to cities. Arku

A. Ebekozien (✉)
School of Housing, Building and Planning, Universiti Sains Malaysia, George Town, Malaysia
e-mail: ebekoandy45@yahoo.com

Bekos Energy Service Nigeria Limited, Ikorodu, Nigeria and Bowen Partnership, Quantity Surveying Consultant Firm, Benin City, Nigeria

© The Author(s), under exclusive license to Springer Nature Singapore Pte Ltd. 2021   219
T. G. Nubi et al. (eds.), *Housing and SDGs in Urban Africa*, Advances in 21st Century
Human Settlements, https://doi.org/10.1007/978-981-33-4424-2_12

[5], Angel et al. [4], Andreasen and Agergaard [3] affirmed that municipal growth has taken the shape of spatial increase on the municipal peripheries. One of the resultant effects of this action is pressure on the infrastructural facilities in the cities, including housing. This enhances slums or informal settlements in cities as a possible outcome to these disparities in the perspective of speedy urbanisation and inadequate governmental social housing development especially in developing countries such as Nigeria. Arroyo [6] opined that informal settlements or slums have been constructed through self-help approaches by the people themselves, lacking proper guide. It is possible that these people cannot afford to pay qualified construction professionals such as Architect and Engineers fees. Ibem [23], Olugbenga and Adekemi [30], Okpoechi [29] are among the few studies conducted regarding middle-come housing in Nigeria. Ibem [23] focused on the role of government organisations in public-private partnerships in Nigerian housing delivery yet the problem remains. Okpoechi [29] centred on determining vital functional requirements for affordable housing design. Olugbenga and Adekemi [30] emphasised on land as the major challenge to housing delivery, and more of a review than an empirical article. Whilst this study focuses on organised self-help in achieving housing sustainability for the masses via exploratory approach. This is because Nigeria's housing deficit is within 17–20 million [10]. Power Lunch [33] reported 12–16 million housing units deficits.

This paper offers an exploration of how organised self-help approach can achieve sustainable housing and enhance homeownership in the middle-income earners. This is because self-help housing development has become prevalent possible answer for the housing needs of the middle-income group in low-income countries with insufficient social housing. United Nations Centres for Human Settlements [40] averred that organised self-help has been executed successfully in many projects by non-governmental organisations and community-based organisations in low-income countries earlier and after the global commitment to the Habitat Agenda in 1996. This was corroborated by Arroyo [6]. The author described organised self-help as a procedure that engages community's active involvement and contribution in decision making during planning, design, construction, and post-construction with the technical support of a facilitating establishment. For this paper, organised self-help is all-inclusive, comprises of sites-and-services, aided self-help homes, state-assisted self-help development, and assisted self-help development. It is a combination of a top-down and bottom-up approach, hence, multifaceted to achieve sustainable housing.

Currently, there is a lack of knowledge regarding this approach in developing countries, including Nigeria where the middle-income earners has been shortened to about 10% of the population from 38% within 2003 and 2009; from an exchange rate of NGN120/US\$1 in 2007 to NGN362/US\$1 in May 2019 [28]. Whilst in 2011, the African Development Bank reported that about 23% of Nigerians make the middle class group with monthly average income within NGN75,000 and NGN100,000 (US\$1/NGN362). Thus, it is important to study current practices to acquire knowledge that can improve planning and homeownership among the middle-income group. This would bring new openings and issues for further research as part of the theoretical contributions and significant advance in knowledge. Therefore, this paper attempts to explore the issues encountered by the middle-income group in an attempt to become a homeowner through organised self-help. And proffer possible policy solutions so that homeownership can be enhanced. The goal of this study will be accomplished via the following objectives:

(i) To investigate the challenges faced by middle-income group in the process of becoming homeowners.
(ii) To suggest possible policy solutions that can enhance middle-income group's homeownership via organised self-help.

This paper is comprised of seven sections. The present section is the introduction section while the next section reviews the findings in current related literature. This includes homeownership, self-help housing provision, social relevance and impact, and theory that informed this research. The paper then described the methodological approach used. This was done in a systematic method. The next section was the presentation of key findings from the analysed data, followed by the discussion of the results in an elaborate and detailed pattern with related findings to previous studies. This includes comparing the findings with existing literature. Next on this paper is the recommendation section and the last section concludes the study. The conclusion section includes the implications of this paper and suggestions for future research.

2 Review of Literature

This section focuses on current literature related to homeownership, self-help housing provision, social relevance, and the theory that supported the paper's framework. Regarding the reviewed literature for the organised self-help housing provision, it will capture present practice in developing nations, institutional methods, and organised self-help housing approach.

2.1 Homeownership

It is the dream of the working class to own a home. Several studies across the globe have shown that homeownership is preferred to renting a home [15, 19, 37] except for mobile professionals. Homeownership is the "dream" of many households and a significant life goal [8]. Homeownership is a major form of wealth creation, distribution and displays the accomplishment of each household [15, 16, 25]. Therefore, how can "organised self-help" enhance the achievement of this goal in the Sub-Saharan Africa city.

A broad body of housing research already partly answers this question, with scholars having attempted to address the issue from a different perspectives; such as state-assisted self-help development, sites-and-services, aided housing, and assisted self-help development. Harris [21] noted that aided self-help development or state-assisted self-help development is a top-down method executed by governments for easing poverty. This method originated from Europe immediately after the First World War. Whilst Abrams [2] asserted that the sites-and-services implies the top-down approach, and mostly used by the United States of America aid in the early 1960s, followed by the World Bank to assist the developing countries in infrastructural provision including housing. Arroyo [6] averred that the assisted self-help development is a bottom-up and family-based method to self-help housing that integrates technological support and micro-credit from facilitating organisations. Many of the organisations involved are NGOs and community-based organisations. For this paper, as earlier stated, using the multilateral approach to enhance homeownership is a new area in the concept of the Sub-Saharan city, hence the need to acquire knowledge that can enhance planning and housing practices. The need for stakeholders to learn how to acquire knowledge regarding organised self-help projects with emphasis in planning and implementation cannot be over-emphasised. This may become a substitute to the architectural and planning practice and sustainable housing. This is in line with Arroyo [6] who observed the need for further studies to advance a theoretical framework for organised self-help housing from a capacity approach. Hence, organised self-help development: current practice in developing nations, institutional methods and the organised self-help housing procedure cannot be over-emphasised. The details would be discussed in the next section.

2.2 Self-help Housing Provision

Rodriguez and Astrand [34], Arroyo [6] identified many issues that make governments cautious in incorporating organised self-help housing within housing policies. First, the quality of self-help housing has been probed but experience from organised self-help housing in the developed nations projects has demonstrated good quality. Second, the construction duration for sites-and-services projects was too long in some instances, but evidence shows that some firms that have enforced organised self-help housing as an on-going learning process have managed to enhance their timing [6, 41]. Centre for Housing Policy [11] identified various forms of organised self-help housing as follows: engaging construction contractors (Type 1), engaging contracting firms (Type 2), engaging housing developers from sites (Type 3), housing provision via a group (Type 4), full development for the housing location via community collaboration (Type 5), and without engaging construction contractor (Type 6).

Arroyo [6] found that the current organised self-help housing practice in developing nations shows that from the past two decades, there are new bottom-up methods and experiences; and more action regarding testing different organised self-help housing than an academic debate. Regarding the institutional approaches to organised self-help housing, it is on records that many non-governmental organisations, community-based organisations, and housing co-operatives have implemented various types of institutional methods. This has been achieved via mutual-help. This is an indication that cooperatives can mitigate slums and enhance homeownership through pro-poor housing policy targeted at the disadvantaged for sustainable housing provision [6]. UN-Habitat [38] reported that assisted self-help housing is affordable and easy way of enhancing sustainable housing.

2.3   Social Relevance and Impact of This Study

Policies and programmes aimed at homeownership can contibute to attaining the Sustainable Development Goals (SDGs) connected with housing before the year 2030. UN News Centre [39] stated that SDGs are the framework to secure an enhanced sustainable future for the people. The homeownership programmes for the masses have a link with four of the 17 identified SDGs (2015–2030). This includes no poverty (Goal 1), good health and well-being (Goal 3), reduced inequalities (Goal 10), and sustainable cities and communities (Goal 11) [17]. These goals interrelate because housing development enhances work opportunity and the outcome would be economic growth with a sustainable livelihood. This may mitigate disparities between the rich and the poor, acting as a form of wealth distribution. This agrees with World Bank Press Release [42] submission. The report showed that an inhabitable environment enhances the well-being of the dwellers. The overall outcome would be sustainable cities and communities across Lagos State and by extension, other parts of Nigeria if the proposed pro-homeownership for the middle-income group's policy solutions are well implemented via organised self-help housing provision. Arroyo [6] avowed that organised self-help housing provision is an enabling shelter and construction strategy for conquering poverty whilst building more sustainable cities across the country.

Policies and programmes can be successfully when the accurate institutional framework, within a corruption-free environment, is established. In this regard, the Land Use Act and access to mortgage will need revisiting. Some of the developing countries that have recorded success in their housing provision, a case in point is Malaysia, had to develop a pro-mortgage institutions for the masses to access housing loans. For example, the Bank Negara Malaysia (BNM) released RM 1 billion (RM4.2/US$1) funds for affordable housing, effective from 2nd January 2019. This is to enable income earners not above RM2300 per month become homeowners [7]. The present study thus contributes to research in this area by focusing on the challenges confronting homeownership and offered possible policy solutions based on other countries' approach that has achieved success in housing provision for the masses. The Central Bank of Nigeria need to imitate her counterpart in Malaysia, set the pro-homeownership standard for the maximum household income to be qualified for the Family Home Funds.

2.4 The Theoretical Considerations

This section discusses the related concept and theory that support the framework of this research to enhance homeownership via self-help in Nigerian cities. One phenomenon that is evidence as reviewed in this study is that housing scarcity is extremely high among middle and low-income groups in cities across Nigeria [17]. This framework intends to proffer possible policy so that homeownership can be enhanced via organised self-help. This is pertinent because housing provision is one of the main pillars of welfare. This study adopted the Public Interest Theory of Regulation. It is also called 'helping hand' theory of regulation [15, 32] and based on two assumptions. First, unhindered markets often fail because of the issues associated with monopoly or externalities. Second, governments are concerned and capable of correcting these market challenges via regulation [20]. Scholars like Hertog et al. [22], Ndubueze [27], Iheme [24], Ebekozien [15], Ebekozien et al. [17] among others have utilised this theory of regulation as a suggestion of what governments should do in a civilian government setting. Many of the pro-poor housing policies in Singapore, Malaysia, and Hong Kong are anchored on this theory.

Several scholars, such as Hantke-Domas [20], Ndubueze [27], Iheme [24], Ebekozien [15] asserted that regulation enhances the general welfare of the masses rather than the interests of well-organised stakeholders. They recognise regulation theory as an optimistic theory. This theory was adopted by Iheme [24], Ebekozien et al. [17] in a similar study; hence appropriate for this research. One of the attributes of this theory is the belief that it is for the "public interest," thus, utilising this theory to enhance homeownership via organised self-help by the provision of affordable housing across the major cities in Nigerian cities via pro-homeownership policy and institutional framework. Figure 1 shows the framework for improved home-ownership and to achieve sustainable housing across the urban cities in Nigeria via organised self-help. The hindrances as reviewed from the current literature, for example, weak institutional framework, rural-urban migration, corruption, increasing population (housing demand higher than supply), relaxed housing policy, inadequate subsidies, high cost of land, and high rejection of housing loans may be addressed via driven pro-homeownership government regulation. The process to accomplish this aim is discussed in the next section.

3 Methodology

This study employed a constructivism type of research philosophy and phenomeno-logical form of qualitative research respectively. Bryman [9] avowed that constructivism or subjectivism is a search for explanations of human action by understanding how the world is understood by persons and theory creation. Whilst phenomenology focuses on the experience of the central phenomenon and the intention is to describe a lived experience of a phenomenon by collecting data from people [13]. This is

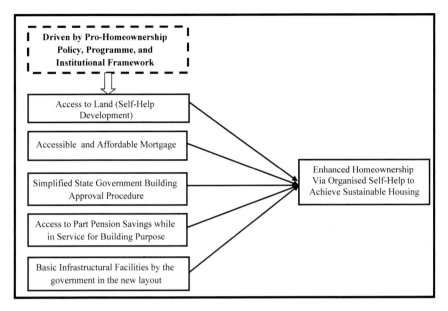

Fig. 1 Proposed framework for enhanced homeownership via organised self-help

because the data are exploratory and descriptive. The need to focus on the intervie-wees' well-informed background and finding possible answers prompted this method [15, 17, 18]. Face-to-face interviews were conducted, followed by validation through secondary sources. To mitigate the impact of the responses of uninformed partici-pants on the data, the paper adopted snowball and purposeful sampling techniques in the participants selection. Snowball sampling involves data which are difficult to access by the investigator, thus, the study participants recruit other participants for a test. Whilst purposeful sampling is used to obtain information from specific target groups that are in the best position to provide the information required in a study [18]. The study combined MAXQDA 2018 with thematic analysis to explain the data. This is in line with a similar study conducted by Ebekozien [15]. The study covered six residential locations, four of which were newly developing peripheral locations, and one from each division (Badagry, Epe, Ikeja, Ikorodu, Lagos Island, and Lagos Mainland). This is in line with Sawyer's [36] study. Table 1 shows the background of the 40 participants' interviewed. The participants were in three categories: 15 homeowners, 15 prospective homeowners, and 10 Government Officials. For confi-dentiality, the names of the organisations, participants, and ranks were masked [12]. The interviewees were well-informed in the issue and with vast experience across the spectrum.

Letters of invitation were sent to intending participants and the face-to-face inter-view took an average of 60 min. It took sometime before the researcher could persuade some of the government personnel with the assurance that their identity would be obscured. Guided by arranged interview themes, the oral questions started on a

Table 1 Summary of participants description and locations

S/No.	Participant	Organisation/location	Number	Code
1	Agency A		3	S1–S3
2	Agency B		3	S4–S6
3	Agency C		3	S7–S9
4	Agency D		1	B1
5	Homeowners (private/public organisation)	Ajeromi-Ifelodun (Badagry division)	2	P1–P2
		Ikorodu (Ikorodu division)	2	P3–P4
		Ibeju-Lekki (Epe division)	3	P5–P7
		Ifako-Ijaiye (Ikeja division)	3	P8–P10
		Lagos Island (Lagos division)	2	P11–P12
		Lagos Mainland (Lagos division)	3	P13–P15
6	Prospective homeowners (private/public organisation)	Ajeromi-Ifelodun (Badagry division)	3	P16–P18
		Ikorodu (Ikorodu division)	2	P19–P20
		Ibeju-Lekki (Epe division)	2	P21–P22
		Ifako-Ijaiye (Ikeja division)	3	P23–P25
		Lagos Island (Lagos division)	2	P26–P27
		Lagos Mainland (Lagos division)		P28–P30
	Total		40	

general note before seeking more explicit questions based on the reactions of the participants. Among the questions were the following:

(i)   What do you think are the main challenges faced by middle-income earners in attempting to become homeowners?
(ii) Do you see land purchase as an issue?
(iii) Do you think is easy to acquire land in rural locations?
(iv) What is your take regarding the Lagos State planning approval process?
(v)   Should the process be reviewed?
(vi) Can you suggest possible solutions that would enhance middle-income group homeownership?
(vii) What role do you think the government can play in this task regarding the National Housing Policy review and enforcement?
(viii) Can organised self-help housing development salvage Nigeria's housing shortage?

The interviews were later transcribed verbatim, for illustrative purposes. In addition to the member checking, the investigator adopted triangulation and researcher's reflexivity within a mixed validity approach [12, 13]. Themeing, emotion, invivo, attribute, and narrative coding strategies were adopted in this paper [35]. Sixty-five codes (highlight coding $= 25$, emoticode $= 03$, and text coding $= 37$) emerged and sorted into six categories based on relationship, reference, occurrence, and frequency. From the six categories, emerged three themes for this study. The results are reported and discussed in detail in the next section.

4 Findings

This section presents the findings according to the stated objectives of the study. As noted already, few scholars have explored the "self-help" approach mechanism for achieving sustainable housing in Nigeria. As such, this paper is an attempt to both to investigate the hindrances faced by the middle-income group in the process of becoming homeowners via organised self-help and proffer possible solutions that can enhance the group's homeownership dreams. Findings show that the desire for homeownership has created the problem of residential mobility among the middle-income earners, leading to the city's expansion across the state. This section is divided into three themes as addressed below.

4.1 Theme One: "Organised Self-help" in Sustainable Housing Provision

In this section on "organised self-help" in sustainable housing provision, the participants' responses to the concept of "organised self-help" regarding enhancing sustainable housing provision for the people are presented. One clear finding from this paper is that the majority of the engaged interviewees agree that a new government pro-homeownership policy is needed to stir-up organised self-help housing development in Nigeria. By extension, this would enhance sustainable housing provision across the state, and salvage Nigeria's housing shortage. Theme Three captured the new government's pro-homeownership policy that emerged in this study. Organised self-help housing provision is all about supporting locally-driven housing solutions for sustainability (S2, S5, S9, B1, P3 and P16). According to Participant 16:

Organised self-help housing provision can only be achieved if there is pro-homeownership policy by the government….Do we have a housing policy that is tailored towards this direction in Nigeria? I doubt….

This is an indication that organised self-help housing provision involves consultations with stakeholders. Participant 8 says:

We should not tailor everything on the government, you and I are the government. Although the government need to engage with people and communities, build the strength of local partnership and the enabling environment, and create a supportive national framework that would be self-sustainable via *agencies and non-governmental organisations......this is what is called organised self-help housing, it is all-inclusive and multifaceted....*

Participants S4, P6, and P17 stressed the need for government to lead in organised self-help housing provision as this would enhance homeownership among the middle-income earners' group and expansion of the Lagos city development, and Nigeria at large. Thus, this would result in sustainable housing for all before the year 2030.

4.2   Theme Two: Challenges Faced by Middle-Income Group

This section presents participants' response to the issues faced by the middle-income group in the process of becoming homeowners via organised self-help as presented in the top section of Fig. 1. Findings show that land and access to the mortgage are the major issues being faced by Lagosian attempting to become homeowners, especially in cities. Viewpoint from Participant P 18 says:

> *Lack of robust land administration and financial system is the biggest hindrances. This is because.....housing-loan repayment costs if managed to secure the loan is expensive for us to repay. I have made several attempt to secure the National Housing Fund but to no avail. So, what should a common man like me do?....*

In the opinion of Participant B1, bridging the housing demand-supply gap requires all-inclusive consultations and action plans, as against the perception of some Nigerians that the government should provide houses for them. The participant says:

> *I agree that Nigeria's mortgage finance sector is still in its process of improvement compared to the developed countries but I feel the sector has not done bad. The sector has attempted to spread the target round against the perception that our emphasis is on high-income earners. We release housing-loan to the middle and low-income earners that meet our stipulated requirements.* (Participant B1).

Participant P29 argued bitterly:

> *How can we achieve organised self-help housing provision when the right housing policy that will drive the concept is missing...?..... Do we have a working housing policy tailored towards homeownership for the middle-income earners? I doubt. This should be addressed first...*

The need to develop a feasible housing framework thus comes to light in this study. Moreover, findings show that government needs to provide a template on how the identified slums can be upgraded to cities in phases. In the few locations where this has been attempted, there is absence of consultations with the host communities, leading to community crisis (Fig. 2).

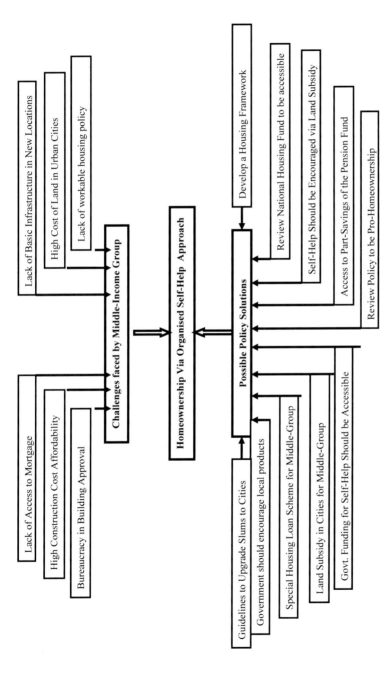

Fig. 2 Thematic network analysis of the homeownership via organised self-help approach

4.3   Theme Three: Possible Policy Solutions

Theme three gives the participants a platform to identify the possible policy solutions to enhance middle-income group homeownership as presented in the bottom section of Fig. 1. Findings show that organised self-help housing development can salvage Nigeria's housing shortage if supported by relevant stakeholders as agreed by the majority of the participants. Participant S2 says:

> Now is the time for concerted efforts by all stakeholders especially the government to come-up with pro-homeownership policy to address the housing deficit not only in Lagos but across the states......I'm aware that the shortage is about 17 million units but there is a need for verification....

P22 says: "*We hope that the NGN100 billion Family Home Fund would be transparent for common middle-income earners like me to access...*" The NGN100 billion Family Home Fund is a yearly contribution to the NGN1 trillion Social Housing Fund. This is an initiative of the Federal Government's social housing project but only for those that can afford NGN30,000 per month contribution. On this issue, a few pertinent questions may be raised. How many middle-income earners can afford to contribute NGN30,000 monthly to qualify for the housing loan, that's if shortlisted? Certainly, this approach is not sustainable. In the opinion of Participants S8, S9, B1, P2, P13, and P19, pension contributors should be allowed access to a portion of their savings while in service, for housing provision. Such a policy would enhance homeownership. Other possible suggestions are developing housing framework (S5), developing guidelines to upgrade slums to cities in phases (S8, P13, and P29), and creating new cities with basic infrastructure (P11 and P26).

Moreover, participants mentioned the need for government to intervene in land administration especially in cities, as a way of sustaining housing provision via organised self-help. Findings show that the cost of land in newly developing peripheral location layout, for example, Ijede in Ikorodu is prohibitively high for the category of many middle-income groups, who often lack additional income with which to buy it; the same applies to other new locations across the state. Participants P5, P13, P17, and P26 suggest land subsidies especially in city locations, while S8 feels that the government can formulate a policy to regulate the price of land within certain locations. Viewpoint from Participant P29 holds the view that the government should open-up new areas and provide basic infrastructure to encourage homeownership.

5   Discussion

The present study findings show that "organised self-help housing" is all-inclusive to enhance housing sustainability via consultation with major stakeholders, community participation and appropriate housing policy that is pro-homeownership. This agrees with Pattison [31], Arroyo [6], Moroni [26], Abd-Elkawy [1]. Arroyo [6] observed that the organised self-help approach is all about responsibility, community participation, technical assistance, solidarity, commitment, self-management, and shared decision making aimed at offering quality housing at lower costs than the usual. Moroni [26] noted that organised self-help is a process involving the preparation, process, and post-project stages. The author stressed the role of the non-governmental organisation and community-based organisation in achieving sustainable housing via organised self-help housing provision for the urban poor all around the world. In the opinion of Abd-Elkawy [1], organised self-help involved planning and building people and government with taking the standards of planning and design into consideration. Therefore, sustainable housing can be achieved via organised self-help; supported by government pro-homeownership policy in Nigeria.

Finding across the board shows that the middle-income group faces hindrance in the process of attempting to become homeowners. This agrees with Ibem [23], Olugbenga and Adekemi [30], Sawyer [36], Ebekozien et al. [14, 17]. Ibem [23] affirmed that a lack of basic amenities and the inability to access the land for housing development is a challenge for housing delivery in Nigeria. Olugbenga and Adekemi [30] identified rising land prices and charges above the official capacity of many middle-income groups, and land value is a major determinant in the housing neighbourhood. Sawyer [36] opined that inadequate attendants to infrastructural facilities such as electricity, road, water among others hindered housing provision. Ebekozien et al. [14] found lax enforcement, bribery and corruption, lack of adequate funding, political interference, inadequate monitoring and evaluation of housing programmes among others as the encumbrances faced by Nigerians in attempt to become homeowners. Whilst Ebekozien et al. [17] found a vulnerable institutional framework, relaxed policy and enforcement, insufficient political will, inadequate funding among others as the root cause of housing policies failure in Nigeria. Finding across the board shows the need to improve on the middle-income homeownership via organised self-help housing provision. This would boost sustainable housing provision in the middle-income group. This agrees with Pattison [31], Arroyo [6], Moroni [26]. Pattison [31] recommended greater capacity for the development of financial models, foster knowledge transfer, promote self-help housing among others. Whilst Arroyo [6] asserted the need for organised self-help housing to be recognised by the appropriate housing authority as a key enabling housing strategy. Moroni [26] found that with the support of a facilitating organisation such as architectural service, organised self-help housing is a good technique of construction applicable in poor urban settlements across the globe.

6 Recommendations

In summary, the key findings are:

1. There are inadequate policies and programmes to address the homeownership of middle-income earners in Nigeria.

2. The high cost of land, bureaucracy in the residential building approval process, lack of access to mortgage, high construction cost, lack of basic infrastructural in new locations among others were identified as the most cumbersome faced by intending homeowners.
3. Organised self-help housing provision can enhance sustainable housing if supported by pragmatic policies and implemented by the government.

In line with the above findings, the following are recommended:

1. There should be more feasible policies with clear-cut direction to enhance pro-homeownership. This will help to open up frontiers for the views of other major stakeholders such as the non-governmental organisation, community-based organisations, religious organisation, etc.
2. The visions of the state should be unified into pro-homeownership policymaking rather than having more regulations that are not feasible at the state level.
3. Also recommended is the promotion of organised self-help housing with focus on developing and sharing sustainable financial models such as contributors to the pension fund. The pension contributors should have access to their pension contributions, and government land subsidies in urban locations.
4. The government should encourage local government authorities and housing associations to build partnerships with a community-based organisation and non-governmental organisation. This will enhance self-help housing for the benefit of all.
5. More importantly, government funding for self-help housing should be accessible to local community groups and other organisations involved in self-help housing provision.

7 Conclusion

This paper shows that sustainable housing can be achieved via organised self-help; supported by government pro-homeownership policy in Nigeria. One of the effects of this will be an increase in homeownership among the middle-income earners; it will also lead to expansion of the city development, in line with the drive to achieve sustainable housing for the middle-income group before the year 2030 based on the 17 Sustainable Development Goals from 2015 to 2030. As highlighted with the three main themes, organised self-help should be supported by all stakeholders in the housing sector, based on government pro-homeownership policies that will stir the vision of sustainable housing for the middle-income earners in Lagos and other parts of Nigeria. Such a framework is currently missing in Nigeia's housing policy. However, its emergence would pioneer the provision of sustainable housing via organised self-help for liveable sustainable communities and cities across Lagos. In this regard, this study opens up new frontiers for further research to explore strategies for supporting organised self-help housing in Nigeria as part of the policy implications for sustainable housing and pioneering the transformation of homeownership for the middle-income earners in the country. Moreover, the study's framework and the supporting theory (Public Interest Theory of Regulation) can be further examined to explain the generic significance of organised self-help housing to sustainable housing via qualitative insight.

Acknowledgements The author would like to thank Prof. Abdul-Rashid Abdul-Aziz and Prof. Mastura Jaafar for their support and supervision during his PhD programme. Thanks are also due to Bowen Partnership, (QS Consultancy Firm in Edo State, Nigeria) for their assistance during the data collection, and to School of Housing, Building and Planning (Universiti Sains Malaysia) for the enabling environment as well as anonymous reviewers and editors for their insightful comments on earlier draft of this paper.

References

1. Abd-Elkawy AAM (2017) Mechanisms of application organized self-help housing program in new Egyptian cities for achieving low-cost housing (Case study: Ebny Betak project-Six October city)
2. Abrams C (1969) Housing in the modern world: man's struggle for shelter in an urbanising world. Faber and Faber, London
3. Andreasen HM, Agergaard J (2016) Residential mobility and homeownership in Dar es Salaam. Popul Dev Rev 42(1):95–110
4. Angel S, Jason P, Daniel LC, Blei MA (2011) Making room for a planet of cities. In: Lincoln Institute of Land Policy (ed) Policy focus report series. Lincoln Institute of Land Policy. Cambridge, MA
5. Arku G (2009) Rapidly growing African cities need to adopt smart growth policies to solve urban development concerns. Urban Forum 20(3):253–270
6. Arroyo I (2013) Organised self-help housing as an enabling shelter and developing strategy. Lessons from current practice, institutional approaches and projects in developing countries. PhD thesis submitted to Lund University, Sweden
7. Bank Negara Malaysia Press Release (2018) Bank Negara Malaysia's special measures for affordable homes. Bank Negara Malaysia Press Release. Retrieved from http://www.bnm.gov. my/index.php?ch=en_press&pg=en_press&ac=4765
8. Beracha E, Johnson KH (2012) Lessons from over 30 years of buy versus rent decision: is the American dream always wise? Real Estate Econ 40:217–247
9. Bryman A (2016) Social research methods, 5th edn. Oxford University, Oxford, United Kingdom
10. Centre for Affordable Housing Finance in Africa (CAHF) (2018) Housing finance in Nigeria. Centre for Affordable Housing Finance in Africa. Retrieved from http://housingfinanceafrica. org/countries/nigeria/
11. Centre for Housing Policy (2015) How local authorities can support self-build, self-build report, York. Retrieved from https://www.york.ac.uk/…/2015/selfbuildreportforcyc2015intern.pdf
12. Creswell WJ (2014) Research design. Qualitative, quantitative, and mixed methods approaches, 4th edn. Sage, Thousand Oaks, California
13. Creswell JW, Creswell DJ (2018) Research design: qualitative, quantitative, and mixed methods approaches, 5th edn. Sage, London, United Kingdom
14. Ebekozien A, Abdul-Aziz A-R, Jaafar M (2017) Comparative analysis of low-cost housing policies in Malaysia and Nigeria. Int Trans J Eng Manag Appl Sci Technol 8(3):139–152
15. Ebekozien A (2019) Root cause analysis of demand-supply gap to low-cost housing in Malaysia. PhD thesis submitted to Universiti Sains Malaysia, Malaysia

16. Ebekozien A, Abdul-AzizA-R, Jaafar M (2019) Housing finance inaccessibility for low-income earners in Malaysia: factors and solutions. Habitat Int 87:27–35. https://doi.org/10.1016/j.hab itatint.2019.03.009

17. Ebekozien A, Abdul-Aziz A-R, Jaafar M (2019) Low-cost housing policies and squatters in Nigeria: the Nigerian perspective on possible solutions. Int J Constr Manag. https://doi.org/10.1080/15623599.2019.1602586

18. Fellows FR, Liu MMA (2015) Research methods for construction, 4th edn. Wiley-Blackwell, West Sussex

19. Gabriel SA, Rosenthal SS (2005) Homeownership in the 1980s and 1990s: aggregate trends and racial gaps. J Urban Econ 57:101–127

20. Hantke-Domas M (2003) The public interest theory of regulation: non-existence or misinterpretation? Eur J Law Econ 15(2):165–194

21. Harris R (1999) Slipping through the cracks: the origins of aided self-help housing, 1918–1953. Hous Stud 14(3):281–309

22. Hertog DP, Broersma L, Van-Ark B (2003) On the soft side of innovation: services innovation and its policy implications. De Economist 151(4):433

23. Ibem OE (2010) The role of government agencies in public-private partnerships in housing delivery in Nigeria. J Constru Dev Countries 15(2):23–48

24. Iheme OJ (2017) Factors for the implementation of affordable federal public housing policies in South-South Region of Nigeria. PhD thesis submitted to University of Salford, Manchester, United Kingdom

25. Kamal ME, Hassan H, Osmadi A, Fattah AH (2019) Government and homeownership: the Penang scenario. Penerbit Universiti Sains Malaysia, Pulau Pinang, Malaysia

26. Moroni A (2015) Organised self-help housing: the role of the facilitating organisation. Retrieved from http://www.hdm.lth.se/fileadmin/hdm/Education/Undergrad/ABAN06_2015/Organized_Self-Help_Housing-Anna_Moroni.pdf

27. Ndubueze OJ (2009). Urban housing affordability and housing policy dilemmas in Nigeria. PhD thesis submitted to the University of Birmingham, United Kingdom

28. Obiaraeri N (2019, May 21) Decimation of Nigeria's middle class. Punch. Retrieved from https://punchng.com/decimation-of-nigerias-middle-class/

29. Okpoechi C (2014) Middle-income housing in Nigeria: determining important functional requirements for mass housing design. Architect Res 4(1A):9–14. https://doi.org/10.5923/s.arch.201401.02

30. Olugbenga E, Adekemi O (2013) Challenges of housing delivery in Metropolitan Lagos. Res Humanit Soc Sci 3(20):1–9

31. Pattison B (2011) Self-help housing. World Habitat. Retrieved from https://www.google.com/search?client=firefox-b-d&ei=NL83XbDNOJm-vATY3YyACw&q=Organised+Self-Help+housing+provision&oq=Organised+Self-Help+housing+provision&gs_l=psy-ab.3..33i160.21592.30760..31847...3.0..0.134.1936.15j6......0....1..gws-wiz.......0i71j33i21j33i22i29i30.lRYHiXIQLCs&ved=0ahUKEwiwt9WKvszjAhUZH48KHdguA7AQ4dUDCAo&uact=5

32. Pigou AC (1932) The economics of welfare. Macmillan, London, United Kingdom

33. Power Lunch (2014) Nigeria's housing shortage. Retrieved from https://www.youtube.com/watch?v=r5n1CW_u8D4

34. Rodríguez M, Åstrand J (1996) Organised small-scale self-help housing. Lund University, Lund

35. Saldana J (2015) The coding manual for qualitative researchers, 3rd edn. Sage, London, United Kingdom

36. Sawyer L (2014) Piecemeal urbanisation at the peripheries of Lagos. Afr Stud 73(2):271–289. https://doi.org/10.1080/00020184.2014.925207

37. Stotz O (2019) The perception of homeownership utility: short-term and long-term effects. J Hous Econ 44:99–111. https://doi.org/10.1016/j.jhe.2018.11.003

38. UN-Habitat (2005) Financing urban shelter: global report on human settlements 2005. United Nations Human Settlements, Nairobi

39. UN News Centre (2018) Sustainable development goals-poverty eradication, inclusive growth focus of UN social development commission's 2018 session. Retrieved from https://www.un.org/sustainabledevelopment/blog/2018/01/poverty-eradication-inclusive-growth-focus-un-socialdevelopment-commissions-2018-session
40. United Nations Centres for Human Settlements (1996) The Habitat Agenda is a global action for adequate and sustainable human settlements. Oxford University Press, Nairobi
41. Viales R (2007) Organised self-help housing: FUPROVI's Model. Course organised self-help housing. (Housing Development & Management, Compiler) San José, Costa Rica
42. World Bank Press Release (2017) World Bank approves new financing to support affordable housing in Indonesia. World Bank Press Release. Retrieved from https://itpcchennai.com/world-bank-approves-new-financing-to-support-affordable-housing-in-indonasia

Exchange Rate and Housing Deficit Trends in Nigeria: Descriptive and Inferential Analyses

John Ogbonnaya Agwu

Abstract Among all possible strategies advanced by policymakers, researchers and practitioners towards resolving housing deficit, macroeconomic strategy remains underexplored. In this chapter, archival data from 1960 to 2019 were used to explore the dynamics between macroeconomic variables and housing development in Nigeria. Specifically, the chapter examined the linkage between exchange rate and rental values. The investigation involved both descriptive and inferential analyses, with the aid of Pearson Product Moment Coefficient in SPSS and Pivot graph in Excel. Results from the descriptive analysis revealed that exchange rate and housing deficit in Nigeria from 1960 to 2019 shared similar trends in stability and volatility. On the other hand, results of the inferential analysis showed a strong positive significant correlation between the naira-dollar exchange rate and housing deficit, at r $(60) = 0.924$, $p < 0.0001$, with exchange rate contributing 92% of the variation in housing deficit. These results agreed with previous studies claiming that aggregated macroeconomic indices influence the real estate sector. Thus, Nigeria should employ exchange rate as a catalyst to stimulate housing development and resolve housing deficit.

Keywords Exchange rates · Housing deficit · Housing development · Macroeconomic variables · Rental values

1 Introduction

In Nigeria, housing deficit, which is simply the shortfall between effective housing demand and supply, is a national problem that has defied many strategies and left millions of Nigerians homeless or sheltered under miserable environmental conditions. Historically, the housing deficit in the country has been associated with population increase, oil boom, land tenure system, housing finance, cost and time required to

J. O. Agwu (✉)
Department of Urban and Regional Planning, Imo State University, Owerri, Nigeria
e-mail: Agwujo@imsu.edu.ng

T. G. Nubi et al. (eds.), *Housing and SDGs in Urban Africa*, Advances in 21st Century Human Settlements, https://doi.org/10.1007/978-981-33-4424-2_13

complete a housing project, legal procedures surrounding property, land procurement and cost of building materials.

However, in recent years, other aggregated macroeconomic factors such as foreign direct investment, migration, exchange rate, technological gap, insecurity and climate change have gained entry into the real-estate performance debate. Unfortunately, researchers, policymakers and practitioners have focused mostly on conventional aggregate microeconomic factors and have neglected the core macroeconomic factors that influence and drive international economic development and national housing supply. One of these neglected core macroeconomic factors is the exchange rate.

In this chapter, we argue that the exchange rate is a critical factor that directly affects consumer spending, capital investment, government spending and net export and that it indirectly affects the interest rate, mortgage ecosystem, disposable income, government policy, inflation and rate of capital flows. As such, it directly and indirectly affects economic development and housing supply, especially in an import-dependent nation like Nigeria. The level of influence exerted on the national housing deficit by the exchange rate can be determined with the basic formula for an economy's GDP, which is expressed as $GDP = C + I + G + (X - M)$. Where;

C = consumption or consumer spending, the biggest component of an economy
I = capital investment by businesses and households
G = government spending and
$(X - M)$ = export minus imports or net exports.

This implies that the exchange rate has both positive and negative effects on domestic economic variables. In a situation where the exchange rate is unstable, households and businesses spend most of their income on imported consumable goods, with less income are available for investment. The same applies to government spending and the national net export, even as foreign investment declines. Therefore, in a situation where the currency is unstable, households and businesses will spend more on food items and less or none on housing projects. Similarly, overall government spending, including housing expenditure will retard while import will exceed export leading to over-reliance on imported building materials for housing projects at high money and time costs. In general, in a situation of unstable currency, housing supply and GDP will be low (this is currently the case in Nigeria today).

Outside substantive problem arising from macroeconomic variables, lack of political will and funding, Feitosa et al. (2015), argued that lack of accurate and reliable data collection approach hinders the performance of housing policies in developing countries. For instance, extant literature suggest that the use of census data such as individual-level sample data (microdata) and universal data with detailed spatial resolution (small areas known as census tracts) is limited by the universality factor, confidentiality (obscuring detailed geographic information) and aggregation. Thus, housing policies based on census data often fail to address the global goal of social equity and environmental justice.

Again, Jakob (2016) observed that economic theory does not clearly articulate how exchange rate regimes can affect economic growth. According to Levy-yeyati and Sturzenegger (2001), studies exploring the nexus between exchange rates and economic growth are few "probably due to the fact that nominal variables are considered to be unrelated to longer-term growth performance" (p. 2). Moreover, results from the few existing studies appear to be contradictory. Some studies report that the choice of exchange rate regimes does not have a significant impact on economic growth and that when it does, the impact differs based on economic region and the financial market (Huang and Malhotra 2004).

Some other studies, such as Gulde et al. [12], found a moderately weak connection between the exchange rate regime and growth of output. Gulde et al. [12] also noted that countries with pegged exchange rates tend to achieve higher levels of investment but attain lower productivity compared to countries with floating exchange rates. In that study, three economic indicators-international trade, international division of labour and low interest rates-were found to be associated with the upsurge in investment and economic growth.

Despite this theoretical linkage and empirical studies linking housing deficit with development problems, research focusing on housing has not effectively explored the empirical relationship between exchange rate and housing deficit. Consequently, the contribution of exchange rate in housing deficit is blurred. To address this knowledge gap, this chapter adopted explanatory approach to explain the empirical relation between exchange rate and housing deficit in Nigeria using historical data between 1960 and 2019.

Previous attempts to boost housing supply using microeconomic variables such as the land tenure system, housing finance, housing regulation, housing development programmes, housing and urban development plans, etc., have not produced the required results. Consequently, the current housing challenge in the country remains alarming. In 2015, the National Bureau of Statics estimated that the housing deficit in Nigeria is over 17 million based on an estimated population of over 174 million. Despite this alarming figure, successive governments have not been able to bridge this gap. With the population projected to increase to over 180 million, skewed mostly in favour of the urban sector, the housing deficit is expected to worsen.

Hence, it is necessary to examine the linkage between exchange rate and housing deficit on one hand and between exchange rate and rental value on the other hand. Besides this introduction, this chapter has three other sections. Section 2 discusses the linkage between economic environment and housing development while Sect. 3 explains the research methods adopted. Section 4 presents the results and discussions, while Sect. 5 concludes the chapter.

2 Linking Economic Environment with Housing Development

The success story of cities such as Singapore and Hong Kong provide glaring evidence linking economic performance and housing development. However, the housing economic relations witnessed in some cities of developing countries such as Lagos,

Jakarta and Metro Manila contradict any assumption that economic development is linked to housing provision. For instance, despite high city GDP per capita figures above the national figures in South Korea, Japan and Nigeria, the proportion of the urban population living in slums remains very high.

The problem is worse in developing countries like Nigeria, where rapid economic development and urbanisation coincide over a short period in a fashion that frequently overwhelms the capacity of government. This usually manifest as poor housing, inadequate urban infrastructural support and lack of or poor social and reproductive health services. In other words, squatter and slum settlements emerge mainly because of the inability of city governments to plan and provide affordable housing for the low-income segments of the urban population [26].

Apart from poor urban housing and sordid environmental conditions that tend to provoke innovative urban policies, today's age of global environmental and social change has necessitated the need for the economic professionals to recognise that healthy environments integrated with inclusive social policies, representative political institutions and fair legal frameworks is the bedrock of successful societies and strong economies [18]. In this regard, it is necessary to rethink the way macroeconomic analysis is conducted. Particularly now that meta-analysis of macroeconomic studies shows that the frequency and magnitude of global crises, the rise of political populism, acceleration of climate change and concentration of research on developed regions, etc., all suggest that existing economic and monetary policies have failed and new approach(s) are required [15].

As observed by Gerlach et al. [8, 11, 30], the frequency and intensity of these failures are mediated mainly by mismanagement of monetary policy and prudential regulations. Other factors include skewed distribution of welfare gains from technological change and globalisation, restriction of social mobility, the maladministration of short-term growth goals at the expense of the environment and long-term sustainability and the overarching focus on developed regions, which are outside where the bulk of the world's population live and where dynamic change is prevalent.

To redeem these flaws, macroeconomic policies need to mediate sustainability. According to [15], three channels have been identified as possible routes: economic stability, distributional equity, and broad social goals such as income security, education, universal health care, adequate housing, infrastructure and basic services. In response, a number of studies have been conducted to capture a broad array of factors that influence these channels. Most of these studies focus on the effects of monetary policy on house prices and other economic variables.

From studies conducted by Williams (2016), Muhammad et al. (2014), there are two ways by which monetary policy can affect the housing market. One relates to increase or decrease in housing demand because of fluctuations in the interest rate, official cash rate, inflation and the exchange index. The other one relates to increase or decrease in housing supply because of fluctuations in the interest rate, official cash rate, inflation and the exchange index. In both cases, it means that monetary policy shocks have effects on the housing market and financial stability. This relationship has been confirmed by studies conducted by Ya-chen and Shuai (2013), Xiuzhi and

Xiaoguang (2006), Jorda et al. (2015) who examined the linkage between RMB Exchange Rate and Real Estate Price, Interest Rate and Housing Price.

Despite such studies focusing on economic matters, no study has combined a measure of economic strength with housing supply in Nigeria, although Dunga et al. [10] attempted to align macroeconomic variables with UN Sustainable Development Goals indicators. In terms of adequate housing, rental value and slum population are central to linking economic and monetary policy with sustainability. In that case, SDG Goal 11 offers the opportunity to examine the relationship between macroeconomic variables and sustainability.

Thus, as one of its objectives, this chapter examines the empirical relationship between the exchange rate and the rental value of single rooms in the capital city of Abuja between 2007 and 2016. The ultimate objective of Goal 11 is to make cities and human settlements inclusive, safe, resilient and sustainable. United Nation through goal 11 expects countries to strive to ensure access for all, adequate, safe and affordable housing and basic services as well as the upgrade of slums by 2030. Meeting this target through economic policy and planning requires strong evidence linking macroeconomic variables with rental value, urban slum population and other parameters portraying housing adequacy.

2.1 Concept of Exchange Rate

The exchange rate is a macroeconomic variable defined as the price of one currency in terms of another currency. It expresses the national currency's quotation with respect to foreign ones. The exchange rate plays a critical role in an economy because imports and exports constitute a large part of the economy. Using Piana [28] illustration, if one US dollar is worth 360 Nigerian Naira, the exchange rate of dollar will be 360Naira. If something costs 360,000 Naira, it will automatically cost 1000 US dollars.

Expressed in terms of purchasing power, a house that costs 3600,000 Naira will cost 10,000 US dollars. Expressed in terms of GDP, Nigeria's GDP of 8 million naira will then be worth 22, 222.22 dollars. Thus, the exchange rate is a price as well as a conversion factor, a multiplier or a ratio, depending on the direction of conversion and perspective ([28], para. 1). As a price, the exchange rate may turn out to be the fastest moving price in the economy, bringing together all the foreign goods with it, if allowed to move freely.

2.1.1 Types of Exchange Rate

Different approaches to classifying the exchange rate exist in the literature and practice, the most common being the nominal exchange rates and the real exchange rates. The nominal exchange rate is defined as the number of units of the domestic currency that can purchase a unit of a given foreign currency or the price of a foreign currency in terms of the home currency [4]. According to Economic-Point.com, it is called nominal because it takes into account only the numerical value of the currencies, not the purchasing power of the currencies. With a high nominal exchange rate, imports become more expensive and all production sectors that depend on import fall.

The nominal exchange rate also influences the flow of capital between countries and the interest rate. It affects investments as well as the economic growth and welfare of the people. The real exchange rates are the nominal rates corrected somehow by inflation measures. For instance, if Nigeria has an inflation rate of 10% and the USA has an inflation of 5%, without any changes in the nominal exchange rate, Nigeria will now have a currency whose real value is 10%-5% = 5% higher than before. The real exchange rate influences the competitiveness of different sectors of a country's economy.

Another classification of the exchange rate is based on the number of currencies taken into account. When only two currencies are considered, what exists is called a bilateral exchange rate, which is usually computed by matching demand and supply on financial markets or in banking transactions. In this latter case, the central bank usually acts as one of the parties to the relationship. Another type of exchange rate under this classification system is the cross exchange rate, which is an exchange rate between two currencies to a third currency, that is, the exchange rate between two currencies expressed in terms of the exchange rate between them and a third currency.

For example, given the US dollar/naira and Japanese yen/naira exchange rates, the dollar/yen exchange rate becomes the cross exchange rate. When the currencies under consideration are more than three, what exists is called a multilateral exchange rate. Multilateral exchange rates are computed in order to judge the general dynamics of a country's currency toward the rest of the world. One takes a basket of different currencies, selects a (more or less) meaningful set of relative weights, then computes the effective exchange rate of that country's currency.

Another way of classifying the exchange rate is based on formal and informal market operations. An exchange rate determined outside formal market operations is called a black market exchange, while an exchange rate determined by formal market operations or central bank regulations is called a real and effective exchange rate. In both cases, there are usually different prices for selling and buying but the central value is usually applied as the official exchange rate.

Some countries impose more than one exchange rate, depending on the type and the subject of the transaction. In such a case multiple exchange rates then exist, usually referring to commercial versus public transactions or consumption and investment imports. This situation always requires some degree of capital controls. In many countries, beside the official exchange rate, the black market often offers the foreign currency at another, usually much higher rate.

2.1.2 Exchange Rate Regimes

An exchange rate regime is the way a particular country manages its currency in relation to other currencies and the foreign exchange market. A country's exchange rate regime governs how much its own currency is worth in terms of the currencies of other countries. According to the IMF's de facto classification, exchange rate arrangements can be classified into four categories: hard pegs or fixed regimes (such as currency board arrangements), soft pegs or intermediate regimes (such as crawling pegs, stabilised arrangements, and craw-like arrangements), floating regimes (such as managed floating and free floating), and residuals [16].

Under the fixed exchange rate, the local currency is pegged against either another currency or a basket of other currencies. The main goal of this system is to achieve stability in the value of currency through fixing it against a stronger and more stable currency (or currencies). The main advantage of this system is that the currency does not fluctuate according to market conditions, thereby creating a stable and predictable business climate for investments and trade between the two currencies. However, the main drawback of pegged exchange rates is that it is very difficult for government to implement independent monetary policies and to liberalise capital markets at the same time [34].

Floating regimes describe a situation where the exchange rate is governed by forces of demand and supply. As the name implies, a floating exchange rate regime fluctuates or floats with market forces. In countries that allow their exchange rates to float, the central banks intervene (through purchases or sales of foreign currency in exchange for the local currency) mostly to limit short-term exchange rate fluctuations [33]. Unlike fixed or pegged regimes, the floating regime puts countries in a better position to operate independent monetary policies [9].

In such countries, the foreign exchange and other financial markets must be deep enough to absorb shocks without large exchange rate changes [33]. In-between the fixed and floating regimes exists a third type known as the soft exchange rate peg, which is obtainable when the exchange rate is allowed to float within a fixed narrow (+1 or −1%) or wide range (up to +30 or −30%). Under this regime the exchange rate can be adjusted over a given time frame. Using the IMF's classification system, the residual category is used when the exchange rate arrangement does not meet the criteria for any of the above three categories. As noted by [14], arrangements characterised by frequent shifts in policies may fall into this category.

2.1.3 Determinants of Exchange Rates

Beyond the forces of demand and supply, numerous factors determine exchange rates, many of which are related to the trading relationship between two countries or the trading relationship between one country and the rest of the world. Discussed below are some of the principal determinants of the exchange rate between two countries.

Inflation

Inflation is typically a rate that measures the general increase in prices and fall in the purchasing value of money in a country. For instance, a report from demonstrated that a country with a long history of lower inflation rate always witness a rise in currency value, as its purchasing power increases relative to other currencies [13]. Countries with higher inflation normally witness devaluation in their currency compared with the currencies of their trading partners [32]. Higher interest rates is usually seen as immediate effect of such economic scenario.

Interest Rates

The interest rate is the percentage of principal charged by the lender for the use of its money. As explained by Central Bank of Nigeria [9], foreigners tend to take advantage of high interests by bringing their money into the domestic economy, thus increasing the supply of foreign currencies and leading to appreciation of the domestic currency. Under such scenario, a country may find it difficult to export goods and services because the appreciation of its currency might make imports cheaper and exports costlier [6]. On the contrary, when interest rates fall, foreigners tend to move their money out of domestic economy [6]. In such case, the supply of foreign currencies will drop and the domestic currency will depreciate. However, the impact of higher interest rates can be mitigated if inflation in a country is much higher than that of their trading partners or if additional factors serve to drive the currency down [32].

Current Account Deficits

The current account is the balance of trade between a country and its trading partners, reflecting all payments between countries for goods, services, interest and dividends. A deficit in the current account shows the country is spending more on foreign trade than it is earning and that it is borrowing capital from foreign sources to make up the deficit. In other words, the country requires more foreign currency than it receives through sale of exports and it supplies more of its own currency than foreigners demand its products. The excess demand for foreign currency lowers the country's exchange rate until domestic goods and services are cheap enough for foreigners and foreign assets are too expensive to generate sales for domestic interests.

Public Debt

This refers to the financial indebtedness of a country to other countries, individuals, corporate bodies and international organisations; the debt is usually incurred to procure or finance public-sector projects. Developing countries such as Nigeria engage in large-scale deficit financing to pay for public-sector projects and governmental funding in the hope of stimulating the domestic economy but such actions often end up encouraging inflation and discouraging foreign investors. Public debt determines the exchange rate in so many ways. It also creates fake economic health by increasing immediate money supply and servicing this debt in future always relies on cheaper dollar values, leading to inflation.

Terms of Trade
Terms of trade refers to a ratio between a country's export prices and its import prices [20]. According to Shaik et al. [32], terms of trade is said to improve when the price of a country's exports rises by a greater rate than that of its imports. Thus, increase in terms of trade suggest greater demand for the country's exports [9]. This, in turn, results in rising revenues from exports and consequent increased demand for the country's currency (and an increase in the currency's value). On the other hand, if the price of exports rises by a smaller rate than that of its imports, the currency's value will decrease in relation to that of a country's trading partners [32]. This means that terms of trade relates to current accounts and balance of payments.

Strong Economic Performance Economic performance is a macroeconomic index that measures country's ability to meet certain defined economic objectives. These objectives could be a long term objective such as sustainable growth and development, or short term objective, such as stabilising an economy in response to sudden and unpredictable economic shocks like Covid-19. As an index, it is usually defined by several factors such as inflation rate, unemployment rate, budget deficit, change in real GDP, terms of trade, productive labour, savings, investment, poverty rate etc. [19].

In determining the strength of economic performance, countries strive to response to two basic questions: are current economic policies working as desired or simply targeting some hot button issue of the day? How is my economy performing relative to our trading partners? In the context of this chapter, the later question is usually answered by measuring the performance of variables that simultaneously influence three primary economic segments: households, firms, and government [19]. These variables include inflation rate, unemployment rate, budget deficit, and change in real GDP.

Countries that have sound economic performance usually attract foreign investors. From available record, countries with positive economic performance usually draw investment funds away from other countries perceived to have low economic performance. By so doing, exchange rate is negatively affected. In addition, social unrest like Boko Haram insurgency in Nigeria can also cause a loss of confidence in a currency and consequently trigger movement of capital to the currencies of more stable countries. In summary, factors that influence exchange rate movements in an economy include real economic variable, market expectations and the sociopolitical climate.

2.2   Concept of Housing Deficit

The housing deficit, otherwise known as housing shortage, is described as a deficiency or lack in the number of houses needed to adequately and comfortably accommodate the population of an area. Based on this description, the housing deficit goes beyond the number of houses required to shelter the population of an area, as it includes the number of houses required to accommodate them at affordable costs and under secured, comfortable and dignifying social and environmental conditions.

For example, Brazil Capstone Team [7] defined housing deficit as the number of shelters that do not have adequate conditions to be habitable, plus the number of housing units that need to be built to shelter all families who currently lack one and, as a result, share a shelter with another household in overcrowded conditions. Quantitatively, the housing deficit is measured by the need for new constructions owing to the number of families sharing the same space (rustic and improvised housing), excessive rent burden and replacement deficit. In other words, housing deficit is the difference between the total number of households in a given area and the total number of housing units that are habitable and affordable.

Based on these definitions, extant literature observed that while the bulk of the housing deficit in urban areas are due to the comfort and precariousness of dwellings, the housing deficit in rural areas is mainly due to the precariousness of dwellings. In Nigeria, there are no accurate data on the nation's housing stock but a number of variables depicting housing characteristics glaringly exist. For instance, it is a known fact that the level of housing production is only two dwelling units per thousand people, compared to the required rate of about 8–10 dwelling units per 1000 population as recommended by the United Nations [35].

Moreover, the magnitude of housing deficit in Nigeria is shaped by income and urbanity. Additionally, the national and urban poverty rates, as well as the housing condition and population growth rates, are known. According to Rosemback et al. [31], there are seven dimensions of adequacy that must be considered in the assessment of housing needs. A combination of these dimensions provides reliable estimate of housing deficit in any particular area (Table 1).

Although housing attributes and statistics differ from state to state, getting the national average based on these dimensions is not difficult. Therefore, despite the confusion as to the number of new additions, it is obvious that a critical gap exists between housing supply and demand. So far, governments have been able to meet only a minute part of their housing targets.

2.2.1   Factors Contributing to Housing Deficit

Some scholars across developing and developed countries seem to be agreed on factors contributing to the housing deficit. For example, [22], Ansah [3], Iwedi and Onuegbu [17], Afrane et al. [2], Ackley et al. [1] agree that factors contributing to the housing deficit cut across demand and supply factors as well as microeconomic

Table 1 Housing needs: dimensions of housing adequacy

S/No.	Dimension	Description
1	Housing cost	The household spending on housing should not severely compromise the total household income
2	Physical suitability of the dwelling unit	Dwellings should be made of materials that permanently ensure weather protection, the health, privacy and security of their residents
3	Dwelling unit suitability to the household	The household density in a building should not be excessively high. Families should not cohabit for lack of choice
4	Environmental safety	Dwellings should not be located in areas of environmental risks, including risks of flooding or landslides, contaminated areas, etc.
5	Legal security	Households must have legal security of tenure
6	Infrastructure and public services	Dwellings should be served by sewage, water supply, electricity network, street lighting, paving, trees, curb, sidewalk, etc.
7	Location and accessibility	The location of dwellings should promote the integration into the city, including appropriate access to employment options, efficient public transportation, health services, school, culture and leisure

Source Adapted from [31]

and macroeconomic factors. Demand factors relate to housing deficit arising from household demographic attributes, including environmental and building conditions, while supply factors relate to policy strategies and the disposition of government towards satisfying housing needs.

As shown on Table 2 above, out of 13 factors contributing to the housing deficit, only four (4) are demand factors while the rest are supply-related factors. In both cases, data are usually generated on an aggregate basis. However, in developed countries such as Australia, the USA and the United Kingdom, there is always a balance between the two sides and data are usually disaggregated. For example, the Australia Senate Committee on Business identified higher incomes, low interest rates, high rent, credit availability, speculative demand and taxation influences, etc., as factors influencing demand for housing.

However, in both developed and developing countries, microeconomic factors are mostly deemed to affect the housing deficit. With the exception of Makinde (2014), macroeconomic factors tend to be aggregated and considered as economic growth in the housing literature. In the economics literature, factors such as interest rate, trade

Table 2 Demand and supply factors contributing to housing deficit

S./No.	Demand factors	Supply factors
1	Rural-urban migrations	Lack of continuity (political instability)
2	Population growth	Unhealthy mortgage system
3	Urbanization	Defective land tenure system
4	High cost of building materials	Lack of political will
5	-	Paucity of fund
6	-	Corruption
7	-	Poor state of infrastructure
8	-	High cost of land procurement
9	-	Cumbersome regulatory framework

Adapted from [17]; Afrane, Bujang, Liman and Kasim (2016)

balance, exchange rate and government spending, etc., contribute to the exchange rate and economic growth.

3 Methodology

3.1 Study Location

Nigeria is situated in the West African region and lies between longitudes 3° and 14° and latitudes 4° and 14°. Nigeria was colonised by the British between the late nineteenth and early twentieth centuries. The country gained her independence on 1st October 1960. Officially, Nigeria is known as the Federal Republic of Nigeria. It has 36 states and a Federal Capital Territory known as Abuja. Nigeria has a land mass of 923,768.00 km² and is bordered to the north by Niger and Chad. It shares borders to the west with the Republic of Benin, while the Republic of Cameroun shares the eastern borders right down to the shores of the Atlantic Ocean, which forms the southern limits of Nigeria's territory.

The 800 km of coastline confers on the country the potential of a maritime power. Land is in abundance in Nigeria for housing as well as for agricultural, industrial and commercial activities. According to current projections Nigeria's population is over 200.96 million, making it the seventh most populous country in the world, with more than 50% living in urban centres (Nigeria Population, 2019-05-12). Based on this projection, Nigeria's average population density is approximately 212.04 persons per square kilometre, making it one of the most densely populated countries in the world. Officially, the country has 277 local governments spread across three major ethnic groups of Hausa, Igbo and Yoruba.

In economic terms, the World Bank [36] classified Nigeria as a lower-middle-income country with rich oil and other mineral deposits. It occupies the seventh position as an OPEC member country Organization of the Petroleum Exporting Countries (OPEC) [27] and accounts for a quarter of Africa's oil production Ogwumike and Ogunleye [24]. Nigeria is currently faced with a chaotic housing situation. Over 100 million of the country's 180 million people lack access to proper housing or live in deplorable housing conditions [23].

The Nigerian housing finance market is organised along formal and informal lines. The formal sub-division has two components: the upper-income groups, whose undertakings are located in the urban areas, and the lower-income groups, which depend on the subsidised National Housing Trust Fund (NHTF) for access to housing. The informal area includes the rotating savings and loan associations, the traditional cooperative system, credit co-operatives, as well as individual and family savings Bichi [5], Okonkwo [25].

The country has two important capital cities: Abuja and Lagos. Abuja is the current administrative capital of Nigeria and it is located at the centre of the country on latitude 9.0765° N and longitude 7.3986° E on 491 m elevation above sea level. According to the 2006 census, the city of Abuja had a population of 776,298, making it one of the ten most populous cities in Nigeria. Lagos used to be Nigeria's capital and remains the country's commercial capital. In 2006, the population of Lagos was officially reported to be 9,113,605, making it the second most populous city in the country, after Kano.

Lagos has a total area of 3345 km^2, with a population density of 3752 persons per km^2. It is located on latitude 6.5244° N and longitude 3.3792° E in the southwest geopolitical zone of the country, where it is bounded by the Bight of Benin on the South (Fig. 1).

3.2 Research Methods

The study involved historical data on the dollar-naira real effective exchange rates, housing deficits and mean room rental values for Abuja, as separately extracted from Central Bank of Nigeria (CBN), National Planning Commission (NPC), National Bureau of Statistics (NBS), Census and Economic Information Centre (CEIC), International Monetary Fund (IMF), International Financial Statistics, and World Bank archives covering 1960 to 2019. Data on slum population and rental value covered only seven years from 1990 to 2014.

With the exception of the 2019 exchange rate values, the central market value of Nigeria's currency to the USA dollar for the year 2019 was used to represent the exchange rate for each year, while the estimated and planned housing unit for each year or plan period was used for the housing deficit. The exchange rate value of 31st

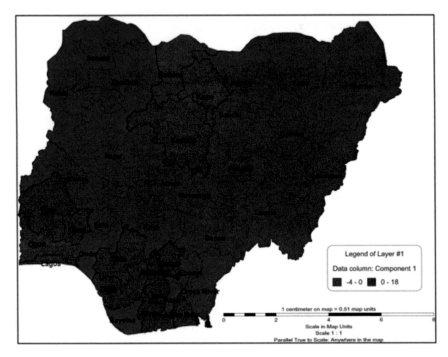

Fig. 1 Map of Nigeria showing housing quality: scores on component Source: Morenikeji et al. (2017)

May 2019 was used as the 2019 exchange rate value. The 60-year data for exchange rates and housing deficits, as well as the seven-year data on slum population and room rental value collected through archival reviews, were analysed using descriptive and correlation statistical techniques with the aid of Statistical Package for Social Science (SPSS).

4   Results and Discussions

4.1   *Exchange Rate and Housing Deficit*

The major argument in this study is that the national exchange rate has influence on the national housing deficit, thus supporting the economic theory that housing supply is a function of economic performance (growth). In addressing this proposition, the study employed both the descriptive and inferential approaches. As shown in Table 3, summary statistics from the descriptive analysis revealed that the study collected 60 years of archival data on the Nigerian exchange rate ($M = 60.93$, $SD = 89.531270$) and the housing deficit ($M = 6.007,967$, $SD = 6,705,123.530$).

Study variables	Mean	Std. Deviation	N
Exchange rate	60.92962	89.531270	60
Housing deficit	6,007,966.67	6,705,123.530	60

Table 3 Descriptive statistics for exchange rate and housing deficit

This represents 60 years of Nigeria's existence as an independent nation. This shows that the mean exchange rate for the period was 60.93 naira for a dollar, while the mean housing deficit for the same period was 6.007,967 units (approximately 6 million). This implies that the minimum and maximum values of the exchange rate and housing deficit strongly deviated from their mean values, thus demonstrating that both the exchange rate and the housing deficit in the country have been highly variable and progressive.

A visual trend analysis of the exchange rate and housing deficit in Nigeria from 1960 to 2019, as shown in Fig. 2, also shows that the sum and percentage distribution of the exchange rate and housing deficit are fairly similar. The trend shows that between 1960 and 1985 there was a fair stability in the exchange rate and housing deficit. A similar trend also occurred between 1994 and 1999, although with higher values. The patterns of increase and decline were also similar. In 1993 and 1994 both variables sharply increased and declined in 1999. Between 2001 and 2003, the exchange rate and housing deficit again witnessed a sharp upsurge, jointly declined in 2007, and later increased between 2009 and 2013.

As shown in Fig. 2, both variables also experienced sharp increases and have been stabilised since 2018 above 250% till date. The two trends descriptively demonstrate association and pose a posture asking for inferential support. To address this, the study hypothesises that there is a significant relationship between the exchange rate and housing deficit. This was inferentially tested using Pearson's product-moment correlation coefficient in SPSS.

As shown in Table 4, a Pearson product-moment correlation coefficient was computed to assess the relationship between the Nigeria-USA real exchange rate and the housing deficit from 1960 to 2019, covering a period of 60 years and comprising

Fig. 2 Exchange rate and housing deficit trends in Nigeria from 1960 to 2019

Table 4 Pearson product-moment correlations of Naira-USA dollar exchange rate and housing deficit from 1960 to 2019

Statistics	Exchange rate	Housing deficit
Pearson correlation	1	0.924[a]
Sig. (2-tailed)		0.000
N	60	60

[a]Correlation is significant at the 0.01 level (2-tailed)

17 different regimes (8 military and 9 civilian). Result of the analysis showed a strong positive significant correlation between the naira-dollar exchange rate and the housing deficit, at r (60) = 0.924, p < 0.0001, with an exchange rate contributing 92% of the variation in housing deficit. This implies that as the dollar gains value against the naira, the housing deficit in the country increases or worsens; clearly this does not occur by chance. The confidence level of this result is 95%, which is considered accurate since the result meets all the basic assumptions that underpin a Pearson correlation.

As already noted, the sample size of 60 was considered large enough since it covered the 60 years of Nigeria's existence as an independent nation. Moreover, the two variables (exchange rate and housing deficit) are continuous. The scatterplot as shown in Fig. 3 depicts a strong linear relationship without any outlier.

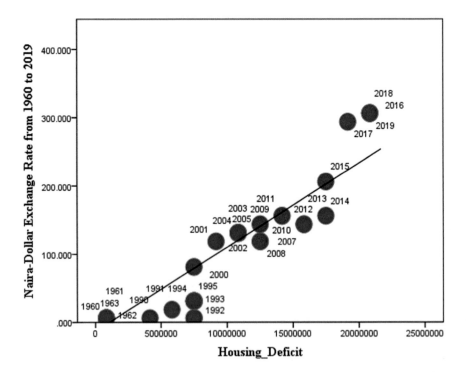

Fig. 3 Scatter graph showing strong linear relationship between exchange rate and housing deficit

Consequently, the results from this study support the proposition that the national exchange rate has influence on the national housing deficit, thus providing empirical evidence that aggregated macroeconomic indices such as exchange rate, capital accumulation and interest rate, etc., influence the real-estate sector and in particular housing production. From the summary statistics, results of the trend analysis agree with an earlier study by Owuru and Farayibi (2016), who demonstrated that exchange rate trends affected export performance in Nigeria from 1970 to 2015.

Similarly, the results of the product-moment correlation coefficient agree with a similar study conducted in 2014 by Qiao and Guo [29], who showed that the exchange rate of RMB was associated with the housing price in China. Their empirical analysis used the pooled regression model of panel data from 2005 to 2010. The results of another study conducted in China by Qiao and Guo [29], Li [21] suggested that housing prices in the cities of Guangzhou and Shenzhen had significant relationships with GDP per capita and per capita disposable income. However, their study also indicated that variability of housing stock had little statistical relationship with its t-value.

The present study thus provides evidence that the exchange rate could be an important factor in efficient housing policy and in forecasting physical development at the national level. Findings from the study also indicate that limiting housing policy to microeconomic variables is no longer fashionable in this age of globalisation. Thus, developing countries such as Nigeria, whose financial markets may not yet be robust enough, should pay attention to macroeconomic variables as catalysts for stimulating physical development and resolving the housing deficit and meeting global agenda like the Sustainable Development Goal (SDGs).

In this regard, this section examines the effect of the exchange rate on cities' sustainability by investigating the relationship between the exchange rate and the rental value of single rooms in Nigeria's capital city Abuja.

As seen in Table 5 above, the descriptive analysis of the exchange rate ($M = 167.8490$, $SD = 52.45095$) and rental value ($M = 14,836.7000$, $SD = 3477.17619$) of single rooms in Abuja revealed that from 2007 to 2016, the mean exchange rate was 167.8490 while the mean rental value for single rooms in Abuja was 14,836.7000.

This implies that the means significantly deviated from the minimum and maximum values of the exchange rate and rental values, thus demonstrating that both the exchange rate and rental in the country's capital for the 10-year period were highly variable. A low standard deviation would have suggested a stable exchange rate or rental value but the inverse situation, as witnessed from the summary statistics, obviously requires the deployment of inferential analysis to ascertain if the variations in rental values during this period were associated with the dollar-naira exchange rate for the same period.

Table 5 Descriptive statistics for exchange rate and rental value	Study variables	Mean	Std. Deviation	N
	Exchange rate	167.8490	52.45095	10
	Rental value	14,836.7000	3477.17619	10

Table 6 Pearson product-moment correlations of Naira-USA dollar exchange rate and rental value of single rooms in Abuja from 2006 to 2016

Statistics	Exchange rate	Rental value
Pearson correlation	1	0.814[a]
Sig. (2-tailed)		0.004
N	10	10

[a]Correlation is significant at the 0.01 level (2-tailed)

As shown in Table 6, a Pearson product-moment correlation coefficient computed to assess the relationship between the Nigeria-USA real exchange rate and the rental value of single rooms in Abuja from 2007 to 2016 showed a strong positive significant correlation, at $r(10) = 0.814$, $p < 0.0001$, with the exchange rate contributing 91% of the variation in rental value. This implies that as the dollar gained against the naira, rental value in Abuja for the study period increased, and this was not a chance occurrence. The confidence level of this result is 95%, which is considered accurate since the result met all the basic assumptions that underpin a Pearson correlation.

However, further studies are required to determine which exchange rate regime is more associated with the housing deficit. A comparative study is also required to examine whether the magnitude and direction of association between the exchange rate and the housing deficit differ within developing economic regions and across developed and developing economic regions. Such studies are expected to cushion the effects of time and population, etc., on the exchange rate and the housing deficit increase.

5 Conclusion

Based on the empirical evidence from this study, which showed that the exchange rate is significantly associated with the housing deficit, it bears restating that housing supply is a social and economic factor. As an economic factor, housing supply involves microeconomic explanations. Experientially, microeconomic indicators such as the land tenure system, population, etc., tend to influence housing supply.

However, under a stable currency regime and a healthy financial market, individuals and businesses will have enough disposable income and will not deny themselves a habitable shelter, since the responsibility for housing delivery would fall to them. No doubt, in this age of globalisation, developing countries like Nigeria with poor financial markets should employ the exchange rate as an economic tool to stimulate and improve housing production in the bid to resolve the housing deficit.

References

1. Ackley AU, Teeling C, Atamewan E (2018) Factors affecting the shortage and or provision of sustainable affordable housing in developing countries—a case-study of Cross River State, Nigeria. J Sustain Architec Civ Eng 21(1):27–38. https://doi.org/10.5755/j01.sace.22.1.20573
2. Afrane E, Bujang AA, Liman HS, Kasim I (2016) Major factors causing housing deficit in Ghana. Dev Country Stud 6(2). Retrieved from www.iiste.org
3. Ansah SK (2014) Housing deficit and delivery in Ghana: intervention by various Governments. Int J Dev Sustain 3(5):978–988. Retrieved 5 May 2019 from 2168-8662. www.isdsnet.com/ijds
4. Arkolakis C (2014, January 20) Exchange rates. New Haven, Connecticut, USA. Retrieved 12 Mar 2020 from http://www.econ.yale.edu/~ka265/teaching/UndergradFinance/Spr11/Slides/Lecture 4–5 Exhange Rates.pdf
5. Bichi KM (1997) Housing finance in the context of vision 2010. J Hous Co-oper Niger Hous Today 30–35
6. Bishop T (2006) Exchange rates and the foreign exchange market: an asset approach. In: Krugman PR, Obsfeld M (eds) International economic theories and policies. Pearson Addison-Wesley, Boston, pp 13–43
7. Brazil Capstone Team (2016) Addressing the housing deficit in São Paulo metropolitan area: perspectives from a social movement. Capstone, Taubman. Retrieved 13 May 2020 from https://taubmancollege.umich.edu/sites/default/files/files/mup/capstones/Addressing-Housing-Deficit-Sao-Paulo_2016capstone.pdf
8. Canuto O, Cavallari M (2013) Monetary policy and macroprudential regulation: whither emerging markets. Policy research working paper; No. 6310. World Bank, Washington, D.C. Retrieved from https://openknowledge.worldbank.org/handle/10986/12175. License: CC BY 3.0 IGO
9. Central Bank of Nigeria (2016) Foreign exchange rate. Educ Econ Ser 4:1–50. Retrieved 14 Feb 2019 from https://www.cbn.gov.ng/out/2017/rsd/education%20in%20economics%20series%20no.%204.pdf
10. Dunga Y, Hardie N, Kelly S, Lawson J (2019, March 25) VOX CEPR policy portal. Retrieved from social capitalism: incorporating sustainability factors into macroeconomic analysis. https://voxeu.org/article/incorporating-sustainability-factors-macroeconomic-analysis
11. Gerlach S, Giovannini A, Tille C (2009, July 17) Are the golden years of central banking over? Retrieved from VOX CEPR Policy Portal https://voxeu.org/article/are-golden-years-central-banking-over
12. Gulde A, Ostry JD, Wolf H (1996) Does the exchange rate regime matter for inflation and growth? (2 No. 1–55775–614–7 Published). Economic Issues. Washington DC. Retrieved from publications@imf.org
13. Ha J, Kose AM, Ohnsorge F (2019) Inflation in emerging and developing economies evolution, drivers, and policies. International Bank for Reconstruction and Development/ The World Bank, Washington D.C. https://doi.org/10.1596/978-1-4648-1375-7
14. Habermeier K, Kokenyne A (2009) Revised system for the classification of exchange rate arrangements. International Monetary Fund, Washington
15. Harris J (2001) G-DAE working paper No. 01–09: "macroeconomic policy and sustainability". Global Development and Environment Institute, Tufts University, Medford
16. International Monetary Fund (2013) Annual report on exchange arrangements and exchange restrictions. International Monetary Fund, Washington, D.C. Retrieved 10 Apr 2020 from https://www.imf.org/external/pubs/nft/2013/areaers/ar2013.pdf
17. Iwedi M, Onuegbu O (2014, December 13) Funding housing deficit in Nigeria: a review of the efforts, challenges and the way forward. Int J Bus Soc Sci 5(13):206–209. Retrieved from http://ijbssnet.com/journals/Vol_5_No_13_December_2014/22.pdf
18. Kelly S (2018, Novermber 29) Scottish financial news. Retrieved 3 June 2020 from 'Social Capitalism': a new measure of national success. https://scottishfinancialnews.com/article/social-capitalism-a-new-measure-of-national-success

19. Khramov V, Lee JR (2013) The economic performance index (EPI): an intuitive indicator for assessing a country's economic performance dynamics in an historical perspective. International Monetary Fund, Russian
20. Kopp CM (2019, April 9) Investopia. Retrieved from economics-terms of trade. https://www.investopedia.com/terms/t/terms-of-trade.asp
21. Li R (2014) Factor analysis of commercial housing prices' fluctuation in the Pearl River Delta Region: a case study of Guangzhou and Shenzhen. In: 2014 international conference on construction and real estate management. American Society of Civil Engineers, Kunming, pp 149–159. https://doi.org/10.1061/9780784413777.149
22. Makinde OO (2013, July 03) Housing delivery system, need and demand. Environ Dev Sustain 16:49–69. https://doi.org/10.1007/s10668-013-9474-9
23. National Bureau of Statistics (2016, Novermber 6) National Bureau of statistics. Retrieved from latest report https://www.nigerianstat.gov.ng/
24. Ogwumike FO, Ogunleye EK (2008) Resource-led development: an illustrative example from Nigeria. Afr Dev Rev 20(2):200–220. https://doi.org/10.1111/j.1467-8268.2008.00182.x
25. Okonkwo O (1999) Mortgage finance in Nigeria. Esquire Press Ltd, Abuja
26. Ooi GL, Phua KH (2007) Urbanization and slum formation. J Urban Health Bull N Y Acad Med 27–34. https://doi.org/10.1007/s11524-007-9167-5
27. Organization of the Petroleum Exporting Countries (OPEC) (2009) OPEC annual report 2009. Organization of the Petroleum Exporting Countries, Public Relations and Information Department. Organization of the Petroleum Exporting Countries, Vienna. Retrieved from https://www.opec.org/opec_web/static_files_project/media/downloads/publications/AR2009.pdf
28. Piana V (2001) Economic web institute. Retrieved from exchange rate http://www.economics webinstitute.org/books.htm
29. Qiao X, Guo H (2014) Research on the effect of the exchange rate of RMB on housing prices based on the VAR Model. In: 2014 international conference on construction and real estate management. American Society of Civil Engineers, Kunming, pp 203–210
30. Rodrik D (2017, July 30) VOX CEPR policy portal. Retrieved from Economics of the populist backlash https://voxeu.org/article/economics-populist-backlash
31. Rosemback RG, Rigotti JR, Feitosa FF, Monteiro AM (2014, November 28) Brazilian Association of population studies. Retrieved from the dimensions of the housing issue and the role of census data in municipal diagnostics: an analysis suggestion in light of the new requirements of the National Housing Policy http://www.abep.org.br/publicacoes/index.php/anais/article/download/2099/2055
32. Shaik S, Baba SK, Shaik H (2019) Forex exchange management and challenges in current global economic environment. In: Prasad MS, Sekhar GV, Prasad MS, Sekhar GV (eds) Currency risk management: selected research papers. Vernon Press, Wilmington, pp 77–86. Retrieved 3 June 2020 from https://books.google.com.ng/books?id = Sq-ZD-wAAQBAJ&pg=PA78&lpg=PA78&dq=Countries+with+higher+inflation+typically+see+depreciation+in+their+currency+about+the+currencies+of+their+trading+partners&source=bl&ots=zI-swp79xW&sig=ACfU3U0ZMmaqy__6p07VrTPGqUB3S9XM
33. Stone M, Anderson H, Veyrune R (2008, March 15) Back to basics. Retrieved from Exchange Rate Regimes. https://www.imf.org/external/pubs/ft/fandd/2008/03/pdf/basics.pdf
34. Thirlwall A (2003) Trade, the balance of payments and exchange rate policy in developing countries. Edward Elgar, Canterbury
35. United Nation (2016) Policy paper 10: housing policies. In: United Nation, Preparatory Committee for the United Nations Conference on Housing and Sustainable Urban Development (Habitat III). Surabaya: United Nation. Retrieved December 12, 2019 from http://habitat3.org/wp-content/uploads/Policy-Paper-10-English.pdf
36. World Bank (2019, January 1) World Bank's data help desk. Retrieved from World Bank Country and Lending Groups Country Classification. https://datahelpdesk.worldbank.org/knowledgebase/articles/906519-world-bank-country-and-lending-groups

Analysing Hernando de Soto's *The Mystery of Capital* in the Nigerian Poverty Equation

Akeem Ayofe Akinwale

Abstract This chapter examines Hernando de Soto's The mystery of capital: Why capitalism triumphs in the West and fails everywhere else, with a view to enhancing understanding of the poverty equation in Nigeria. The following research questions were addressed: How suitable are de Soto's ideas in addressing the socioeconomic situation of the poor in Nigeria? What lessons can be derived from the poverty equation in Nigeria? The chapter undertakes a systematic review of the relevant literature on de Soto's ideas concerning poverty alleviation against the backdrop of Nigeria's capitalist ideologies and anti-poverty programmes. Weick's Sensemaking Theory was used to explain de Soto's ideas concerning the role of property rights and access to capital among the poor in developing countries. The theory was also used to explain the suitability of de Soto's ideas in addressing the poverty equation in Nigeria. Issues such as legal protection of property rights and poverty alleviation through establishment of limited liability companies constitute the premises of de Soto's ideas in The mystery of capital. However, De Soto's ideas appear inappropriate for poverty alleviation in Nigeria. As such, possible remedies for the poverty situation in Nigeria are presented here.

Keywords Capitalism · de soto · Informal economy · Poverty equation · Property rights

1 Introduction

In the year 2000 Hernando de Soto published the well-received book *The mystery of capital: Why capitalism triumphs in the West and fails everywhere else*. The book generated controversy in the academic world and outside of it, hence the emergence of several studies on its subject matter—both positive and negative. However, none of these studies has adequately addressed the poverty equation in Nigeria. This chapter

A. A. Akinwale (✉)
Department of Employment Relations and Human Resource Management, University of Lagos, Lagos, Nigeria
e-mail: aakinwale@unilag.edu.ng

© The Author(s), under exclusive license to Springer Nature Singapore Pte Ltd. 2021   257
T. G. Nubi et al. (eds.), *Housing and SDGs in Urban Africa*, Advances in 21st Century Human Settlements, https://doi.org/10.1007/978-981-33-4424-2_14

is therefore an attempt to fill the gap by providing the reasons why de Soto's prescriptions may not work for the landless poor in Nigeria. The chapter also addresses lessons from the poverty equation and how to solve it in Nigeria. The issues addressed in the chapter can serve as a guide in the efforts to achieve the first sustainable development goal: poverty reduction. As shown in the report by the United Nations Development Programme (UNDP), extreme poverty remains a key challenge in a world where more than 700 million people globally live on less than US$1.90 per day, where inequality is growing within countries, where levels of unemployment and vulnerable employment are rising, and where more than 200 million people are unemployed globally [34].

A close observation of available records shows that poverty is not simply a lack of adequate income but a multidimensional problem in terms of the disadvantages that poor people experience across different areas of their lives, including education, health, and living standards. In this regard, literacy rates, life expectancy, quality of housing and dependency ratio, etc., are strong indicators of poverty. According to the Global Multidimensional Poverty Index (MPI), 1.6 billion people in 108 countries, home to 78% of the world's population, are identified as multidimensionally poor, with 81% of them living in households with inadequate sanitation [34].

These facts call attention to the link between extreme poverty and quality of housing. A visit to slum communities in Nigeria will reveal not only the harrowing scenes of poverty and informal housing but also the challenges of bringing the multidimensionally poor out of poverty. The socioeconomic conditions of the poor in Nigeria deserve urgent attention if the first sustainable development goal must be realised. Nigeria has the world's largest number of people living in extreme poverty, with an estimated 87 million persons thought to be living below the poverty line of less than $1.90 a day [1]. Similarly, Dauda [10: 7] showed that "out of the world's 736 million extreme poor (those living on less than $1.90 a day) in 2015, half of the total (368 million) live in just five countries: India, Nigeria, Democratic Republic of Congo, Ethiopia, and Bangladesh." Altogether, these countries account for 85% (629 million) of the poor in the world [43].

From a multidimensional perspective, Dauda [10: 13] equated poverty with deprivation, which she described as "a multidimensional view of poverty that includes hunger, illiteracy, illness and poor health, powerlessness, voicelessness, insecurity, humiliation, and a lack of access to basic infrastructure." This chapter adopts Dauda's description of poverty for its wide coverage and relevance to the poverty equation in Nigeria, where the use of the poverty line based on living on less than $1.90 a day is not adequate to explain poverty.

Consistent with the above-mentioned facts, Nigeria has a low ranking (157th position out of 189 countries) on the 2018 Human Development Index, which shows a harrowing picture for Nigeria on different dimensions of human development such as life expectancy, access to education and standard of living [35]. The current life expectancy in Nigeria is 53 years [42]. Any work that promises a solution to poverty is therefore relevant in the country.

It is against this backdrop that this chapter examines de Soto's ideas on capitalism and poverty alleviation as presented in *The mystery of capital*, with a focus on the following research questions: How suitable are de Soto's ideas in addressing the socioeconomic situations of the poor in Nigeria? What lessons can be derived from the poverty equation in Nigeria? The following is the rationale for focusing on de Soto's work: The crisis of poverty is getting out of hand in Nigeria, whereas de Soto has only worked in Asia, the Middle East and Latin America on the practical implementation of measures for bringing the poor into the economic mainstream as suggested in *The mystery of capital.*

There is considerable praise from individuals and groups for de Soto's work. For instance, on the cover page, the *Daily Telegraph* describes it as "a hugely persuasive and important book." A former Sectary-General of the United Nations described the book as "a crucial contribution and a new proposal for change that is valid for the whole world." *The Economist* described de Soto as the second most important thinker in the world, while *Time* magazine listed him as one of the five leading Latin American innovators of the twentieth century.

Unfortunately, there is paucity of data on the extent to which de Soto's ideas can be adopted in addressing the crisis of extreme poverty and proliferation of informal housing in Nigeria's slum communities, where rapid urbanisation and the inability of the state to provide affordable formal housing units for the low-income groups are the major reasons why the urban poor have resorted to building poor-quality houses in slums [4]. Alabi [4] has observed that most of the slums emerged as cases of planning contravention, as only 30% of houses in Lagos have approved building plans, with 73% of residents occupying one and two rooms, occupancy rates in some cases being up to eight persons per room and most rooms being not more than 9.3 m².

2 Materials and Methods

The data presented in this article were derived from a systematic review of textbooks, journal articles and other important documents obtained from libraries and online databases, including Google Scholar, JSTOR and EBSCO Host. The literature comprised de Soto's work and other more recent materials showing the strengths or weaknesses of the capitalist ideologies and prescriptions that cannot work for economies of the global south including Nigeria. The technique adopted for the review of the literature was in accordance with Hart's [17: 2] description of the quality of a review of literature in terms of "appropriate breadth and depth, rigour and consistency, clarity and brevity, and effective analysis and synthesis."

The research technique adopted was based on critical realism and sociological imagination, which shape both the questions and the answers presented in the chapter. This is in line with the virtually universal support for critical thinking: a process of looking beyond appearances, understanding root causes and asking relevant questions [9]. It is this process that Berger [8] called the "debunking" tendency of sociological consciousness.

In fact, critical thinking and sociological imagination make clarification of complex issues possible for better understanding of a given social reality. Historical analysis of a given subject matter is a crucial part of the sociological imagination [20]. In this regard, available records of capitalist ideologies and anti-poverty programmes in Nigeria were juxtaposed with de Soto's ideas in *The mystery of capital*. Applicability of de Soto's prescriptions for poverty alleviation in Nigeria was considered alongside the lessons from the poverty equation in Nigeria. Several cases of commendations and criticisms against de Soto's ideas in the book were also presented and analysed.

3   Theoretical Framework for Analysing de Soto's Ideas in *The Mystery of Capital*

The theory of sensemaking was deployed towards understanding of de Soto's ideas on capitalism and poverty alleviation in developing countries. The theory was chosen based on the submission that plausibility rather than accuracy is the ongoing standard that guides learning [38]. It is also noted that:

> People do not need to perceive the current situation or problems accurately to solve them; they can act effectively simply by making sense of circumstances in ways that appear to move toward general long-term goals. The important message is that if plausible stories keep things moving, they are salutary [38], p. 415)

In this regard, it can be argued that the plausibility rather than the accuracy of the ideas made de Soto's ideas reverberate across the world. The concept of sensemaking simply means the process of using a reasonable explanation to make sense of a situation. In discussing the substance of sensemaking, Weick [37: 106] noted that sense is "generated by words that are combined into sentences of conversation to convey something about an ongoing experience." Sensemaking starts with three elements, i.e., frame, cue, and connection, with frames tending to be past moments of socialisation and cues being present moments of experience (Weick, [37: 111]. Explicit efforts at sensemaking tend to occur when the current state of the world is perceived to be different from its expected state. Sensemaking is about the interplay of action and interpretation [38]. It involves a retrospective development of plausible images of human behaviour, although what is plausible for one group (e.g., managers) may prove implausible for another group (e.g., employees).

Moreover, sensemaking deals with the use of language and communication for an explicit description of a given situation, organisation or environment. Discussions of sensemaking often include words like "construct," "enact," "generate," "create," "invent," "imagine," "originate" and "devise" [38: 417]. The implications of de Soto's [32] use of words in his book can be understood and evaluated in this context. The book contains words and phrases like "mystery," "capitalism," "triumphs," "fails," "dead capital," "extralegal economy" and "the poor." His interpretations of these words have produced mixed reactions from individuals and organisations in different countries. In his effort to make sense of de Soto's concept of capital, Ahiakpor [3: 60] observed as follows:

When a businessperson says he is looking for some "capital" to invest or set up an enterprise, he has in mind a sum of money (funds), typically to be borrowed at interest. Some of the funds may be used to purchase or rent machinery or equipment, rent space for the enterprise, purchase raw materials, and pay for the services of workers.

This description of capital is directly applicable to a type of capital known as economic capital, which is closely connected with savings and interest rates. Other types of capital include human capital, social capital and cultural capital. Describing these types of capital is beyond the scope of this article. With regard to the use of capital as a factor of production, other forms of capital such as fixed capital, working capital and nominal capital may be considered. De Soto summarised his agenda in *The mystery of capital* as follows:

In this book I intend to demonstrate that the major stumbling block that keeps the rest of the world from benefitting from capitalism is its inability to produce capital. Capital is the force that raises the productivity of labor and creates the wealth of nations. It is the lifeblood of the capitalist system, the foundation of progress, and the one thing that the poor countries of the world cannot seem to produce for themselves, no matter how eagerly their peoples engage in all the other activities that characterize a capitalist economy. [32: 5]

It is important to note that since de Soto also mentioned capitalism in the subtitle of his book, his description of the concept of capital, to avoid ambiguity, can be examined from the perspective of economic capital or capital as a factor of production. De Soto also refers to the concept of dead capital while noting that capital has an invisible existence like television or radio waves, which cannot be seen or touched. For him, capital is congealed in the residential houses and lands that the poor occupy. In this regard, de Soto argued that the problem of the poor in the Third World is a failure to recognise or understand how to convert their savings into capital. He noted that the quantity of saving by the poorest sectors of society is enormous but "most of it is dead capital" [32: 12]. According to de Soto,

Since the nineteenth century, nations have been copying the laws of the West to give their citizens the institutional framework to produce wealth. They continue to copy such laws but most citizens still cannot use the law to convert their savings into capital. Why this is so and what is needed to make the law work remains a mystery. [32: 13]

The next section analyses de Soto's ideas in *The Mystery of Capital*.

4 The Main Ideas in de Soto's *the Mystery of Capital*

De Soto's [32] thesis is that the poor in developing countries often have many assets— homes, informal businesses, and plots of land but what they lack is formal property rights to these assets [31]. Moreover, de Soto's [32] main argument is that informality and dead capital are obstacles to the development of capitalism, since the majority of the population of developing countries lack access to credit because they secured ownership of their property informally, hence their inability to use their property as collateral to borrow money that would help them escape poverty [3, 15, 40].

Based on his belief that clearly defined property rights are essential to capital formation and ultimately to economic growth and poverty alleviation, de Soto [32] called for the transformation of "dead capital" into "live capital" through the development of a legal property system and formalisation of titles [16, 19]. He argued that property ownership is the key to ending poverty but that it will work only if the poor can use their property to generate further wealth. This is consistent with his belief that despite their low incomes, the poor of the world have a surprisingly large amount of property, although the property is not legally regarded as theirs.

He also argued that the failure of capitalism outside the west is not due to the usual reasons such as cultural differences, lack of enterprise, religion, fecklessness or laziness. He rightly observed that the developing world comprises people with hard work, entrepreneurial skills and ingenuity. Unfortunately, de Soto did not consider the possibility of a direct alleviation of poverty through hard work, entrepreneurial skills and ingenuity. The Yoruba concept of work as antidote to poverty is relevant here. The traditional belief among the Yoruba is that laziness is a major cause of poverty, which can be overcome through work and dedication to a given occupation.

4.1   An Appraisal of the Main Ideas in de Soto's Mystery of Capital

Many researchers have either commended or criticised de Soto's ideas on poverty alleviation. However, none of the existing commendations or criticisms addressed the poverty equation in Nigeria, despite the high rates of poverty in Nigeria and the fact that Nigeria is the most populous country in Africa and the seventh most populous country in the world. De Soto's ideas do not in fact capture the poverty context in Nigeria where the majority of the poor do not have land that can be converted into living capital. For instance, in the case of the demolition of Maroko in Lagos State, Nigeria, out of the 300,000 inhabitants of the community, only 18,000 were property owners while the remaining 282,000 inhabitants of the community were low-income tenants [23].

Dispensing with the informal economy is also not a likely solution to poverty in Nigeria, given that the majority of the extreme poor in Nigeria are persons with little or no formal education who find it easier to survive in the informal economy than relying on the Nigerian government or corporate organisation for anything. The informal economy employs the larger proportion of Nigeria's economically active population.

Employment in the informal economy comprises more than one-half of non-agricultural employment in most regions of the developing world—82% in South Asia, 66% in sub-Saharan Africa, 65% in East and Southeast Asia, and 51% in Latin America [36]. A major reason for the large share of the informal employment is the inability of the private and public sectors of the formal economy to absorb the growing labour force.

4.2   *Commendations for* **The Mystery of Capital**

Prominent persons and organisations have endorsed de Soto's [32] ideas in *The mystery of capital*. The late Margaret Thatcher noted that the book has the potential to create a new revolution, as it addresses the single greatest source of failure in the Third World and ex-communist countries. Francis Fukuyama described the work as one of the few and genuinely promising approaches to overcoming poverty. Moreover, the World Bank endorsed de Soto's idea that the creation of individual property in housing and land will revive "dead capital" and enable the poor to emerge from abject poverty [24].

As an indication of recognition of his ideas, de Soto has served as an advisor to governments on economic development policies through legal and institutional reforms in more than 100 countries [31]. Several governments in the Third World have invited him to advise them on how to unlock the hidden wealth in their countries. According to Gilbert [15],

> Governments in Egypt, Peru, Philippines and Tanzania have launched titling programmes that, some claim, have produced impressive outcomes. The Commission for Legal Empowerment of the Poor (CLEP) was set up in 2005 – an independent global body that seeks to reduce poverty by expanding legal protection and economic opportunities for the poor. Also, various consultants have been busy pushing his ideas in poor countries [15: 5].

De Soto's proposals for granting formal titles to holders of informal real estate have been carried out in Peru and elsewhere with some success [16]. As observed by Home [18], *The mystery of capital* has transformed the previously obscure topic of land titling with a focus on the relationship between sustainable capitalist economic development and the need of the Third World poor for secure land tenure. From de Soto's call for integrated property systems, Home [18] provides an illustration of specific themes with examples from different countries, e.g., cadastral reform in Southern Africa, adverse possession in Israel, usucapio in Brazil, land tenure systems in Botswana, and land assembly and infrastructure provision for urban development and land readjustment in Japan and India.

De Soto used the Japanese experience of land reform to buttress his ideas. The next paragraph shows a transition from feudalism to land reform in Japan. When MacArthur arrived in Japan in 1945 he worked with a few American intellectuals there. Their primary goal was to ensure that Japan did not attempt military expansion again. They observed that since the feudal class had financed the military, the best way to prevent another military expansion would be to eliminate the feudal class and the best way to go about this was to take away all the feudal land, convert it into private property and give it to the people who worked on it [31].

In Japan of 1946, the farmers and people living in the cities were organised into neighbourhood associations called Burakus. There were 10,900 Burakus in Japan at that time and the authorities legalised them and converted them into land commissions [31]. The property law instituted during the Meiji restoration was discarded because it protected only the interests of the feudal lords. This was how Japan was transformed from a feudal system to a property system that has allowed the country to become much more prosperous than any of the systems in Latin America.

Like the Japanese, the Koreans and the Taiwanese were also able to achieve the transition from a feudal system to a property system. China's Mao Zedong also created property systems that he put under the people's control. De Soto reasoned that if Japan could do it, then Peru could also do it since Peruvians used to be wealthier than the Japanese.

4.3 Criticisms of **The Mystery of Capital**

There have also been criticisms of de Soto's ideas in *The mystery of capital*, including by Ahiakpor [3], Fernandes [14], Gilbert [15], Gonzales [16], Home [18], Obeng-Odoom [24], Otto [28], Smith et al. [31], Williamson [39], and Woodruff [40], among others. Ahiakpor [3: 60] noted that "the poor in the Third World typically are not property owners."

Woodruff [40] observed that the amount of capital that may be unlocked from the poor's assets might be only a small percentage of the US$9.3 trillion de Soto suggested. Woodruff [40] also noted that the owners of untitled lands were as likely to receive credit from banks and non-bank sources as those with titled land in Thailand. Similarly, Gilbert [15] reported that the granting of titles in Bogota, Columbia has made little difference in people's access to credit. Another criticism also stated that:

> Rather than the lack of titles to property, the problem is the inadequacy of their domestic savings to finance investment. Poor people in these countries hardly own assets, the absence of whose formal titles impairs their ability to borrow funds, or "capital," for investment. Thus, de Soto's suggested solution of a massive titling programme by the governments of these countries would be a wasteful diversion from what needs to be done in them to promote their economic prosperity. [3: 58]

In his criticism of de Soto's observation of the triumphs of capitalism in the West and failures everywhere else, Ahiakpor [3] noted that indeed capitalism has triumphed in Hong Kong, Singapore, South Korea, Taiwan, and the Bahamas, showing that the per capita incomes of Singapore and Hong Kong then exceeded those of Portugal, Greece, New Zealand, and Spain, which are capitalist countries in the West.

Consistent with Ahiakpor's [3] observation, an earlier study by Spinner [33] focused on de Soto's idea on how property rights can be used to create capital as the driving force of capitalism. Spinner [33: 334] however noted that de Soto's grouping of "the West versus the rest" is very problematic in many respects and showed that if de Soto had extended his field trips to Central and Eastern Europe he would probably have avoided this oversimplification and realised a need to differentiate. To buttress his point, Spinner [33: 334] asked, "Can the history of the USA or the Western hemisphere in general be repeated or copied?" The foregoing criticisms appear to negate the subtitle of de Soto's book, which is "Why capitalism triumphs in the West and fails everywhere else."

De Soto's assertion that informality is the main obstacle to development has become controversial in academic and political circles. The major limitation is that his proposal that capitalist development in the twenty-first century could occur simply through the surplus generated by the formalisation of property rights is more ideological than practical [16].

His argument that the informal economies have little chance for growth and development without integration into the formal economy through titles is an illusion, given the sociocultural factors that reinforce the development of informal economies in Nigeria and other developing countries.

This observation is in line with UNESCO's prediction that 80% of jobs in developing countries will be created in informal economies. Below are further critical assessments of de Soto's work:

> I read Hernando de Soto's book, *The Mystery of Capital*, while working in Washington DC, in the last quarter of 2000. I was shocked to see how much publicity the book was receiving and how the development banks, and particularly, the Inter-American Development Bank where I was based, were responding to it. I could not understand why a book based on so little real evidence was being given so much attention by serious professionals and policymakers [15: 5]
>
> Most case studies showed that for de Soto's plans to be effective, a range of conditions had to be fulfilled. As long as these conditions are not fulfilled – which happens to be the unfortunate reality throughout the developing world – his plans will not work. Unfortunately, his proposals then are based on a whole set of unrealistic assumptions. [28: 173]

Joireman [19] traced two impediments to the clear definition of property rights in the African context: customary law and the status of women. Both of these issues interfere with the attempt of African countries to rearticulate property law with the goal of capital formation. De Soto also failed to address the problem of enforcement of property rights in under-resourced environments where changes may not be welcomed [19].

Benjaminsen et al. [7] examined three cases of research carried out in Mali, Niger and South Africa in light of the debate about formalisation of land rights. The Malian case shows that lack of a broad access to formalisation processes in high-pressure areas may play into the hands of those with power, information and resources. The experience in Niger demonstrates how impending formalisation led to a scramble for land and increased conflicts. The South African case shows that the very process of surveying and registering rights may also change the rights themselves.

Obeng-Odoom [24] observed that de Soto's assumption that the poor possess some economic agency is sound but with insights from Joseph Schumpeter, Karl Polanyi and Henry George, an application of de Soto's ideas through policy would be ineffective in curbing urban poverty. Regarding these criticisms, the concept of entrepreneurship and recognition of talents among the poor are aspects of de Soto's ideas in *The mystery of capital* that may be useful for poverty alleviation in Nigeria but they were not developed in the book. For instance, de Soto [32: 4–5] recognised the fact that

> The cities of the Third World and the former communist countries are teeming with entrepreneurs. The inhabitants of these countries possess talents, enthusiasm and astonishing ability to wring a profit out of practically nothing. They can grasp and use modern technology.

Instead of using the abovementioned issues as the basis for his prescriptions for poverty alleviation in Third World countries, de Soto veered into the issue of capital and declared that the inhabitants of these countries lack the ability to produce capital, the lifeblood of the capitalist system. He used the concept of dead capital to describe all the informal resources and assets possessed by people in Third World countries. His justification for use of the concept of dead capital is presented below:

> They hold these resources in defective forms: houses built on land whose ownership rights are not adequately recorded, unincorporated businesses with undefined liability, industries located where financiers and investors cannot see them. Because the rights to these possessions are not adequately documented, these assets cannot readily be turned into capital, cannot be traded outside of narrow local circles where people know and trust each other, cannot be used as collateral for a loan and cannot be used as a share against an investment. [32: 6]

5   Lessons from the Poverty Equation in Nigeria

There is no consensus on the phenomenon of poverty in Nigeria, where different categories of the poor may include destitute persons or street beggars, refugees and dwellers in slums with inadequate sanitation. The majority of the poor in this category do not have the type of assets that de Soto [32] described in his book. Yet, if the socioeconomic conditions of the poor in this category do not improve it will be difficult to achieve the first sustainable development goal. The poor in this category are part of the base of the pyramid. Prahalad [29] mentioned the concept of base of the pyramid in his description of the poorest people in the world, with a focus on poverty eradication through profits. Surprisingly, the poor at the base of the pyramid in Nigeria find it more convenient and beneficial to rely on an informal network of support than on social service from Nigerian governments.

However, the current situation is not sustainable because the rising level of poverty in Nigeria suggests that the existing informal network of support for the poor is not adequate. Therefore, there is need for a fundamental change in the relationship between the Nigerian government and the extreme poor, who do not have property to convert into living capital. Nigerian governments need to earn the trust of the poor by genuinely showing considerable interest in the provision of social service in poor communities. In this context, there is need to reconsider the policy of evicting the poor from slums owing to its deleterious consequences. Rather than evict the poor from informal housing, government needs to provide an inclusive social service for them. The case of demolition of Maroko community in Lagos State, Nigeria is relevant here for an illustration of the plight of the extreme poor in Nigeria.

With the demolition of Maroko in 1990, about 300,000 persons were forcibly ejected and the area was transformed into a condominium known as Oniru Private

Housing Estate, which is way beyond the reach of the urban poor [23]. This practice further impoverishes the urban poor by consigning them to a lower quality of life [23]. Maroko had a mixture of people in the formal and informal sectors. It was a service centre to the adjacent Victoria Island and Ikoyi. The majority of the residents of Maroko were low-income earners, being mostly fishermen and semi-skilled self-employed artisans such as mechanics, technicians, etc. [23]. The women were mainly traders. Immigrants from Togo and Ghana also resided in Maroko, with some of them owning houses in the area [23]. The housing stock was spontaneous (makeshift plank building), so it was easily flooded during the raining seasons [23]. Residents of Maroko had no access to pipe borne water and electricity as well as proper drainage and sewage disposal systems. Residents relied on bucket toilets and an average of eight persons per household occupied between one and two rooms [23].

Maroko was demolished in phases: the first in the early 1980s and the second in 1985. In February 1990, the President of the Federal Republic of Nigeria visited the community and promised to improve the living conditions of the residents to no avail until the third demolition in July 1990.

The major reason offered for the demolition was that Maroko's environment was prone to epidemics and therefore not suitable for human habitation [23]. Another reason was that the residents were squatters on land the government had acquired since 1972. The displaced residents of old Maroko moved to Aja, Ikota, Ilasan, Maroko-Beach in Eti-Osa and Okokomaiko in Ojo. The Ilasan and Ikota housing estates are government estates along the Lagos-Epe Expressway. Aja, Maroko-Beach and Okokomaiko are private neighbourhoods with relatively old and unserviced structures.

Only about 2000 out of over 10,000 former Maroko house owners were resettled [6]. The remaining 8000 former Maroko house owners were neither assisted to find alternative accommodation nor offered compensation. The issue of which family was the genuine owner of Maroko had been a tussle between the Oniru and the Elegushi chieftaincy families of Lagos since 1965. The suit was filed and settled in court, which found and declared the Oniru Chieftaincy Family as lawful owners of the land.

However, the Lagos State Government was able to prove that Maroko land belonged to it and acquired the land. The Oniru Chieftaincy Family thereafter urged the government to leave some portion of the land for their family use. Government considered this request mainly to prove that the acquisition was not done in bad faith. A portion of the land was conceded, duly derequisitioned and published in the Official State Gazette in 1977. The State Government paid a total sum of N 6.8 million as compensation to the Oniru Chieftaincy Family for both the land and the structures on it. What was left to the Oniru family was 732 ha out of Maroko's 11,425 ha. Some of the land was sold and some of it (200 plots) was given to Oniru family members.

This Maroko case shows the need to prevent social exclusion in Nigeria. The processes of social exclusion—driven by multiple economic, social, political and cultural factors—continue to play a major role in perpetuating poverty and entrenching inequalities of outcomes and opportunities [34]. Social exclusion denies many—including the urban and rural poor, indigenous people, ethnic minorities, people living with disabilities, women, and youth—the opportunities and capabilities that they need to improve their lives. Until the issue of social exclusion is properly addressed in Nigeria, the poverty equation will remain and this will affect efforts to achieve the sustainable development goals.

In the light of the foregoing, Dauda [10: 26] noted that "poverty reduction and increasing social inclusion are impossible without gender equality." This shows the need for women empowerment programmes in poor communities in Nigeria. The starting point of the empowerment programmes should be provision of opportunities for enrolment in formal education and vocational training. Investment in human capital, particularly formal education and vocational training of women and girls, is a critical element for efficient small and medium enterprises that hold the prospect for poverty reduction, inclusive growth and sustainable development in Nigeria [10].

Moreover, proffering a solution to poverty in Nigeria should begin with an understanding of the systems of production and reproduction, which lead many Nigerians into the poverty trap, especially through social exclusion. A close observation of the system of production in Nigeria shows that the majority of the poor earn a living from the informal economies, including agriculture, crafts and trading. In this regard, establishment of vocational training programmes for the poor and their children will provide a basis for poverty alleviation through self-employment and savings for further investment.

Closely related to the production system is the need to enlighten the poor about the importance of child and maternal health, which can be achieved through community-based family planning. This is consistent with Dauda's [10: 85] prescription that population control policies must be implemented passionately and sincerely if economic growth and sustainable development must be achieved: "The rural poor in Nigeria are largely characterised by the prevalence of large household size, illiteracy, low level of capital, lack of access to land, low self-esteem and economic status, inaccessibility to infrastructural facilities, and higher dependency ratios" [10: 48].

There is no doubt that poverty has become endemic in Nigeria, with most poverty-eradication efforts having failed. For instance, from 2000 to 2001, the government set up the Poverty Alleviation Programme (PAP) and the National Poverty Eradication Programme (NAPEP) to create jobs for unemployed youths and to eradicate poverty, respectively. In their evaluation of these programmes, Nmadu et al. [22: 126] noted that "the poor are not better off than they were in the past and there is no reasonable hope even for the future if things continue the way they are presently."

Furthermore, Dauda [10: 48] noted that "development efforts of many countries have failed to make any meaningful impact on the growing incidence of poverty, inequality, unemployment, and dependency. Extreme dependency renders development unsustainable." Dauda added:

> Attempts by successive governments to ameliorate the suffering of the rural poor have only worsened their plight. This is most especially in the areas of the provision of social services, transferring physical assets such as land to the poor, and empowering the poor to design and implement policies to generate and increase their incomes. [10: 48]

The above analysis is at variance with the assessment of anti-poverty programmes in other developing countries such as Brazil, China and India. Brazil has celebrated 10 years of successful implementation of a poverty eradication programme called "Bolsa Familia." As shown in the World Bank's [41] report, Brazil's *Bolsa Familia* poverty eradication programme has lifted 36 million Brazilians out of extreme poverty and has contributed significantly to the reduction of the poverty rate in Brazil from 9.7 to 4.3%.

6 Implications for Housing Development in Nigeria

The main issues addressed in this chapter indicate that extreme poverty has provided an impetus for the proliferation of informal housing in Nigeria. Moreover, poverty is driven by different factors such as rural-urban migration, mass unemployment, neglect by the government, illiteracy, and marginalisation. Evidence shows that

> The spread of squatter settlements and slums is hardly surprising given the large numbers of poor people, increasing rural-urban migration, and lack of government concern. Most people migrated in search of employment opportunities which are often not available. They, therefore, engage in irregular, low-income employment and they live in sub-standard housing. [23: 164]

The question of ownership or access to land or other property is important in this regard. Land is a major component of housing provision and delivery. However, access to land for housing development has become an almost insurmountable challenge due to rapid urbanisation and the fact that many low- and middle-income earners cannot afford the exorbitant cost of land in Nigerian cities, as noted by Opoko and Oluwatayo [26]. The authors further noted that

> The problem of affordable housing in Nigeria is further exacerbated by the constraints imposed by the Land Use Act, a moribund and repressive Act that hinders mortgage financing and creates enormous obstacles to private sector involvement in the housing industry and which has constrained the transfer of titles and made mortgage finance extremely difficult. [26: 23].

A total of 3.5 million new housing units were required between 1985 and 2015 in the bid to provide adequate housing for residents of Lagos, where many poor households have been constrained to resort to informal housing procurement processes [26]. A major feature of the Lagos urban landscape is the proliferation of slums and informal settlements because of government's inability to provide adequate housing for the teeming population. Studies show that about 60% of Lagos residents are tenants who spend 50–70% of their income to pay rent [27]. Construction of houses has not kept pace with rapidly expanding urban populations, leading to severe overcrowding and congestion in slums. In some areas of Lagos, the cost of living has forced residents to live in low-quality slums and shanty houses [26].

Life in slums is characterised by serious problems of environmental pollution, lack of access to basic social services, poverty, deprivation, crime, violence, general insecurity and life-threatening risks and diseases [12]. Adedeji and Olotuah [2] showed that the level of accessibility of low-income earners to housing finance in Nigeria remains minimal despite the intermediation of private developers and cooperative societies in sourcing funds for housing development. Thus, many people have resorted to personal savings, funds from family and friends and other sources of funds for housing development. Murtala [21] opined that the clients, stakeholders and promoters of low-cost housing have limited access to formal procurement systems. This situation is responsible for the high incidence of informal housing procurement in developing countries.

Egidario et al. [13] observed that urbanisation has exacerbated housing challenges among low-income earners by promoting the proliferation of urban slums. The pathways to housing low-income urban residents in Nigeria will depend on site services and settlement upgrading strategies, which require active participation by governments, non-governmental organisations and investors. Unfortunately, the Nigerian government appears to be hostile to the extreme poor, who reside in informal housing such as slums. A recent experience of slum demolition in Lagos State is paraphrased thus:

> The Lagos government has been carrying out forcible demolition of unplanned settlement in the name of regeneration and vision of creating a 21st-century mega-city. Makoko, a residential enclave in Lagos, is an example of recent forced eviction. Public good was invoked to justify the eviction and demolition. This eviction was carried out without prior consultation, compensation or alternative accommodation. [5: 400]

Slum households are closely associated with the informal economy and extreme poverty, and the implications of forced eviction can be observed in this context. Forced evictions may exacerbate a poverty trap and a cycle of destitution [30]. Moreover, forced evictions violate the provision of international treaties for the protection of residents of informal housing.

Considering the human right to housing and informal settlements, studies have shown that informal dwellers are mostly victims of human rights violations [11, 25]. The "public interest" justification often proffered for forced evictions by African governments is a myth, and the escalating urban and rural land crises in Africa violate human rights with grave implications for human development and regional security [25: 173], e.g., as in the recent demolition of Otodo Gbame in Badagry Local Government Area of Lagos State.

7   Conclusion

This chapter has shown that in his accounts of the triumphs and failures of capitalism in the West and everywhere else, Hernando de Soto was unable to demonstrate clear understanding of how to deal with the poverty equation in Nigeria. Other studies that either commended or criticised de Soto's work were considered with a focus on the issue of poverty and how to solve it sustainably in Nigeria. In this regard, the chapter proposes likely solutions to the poverty equation in Nigeria, in line with the UNDP's [34] position that the multiple dimensions of poverty should be addressed through integrated, coordinated and coherent strategies at all levels.

References

1. Adebayo B (2018) Nigeria overtakes India in extreme poverty ranking. www.cnn.com. Accessed 26 May 2019
2. Adedeji Y, Olotuah A (2009) An evaluation of accessibility of low-income earners to housing finance in Nigeria. Eur Sci J 8(12):80–95
3. Ahiakpor JCW (2008) Mystifying the concept of capital: Hernando de Soto's misdiagnosis of the hindrances to economic development in the third world. Independent Rev 13(1):57–79
4. Alabi M (2018) Political economy of urban housing poverty and slum development in Nigeria. East Afr Soc Sci Res Rev 34(2):59–79
5. Amakihe E (2017) Forced eviction and demolition of slum: a case study of the Makoko slum in Lagos, Nigeria. J Urban Regeneration Renewal 10(4):400–408
6. Amnesty International (2006) Nigeria: making the destitute homeless – forced eviction in Makoko, Lagos State. http://web.amnesty.org/library/Index/. Accessed 2 June 2019
7. Benjaminsen TA, Holden S, Lund C, Sjaastad E (2009) Formalisation of land rights: some empirical evidence from Mali, Niger and South Africa. Land Use Policy 26(1):28–35
8. Berger P (1963) Invitation to sociology. Doubleday, New York
9. Buechler S (2008) What is critical about sociology? Teach Sociol 36(4):318–330
10. Dauda ROS (2019) No simple answers, no easy rides: the economist in the pursuit of development. University of Lagos Press, Lagos
11. Davy B, Pellissery S (2013) The citizenship promise (un)fulfilled: the right to housing in informal settings. Int J Soc Welf 22(S1):S68–S84
12. Dung-Gwom JY, Oladosu RO (2004) Characteristics and physical planning implications of slums in Jos. J Environ Sci 8(2):118–127
13. Egidario B, Patrick A, Eziyi O (2016) Urbanization and housing for low-income earners in Nigeria: a review of features, challenges and prospects. Mediterranean J Soc Sci 7(3):347–357
14. Fernandes E (2002) The influence of de Soto's the mystery of capital. Land Lines 2(1):5–8
15. Gilbert A (2012) De Soto's the mystery of capita: reflections on the book's public impact. Int Develop Plann Rev 34(3):5–25
16. Gonzales E (2001) Hernando de Soto's mysteries. SAIS Rev 21(1):275–282
17. Hart C (2018) Doing a literature review: releasing the research imagination, 2nd edn. Sage Publications Ltd., London
18. Home R (2004) Land titling and urban development in developing countries: the challenge of Hernando De Soto's the mystery of capital. J Commonwealth Law Legal Educ 2(2):73–88
19. Joireman SF (2008) The mystery of capital formation in sub-Saharan Africa: women, property rights and customary law. World Dev 36(7):1233–1246
20. Mills CW (1959) The sociological imagination. Oxford, New York
21. Murtala A (2001) A framework for cost management of low cost housing. In: International Conference on Spatial Information for Sustainable Development, Nairobi, Kenya
22. Nmadu JN, Yisa ES, Simpa JO, Sallawu H (2015) Poverty reduction in Nigeria: lessons from small-scale farmers of Niger and Kogi states. Br J Econ Manage Trade 5(1):124–134
23. Nwanna CR (2012) Gentrification in Lagos state: challenges and prospects. Br J Arts Soc Sci 5(2):163–176
24. Obeng-Odoom F (2013) The mystery of capital or the mystification of capital? Rev Soc Econ 71(4):427–442
25. Ocheje PO (2017) In the public interest. Forced evictions, land rights and human development in Africa. J Afr Law 51(2):173–214

26. Opoko AP, Oluwatayo A (2014) Trends in urbanisation: implication for planning and low-income housing delivery in Lagos, Nigeria. Archit Res 4(1A):15–26
27. Oshodi L (2010) Housing, population and development in Lagos, Nigeria. https://www.Osh lookman.Worldpress.Com/2010/11/24/Urban. Accessed 25 May 2019
28. Otto JM (2009) Rule of law promotion, land tenure and poverty alleviation: questioning the assumptions of Hernando de Soto. Hague J Rule Law 1(1):173–194
29. Prahalad CK (2004) The fortune at the bottom of the pyramid: eradicating poverty through profits. Wharton School Publishing, Upper Saddle River
30. Roberts RE, Okanya O (2018) Measuring the socio–economic impact of forced evictions and illegal demolition: a comparative study between displaced and existing informal settlements. https://www.sciencedirect.com. Accessed 25 June 2019
31. Smith B, Mark D, Ehrlich I (2008) The mystery of capital and the construction of social reality. Carus Publishing Company, Chicago
32. de Soto H (2000) The mystery of capital: why capitalism triumphs in the west and fails everywhere else. Bantam Press, London
33. Spinner M (2003) The mystery of capital: why capitalism triumphs on the west and fails everywhere else by H. de Soto. Acta Oeconomica 53(3):330–334
34. United Nations Development Programme (2016) UNDP support to the implementation of sustainable development goal 1: poverty reduction. UNDP, New York
35. United Nations Development Programme (2018) The 2018 human development report. https://ianiworld.com/ranking-human-development-index-2018. Accessed 25 June 2019
36. Vanek J, Chen MA, Caree F, Heintz J, Hussmanns R (2014) Statistics on the informal economy: definitions, regional estimates and challenges. WIEGO Working Paper, 2, pp 1–41
37. Weick KE (1995) Sensemaking in organisations. Sage, Newbury Park, CA
38. Weick KE, Sutcliffe KM, Obstfeld D (2005) Organising and the process of sensemaking. Organ Sci 16(4):409–421
39. Williamson CR (2010) The two sides of de Soto: property rights, land titling, and development. In: The Annual Proceeding of the Wealth and Wellbeing of Nations, pp 95–108
40. Woodruff C (2001) Review of de Soto's the mystery of capital. J Econ Lit 39(4):1215–1223
41. World Bank (2014) How to reduce poverty: a new lesson from Brazil for the world? World Bank, Washington DC
42. World Bank (2018) World development indicators 2016. https://databank.worldbank.org/data/reports. Accessed 25 June 2019
43. World Bank (2019) Half of the world's poor live in just five countries. https://wrld.bg/km0y50 tjUfO. Accessed 25 June 2019

Beyond a Mere Living Space: Meaning and Morality in Traditional Yoruba Architecture Before Colonialism

Akinmayowa Akin-Otiko

Abstract Houses are built according to purposes and designs that express the mind of the builder. This was particularly the case with the way the Yoruba people built their homes before the encounter with colonial ideas and the eventual effects of urbanisation. With colonialism came a particular architecture that blended the European (nuclear family) design with master-servant-relationship designs. These new designs obliterated the traditional Yoruba architecture that was embedded with meaning and morality. In Nigeria, at different times and levels, governments have embarked on programmes that adopted European and master-servant–relationship architectural designs in solving housing problems. First, this study examines and compares pre- and post-colonial architectures with the aim of indicating the point of departure from tradition. Next, the study shows how postcolonial housing responses departed from traditional designs, meanings and morality. The discussion section then highlights the implications of the departure from traditional designs on attempts to solve the housing problem in Nigeria. The postcolonial responses to the housing problem have for the most part manifested as mere living spaces lacking traditional meaning. In solving Nigeria's housing problem, it would be necessary to reflect cultural meaning and morality in architecture.

Keywords Architecture · Colonialism · Living-space · Morality · Yoruba

1 Introduction

All over the world there have been efforts to preserve and showcase traditional architecture, and whenever the occasion has presented itself, aspects of old architecture have been explained to show the purposiveness in style and design. This is because "Universally, architecture is dependent on culture, which, in simple terms, embodies the way of life of a people. Even though factors such as climate, materials and

A. Akin-Otiko (✉)
Institute of Africa and Diaspora Studies, University of Lagos, Lagos, Nigeria
e-mail: pakin-otiko@unilag.edu.ng

© The Author(s), under exclusive license to Springer Nature Singapore Pte Ltd. 2021
T. G. Nubi et al. (eds.), *Housing and SDGs in Urban Africa*, Advances in 21st Century
Human Settlements, https://doi.org/10.1007/978-981-33-4424-2_15

273

methods directly influence building practices, they submit to the common denominator of culture." Adeosun [1] A typical example is the explanation for the design of lifts in front of houses along the riverbanks of Amsterdam. It was important to have the lifts as a means of moving furniture into the buildings, as the staircases were very narrow. This historical reality is peculiar to every culture, given the context and cultural realities of the locality. The south of Nigeria and urban Lagos in particular has not done well in terms of preserving the architecture that represented the cultural realities of the Yoruba, who lived and built houses in Lagos metropolis.

According to Adegoke [2], "before the advent of Brazilian Architecture early in the twentieth century, Yoruba traditional architecture was spontaneous," being meant to fulfil the central and basic purpose of architecture in providing shelter, protection, and accommodation for the physical activities of man." Martinet [3] notes that a building communicates its ability to shelter, protect and accommodate and that it should in fact shelter, protect and accommodate occupants. Adegoke [2] further notes

> The traditional Yoruba house is mainly designed to protect people from the rain. The scorching heat of the sun is treated as secondary. Therefore, every effort is made by the people to make their houses rain-proof. This was the driving force behind the early traditional Yoruba architecture before its flow into the vernacular style following the impact of the colonial masters and returnee slaves.

Moreover, Prucnal-Ogunsote [4] observes that in the traditional setting, "the populace lived in compounds each of which had a large house set in a square-shaped space bounded by a high wall... There was only a single entrance. Inside, the compounds were divided into numerous rooms".

Differing with the view that African architecture is spontaneous, Osasona [5] asserts that the process of documenting the architecture, meaning and use of space in Africa is far from being well established by African architects. The author therefore queries the existence of an 'African' or a national type because of the myriad of cultures and influences on the continent.

This chapter is an ethnographic study to show how well thought-out architecture was among the Yoruba before internal (government and urbanisation) and external (foreign) influences affected the role that culture and meaning play in the designing and assigning of spaces in households. It can be said that housing considered the well-being and health of those for whom houses were designed and built. The key elements that were responsible for traditional architecture and the meaning that inspired the designs of Yoruba architecture will be examined, as well as the things that have changed in the contemporary architecture. The architecture of the Yoruba reflected the intent, morality and function that are woven into each design, space and arrangement of facilities, as viewed within the context of how urbanisation and modernity have displaced the cultural reality of Yoruba architecture in Lagos. Particular attention will be paid to the design and positioning of the frontage, rooms, corridor and backyard, as compared to what has been contemporary in Lagos since the Millennium. This is done with a view to highlighting how Yoruba architecture has made provision for SDG 11, which is to "Make cities and human settlements inclusive, safe, resilient and sustainable."

2 Literature Review

In defining architecture Vitruvius, the Roman architect of the early first Century AD, observed that "a good building should satisfy the three principles of *firmitas, utilitas, venustas*," commonly known in the original translation as firmness, commodity and delight [6]. The definition in modern English would mean that a house should possess a number of qualities. One, a house should be durable, that is, a building should stand up robustly and remain in good condition. Two, a house should have utility, that is, it should be suitable for the purposes for which it is used. Three, a house should be aesthetically pleasing. All these speak to resilience, as contained in SDG 11.

The nineteenth Century has witnessed definitions of architecture that exclude purpose as stated by Vitruvius; a good instance is found in English art critic John Ruskin's *Seven Lamps of Architecture*, published in 1849. For Ruskin, architecture was the "art which so disposes and adorns the edifices raised by men... that the sight of them" contributes "to his mental health, power, and pleasure" [7]. By the twentieth Century, there was a debate as to whether architecture was just about beauty or not. Focusing on the difference between the ideals of architecture and mere construction, the renowned twentieth Century architect Le Corbusier wrote, "You employ stone, wood, and concrete, and with these materials you build houses and palaces: that is construction. Ingenuity is at work. But suddenly you touch my heart, you do me good. I am happy and I say: This is beautiful. That is Architecture" [8].

Contemporary architecture is multidisciplinary and goes beyond the narrow definition of architecture. This new understanding is reflected in the definition given by the Architects Registration Council of Nigeria (ARCON):

> The art and science in theory and practice of design, erection, commissioning, maintenance and management and coordination of all allied professional inputs thereto of buildings or part thereof and the layout and master plan of such buildings or group of buildings forming a comprehensive institution, establishment or neighbourhood as well as any other organized space, enclosed or opened, required for human and other activities.

This chapter is concerned with the second element of architecture, i.e. commodity: A building should be suitable for the purposes for which it is used as stated by Vitruvius. The other two elements of architecture—durability and beauty—are presumed once commodity is satisfied. The Yoruba had an architectural pattern that was well thought-out and had a purpose; this is different from what is obtainable in the early nineteenth Century. Structures such as the Brazilian houses, which are more than one-room deep or more than one-story tall, were a striking deviation from traditional Yoruba building practices [9].

The typical Yoruba home had some basic features that ensured wellness and community living. Before the colonial era, housing provided opportunities for bonding as well as for instilling and correcting morals. According to Adeosun [1]

> The Yoruba home could take one of two forms: The traditional compound, built around one or more courtyards, or the rooming house, famously called 'face-me-I-face-you'. The rooming house became popular in the 1930s, but the courtyard design is the root architecture of the Yoruba people, inspired by a culture of honouring family.

In Yoruba architecture, spaces consisted of designed structures that bore meaning; this suggests that planning and purpose went into the architectural intention of buildings before the influence of Europeans. This changed first with the return of slaves from Brazil. According to Adegoke [2]

> The clearest indication of Afro-Brazilian influence is reflected by the many two-story houses, which were referred to as *ilé pètésì* or "upstairs house" in the 1970s in addition to other bungalows trimmed with moulded stucco facades. These dwellings first appeared in Lagos in the 1850s and became increasingly more commonplace in the last quarter of the nineteenth century.

Scholars have examined and interpreted the art and design on buildings not only as meaningful but also as measures of status. Adegoke observed that "Every aspect of architecture inevitably contains within itself meaning or intention, for the act of design itself is an act of intention" [2]. For the Yoruba, the function that a space was to play defined how it was planned. To ignore the function is to create a shift from the cultural past to the context of the imported designs. The new designs that have appeared due to international influence, government interventions and the need to provide housing for the huge population in Lagos have shifted attention from the purpose of space to status and class. Adeosun noted that "the architecture of the ancient Yoruba of Southwest Nigeria was a communal endeavour and the house was a statement of ideological, economic and social position in the larger urban context" [1].

Pre-colonial Architecture in Lagos

According to Adeokun [10], there exists "a small but growing body of work focussed on in-depth analyses of traditional (Nigerian) domestic architecture from a specifically morphological perspective [11–13]." What is outstanding is the fact that pre-colonial architecture was designed for the well-being and health of dwellers. "The pre-colonial architecture of the ancient city of 'Eko' (War Camp), as Lagos was initially known by its Bini and then Awori colonists, was largely of the type that characterised the Yoruba namely: Rectangular houses with central inner court-yards, and in well-planned areas potsherd tiled pavements" [14]. According to Adeosun [1],

> To accommodate an extended family, the house would be a rectangular, open-plan compound, with one entrance gate and rooms opening onto one or more courtyards. Between the rooms and courtyards, there would be porticos of lean-to roofs with timber columns for support. A segment of the compound would belong to a lineage or, in the case of traditional rulers and, perhaps, the wealthy, a wife and her children. This system allowed for much personal contact, which contributed to the unity of the community microcosm that is the family. The image of co-wives connecting and gossiping in a courtyard was, in fact, a cliché.

For the most part, scholars have used the architecture of the Yoruba to explain the unity of the family. According to Crooke [15] and Marris [16], most floor plans of Brazilian houses in Nigeria feature a hallway, usually in the middle, which is flanked by rooms on both sides. The arrangement is commonly manipulated to make the rooms to one side of the hall larger than those on the other. Adegoke agreed with Crooke and added that some of the rooms may also be of "equal size ranging from four to ten rooms facing each other" [2]. The traditional Yoruba designs use the hallway for far greater purposes; cooking and moral lessons among other things took place in the hall ways. Beyond the hallways, the traditional architecture of the Yoruba has changed with the influence of returnee slaves from Brazil. Adeosun noted that:

> In 1842, when Reverend Henry Townsend laid the foundation of the White House in Badagry, he rerouted a country's architecture. The Badagry building was the first storeyed building in Nigeria and its construction marked the point at which Yoruba traditional architecture started aspiring to modernism. Later, 85 Odunfa Street, built by a Sierra Leonean in 1914, became the first three-storey building in Lagos. Christened Ebun House, it is in the theatrical Baroque style of sixteenth century Italy. Soon after, storeyed structures became the new measure of wealth. Bungalow owners began to deck their old houses and new buildings would not stop after the first floor. When the owner could not afford more than one floor, he decked his house and hoped to complete it later.

3 Statement of Problem

Most buildings today are defined by contemporary understanding of architecture, which is multidisciplinary and pays more attention to managing spaces for the sake of comfort, aesthetics and sanitation. There seems to be little or no consideration for the holistic well-being of inhabitants and the issues that were germane to ancient builders, who allowed for the cultural needs of the people, e.g. the need for moral formation, health, family arrangement and interaction, to determine the way space and structures were put together. In contemporary architecture, the needs of government and house owners defines space; it is what the government or house owners provide that people adjust to. Spaces are now defined mostly by the desire for aesthetics and status, at the expense of morality, communal living, health and wellness. This new situation has impacted on the morality and communal spirit of Africans. Adeosun [1] further notes that "The standardisation of Yoruba vernacular architecture started when European missionaries arrived armed with the paraphernalia of change. It progressed when ex-slaves returned home from Brazil with a newfound style and reached a peak after Portland cement became popular." Contemporary architecture has departed from tradition and has left Nigeria to deal with the consequences of the lost values that spaces provided in the homes.

Methods and Data

To achieve the purpose of this ethnographic study, the researcher had three Focus Group Discussion (FGD) sessions. The first two FGDs had six members each that were selected from among house owners in the selected areas. The first FGD was in Ebute Meta and the second took place in Isale-Eko. Both FGDs were conducted among house owners who were over 70 years. They were asked questions based on their experiences regarding the meaning of traditional architectural spaces and how the traditional spaces have changed and influenced the plan of their own houses.

These two sites were selected because they reflect the old parts of Lagos. A third FGD was conducted with architecture scholars in the University of Lagos. The third group was chosen to identify contemporary perspectives on space utility in the architecture of buildings in Lagos, taking into consideration government's plans in the area of housing. Findings/responses were transcribed according to FGD responses to reflect the understanding of traditional and contemporary spaces. The discussions highlighted the study objective on how contemporary designs have obliterated the meaning and lessons of traditional Yoruba architecture owing to urbanisation and foreign culture especially with regard to population. The discussions also examined the points of departure from tradition, meaning and morality.

4 Results

The architecture of a house says a lot about both the designer and the owner of the structure; it also reflects the culture and custom of the owner of the structure. The findings of this study are grouped into two parts: the traditional, which reflects the cultural heritage of the Yoruba, and the modern, which reflects the current situation in southwest Nigeria, especially Lagos. The traditional architecture reflects the period when cultural orientations reflected in the construction and design of buildings. This is true as

> The architectural characteristics of the traditional residence of the Yorubas were basically the proper putting together of the rectangle or square buildings to achieve a courtyard system. The buildings were erected in such a way that they joined together at each end. The courtyards were surrounded by the verandahs with doors leading to different rooms. These verandas were covered by cantilevered roof eaves which were supported by carved wood columns, caryatides in form of human figures and mud columns [17].

On its part, contemporary architecture represents designs that reflect modern-day tastes and needs at the expense of cultural heritage.

Regarding the old view, the FGDs from Ebute-Meta and Isale-Eko generally noted that "a typical house was designed to reflect utilities that spaces were meant to serve among the Yoruba." A typical household building was referred to as *Agbo-ilé* (household). An *Agbo-ilé* was designed in such a way that there was a sit-out or frontage called the *gbàgede*, which overlooked the streets. Next to the *gbàgede* was the first room in the house, which usually belonged to the head of the nuclear family (*Iyèwù baálé ilé*). Since the typical Yoruba household was polygamous, there were other rooms for the wives, called *Iyèwù*, which were arranged according to seniority. Then there was the *Orúwá* (the corridor), through which each of the rooms in the house could be accessed. There was also the *àgbàlá*, which was a space behind the house). Each of these spaces was well thought-out and aimed at creating and maintaining the health and well-being of each member of the family.

The *gbàgede* or frontage: Houses were built in such a way as to provide shade in front for the owner of the house and his visitors to sit in the evenings, after work. Where possible, a big tree served the purpose of the *gbàgede*. The front of the house had four basic functions: social, economic, moral, and security. It was a social space where fresh air was enjoyed, general stories were told, friends met and got entertained if it was just a casual visit. As an economic space, sales of minor farm produce, hunted game or homemade items took place there. The moral function became visible when elders sat in that space to address the ills of the day. There, children were discouraged from mingling with wayward friends. Finally, it was a space where security was enforced, that is, the first security post any visitor has to pass through before entering the house. It was where visitors were welcomed into the house by the eldest person or the owner of the house. Information and news was also passed from one person to the other, with strange movements and persons identified. The town crier could be heard and seen from the frontage of the house as well.

The *Iyèwù baálé ilé*) or first room in the house: The man's room was respected and was accessed only on invitation or during the discharge of an errand. Sometimes a powerhouse adjoined it, with the profession or title of the head of the house determining the nature of the powerhouse. Adeosun [1] notes: "The traditional Yoruba man was a polygamist, counting his wives and children when numbering his properties—and his lifestyle fed into his building. Yoruba architecture is a family panegyric—it shouts the glory or misfortune of a family in clear structural language." The man's room was strictly his room. He was free to visit other wives or invite them over to his room as he felt convenient.

The *Iyèwù*) or wife's/wives' room: It was a place of shelter but also a place where secrets were kept. Rooms were built according to the number of existing wives or the number of envisaged wives. Again, the profession, heritage or title of the head of the family determined the number of rooms that he had in his house. *Iyèwù* was ordinarily occupied according to the length of time a woman had been married to the man. The earliest wives lived closest to the man; however, this arrangement can change to suit the head of the family, such that the room of the youngest wife may be closest to his.

The room served different functions. First, it was a place of rest for the woman and her children. Second, it was her abode where secrets were kept. She was free to arrange the room to her taste and keep her treasures there. Third, and very important, the room was the place where a good mother would first reprimand her child before the child got into the bigger view. A mother/wife would try to mould a child's character so that child would be seen as good in the family. The mother gave advice on etiquette, brotherliness, hard work, following of house roster, etc., in the room. The room was also a place to recoil to after a bad experience. Children saw *Iyèwù* as a storeroom for good things if the wife was a friendly one, otherwise it was a place to be avoided if the wife was not friendly.

The *Orúwá* or corridor: The next significant space in the traditional Yoruba household was the *Orúwá* or corridor). This was a purposefully designed space. Bigger than a regular corridor, the *Orúwá* linked all the rooms together; sometimes, for privacy, it may not open directly to the *Iyèwù baálé ilé*. The corridor performed different intended functions. First, each of the rooms had the entrance opening into the corridor; this was not for lack of creativity but it was intended to prevent the keeping of secrets. The corridor was meant to encourage communal life and not secret or individual life. It was believed that communal life was better ensured when there were no secrets among members of the family.

The second function of the open corridor was to ensure that there was little or no partial treatment among the wives and children; the corridor was a leveller where each child or wife got to perform assigned functions. It exposed what happened and who entered the other bedrooms. The corridor helped in the training of children; it was a shared corridor that served as space for domestic chores and relaxation. It usually contained small stools for the wives and elders and there was always a big mat for the children to sit on.

The design (open space) made it almost impossible for anyone in the house to keep a secret. Everyone knew all that went in and out of the rooms. The corridor was where food was cooked and pepper ground on stones, as well as where members rested during the day and where constructive gist about members of the family took place. Feedbacks from the day's activities were shared and disagreements were also resolved. As the family bonded, the process of discernment went on as the family got to know the manners and disposition of each child. Talents were celebrated and support was given to the weak.

The *Orúwá* was generally a moral space where children who had erred were reprimanded in order to deter others from doing the same. It was a space for discipline or training in areas such as cooking, proper sitting positions, sweeping, etc. A child was reprimanded irrespective of who the mother was. Children were corrected or punished whenever found wanting. Punishments were given in the open space so that other children learnt from the incident; thereafter, the mother of the child or children in question took over in the secret of the room to either further punish or give palliative measures. Punishments were given and carried out in the corridor in order to deter other children, unless it was a unit punishment, where the mother handled the matter in her room.

The *Orúwá* was a space for training, as all the wives and elders in the house had a common duty to train children in the household; it did not matter who the biological mother was. This was where children learnt basic home training, such as respecting, greeting and showing regard for older people. Gender roles were defined and attendant skills acquired. Each child's capacity was identified and developed and general training in culinary and other skills were learnt in that space. With the *isasùn* (cooking pot), mothers took turns to cook for the family and children watched them do this, before they eventually took over a substantial part of the family cooking. Skills such as hair and body beautification were also learnt.

Family trust was also built within the *Orúwá*. There was usually one drinking pot (*àmù*) on the corridor from which everyone drank. The pot had a common cup (*ìkéèmù*), which everyone in the household drank with. As children ate and drank together, trust was built. It was a space to bond the family; every woman got the support of all the children in the household. In the case of a barren wife, communal life made house chores easy and possible, as the children were assigned to chores according to their age in the family. Bad habits were addressed through stories and family legend, which were usually told by the elders. This was an important part of family development and engagement.

Àgbàlá (backyard): This was typically a space where the sanitary rooms were built. This space served two basic purposes. It was the space for convenience, designed with or without a physical structure, where members of the household could ease themselves. Àgbàlá also served as the health garden for the household and close neighbours. Simple herbs were left to grow or planted in the garden. The space was needed to house rare species of plants, vegetables for immediate use, as well as herbs needed for first aid.

Contemporary Influence on Lagos Architecture

It is almost impossible to talk about contemporary Yoruba architecture, as spaces have assumed new and very different meanings. With colonialism in the 1800s came one of the significant causes of the far-reaching, permanent modification of traditional architecture in Lagos. Moreover, with specialisation and

> the emerging knowledge in scientific fields and the rise of new materials and technology, architecture and engineering began to separate, and the architect began to concentrate on aesthetics and the humanist aspects, often at the expense of technical aspects of building design. There was also the rise of the 'gentleman architect' who usually dealt with wealthy clients and concentrated predominantly on visual qualities derived usually from historical prototypes [18].

The ideas identified above have been reflected in government housing interventions, which are utilitarian in nature and provide the basic comfort for as many as possible. Such housing estates ignored the fundamental culture of family well-being, morality, support, etc. The structures had just the basic facilities that met government's expectations: sitting room, two rooms, kitchen and convenience.

In contemporary architecture, aesthetics have been given more value than meaning. And because bigger and more complicated buildings are being built, there is little concern for traditional spaces and the meanings attributed to such spaces. It must also be noted that attention is given to the increasing nuclear family and monogamous family structures as against the extended and polygamous family structures. An average architect is mostly concerned about what works and how many people can gain from the space that is designed. There is often the need to mass-produce houses in order to meet the demands of the government and not necessarily the demands of those that are to live in the houses.

Cultural ideals are put aside for the sake of government plan and housing programmes. This idea has also impacted on individuals who build their houses for either personal or commercial purposes. Structures for households are essentially functional, Smaller houses are built because of the number of people that government plans to take care of. The space that is available and the category of people that government wants to care for also affect the way architecture is currently defining spaces and the meaning or purposes that such spaces have. This new development has led to some traditional spaces disappearing and the ones that are being kept are been assigned modified functions. New spaces have also been created in view of SDG 11, to "make cities and human settlements inclusive, safe, resilient and sustainable."

In contemporary designs, the seven architecture scholars who made up the FGD in the University of Lagos explained as follows.

Gbàgede or frontage: The frontage has disappeared in contemporary architecture, which does not allow for houses to open up directly onto the streets. With houses built into compounds, this defeats the purpose of the frontage. In contemporary times, people are not encouraged to sit in front of the house. When they do, it is clear that the traditional function of a frontage is not intended. There are many other recreational activities that people can engage in inside the main building or in an entirely different space. All the functions of the frontage have been taken over by other departments; instead of having the head of the house watching out for security threats, there is a security post manned by guards.

The room of the head of the house or *Iyèwù baálé ilé*: In the ancient architecture, rooms were arranged in such a way that the first room was reserved for the head of the family. This room was semi-detached from the main building. It was significantly different from the other rooms in the house. In contemporary architecture, the *Iyèwù baálé ilé* is gone, replaced by the master' bedroom. One may argue that the master's bedroom serves the same purpose as the *Iyèwù baálé ilé* but a closer consideration shows that the idea of the master's bedroom is different from the *Iyèwù baálé ilé*. The master's bedroom is usually designed to accommodate both parents, unlike the *Iyèwù baálé ilé* that was specifically meant for the head of the house, with others allowed in only on invitation. The master's bedroom is not always exclusive to the head of the family, as children also share it when they have a need or on special occasions. In any case, the secrets of the family head can no longer be left in the master's bedroom.

Wife/wives' room (*Iyèwù*): The ideal today is for there to be no space called *Iyèwù*, since there is usually one wife who shares the master's bedroom with the husband, or for there to be another room that the woman occupies but that does not serve the basic functions of *Iyèwù*. It is not a secret place to hide from other wives or a place where children get reprimanded. The head of the family does not need to go there to gain access to the wife, since the house is usually smaller and facilities are shared in common.

The *Orúwá* or Corridor: Contemporary buildings do not have the idea of the corridor as was conceived in the ancient architecture. The semblance of corridors in contemporary buildings is found in the multi-tenanted buildings, described in the local parlance as 'face-me-I-face-you.'. These corridors have assumed the opposite value of *Orúwá*; unlike a place where disputes were settled, it is now a place where more quarrels are experienced because 'face-me-I-face-you' houses are usually occupied by different families who may not get along well with one another. Unlike the old function of *Orúwá*, corridors destroy the value of living together. Trust is usually destroyed and fighting often occurs, since the sense of sharing things in common is not promoted because strangers live together.

The *Àgbàlá* or backyard: This space has an entirely different meaning. It no longer houses the sanitary rooms, as these are usually incorporated in the building. Where there is a backyard at all, it is only semi-detached. The *Àgbàlá*, as part of the old architecture, had the herbs and first-aid plants; however, in contemporary architecture, there is no space for such knowledge. In the traditional setting, "the courtyards within the palaces and the compound were used for relaxation after the day's work. The private spaces like the front yard (*Ojuode/gbàgede*) and the back-yard (*Ehinkule/Àgbàlá*) were used for boxing and other recreational activities." The *Àgbàlá* (backyard), if it exists in the contemporary architecture, will accommodate what is called the boys' quarters, as was introduced by the Europeans, otherwise it is just an open space that is hardly utilised.

5 Discussion and Conclusion

5.1 Discussion

The findings make it clear that "it is the morn of the twenty-first century and history is crumbling under the weight of the future. The root architecture of the Yoruba people is fast disappearing. Ancient houses are being torn down and modern buildings grow in their place" [1]. Traditional spaces are giving way to new and foreign spaces. The frontage is gradually giving way to balconies that are restricted in their purposes, unlike the frontage in the traditional design that served four basic functions: social, economic, moral, and security. New spaces have also evolved, such as the security post, which does what the head of the family used to do while sitting in the front of the house. Some new spaces have taken over some of the functions of the corridor. Contemporary architecture now has laundry rooms, self-contained rooms, separate kitchens and multiple sitting rooms where guests are entertained. The corridor used to serve the purpose of laundry and sometimes kitchen. All of these developments are happening because urbanisation and population explosion are putting pressure on the government to provide housing, and this has become a strong part of the SDG 11, which is to "Make cities and human settlements inclusive, safe, resilient and sustainable." Different governments have attempted to solve this problem of housing by embarking on alleviation programmes. Many of these programmes have adopted European and master-servant–relationship designs in solving housing problems. The different attempts at resolving the difficulty of housing have resulted in obvious "social changes, particularly the shift from traditional community, family values to more western ways of life, have had variegated impact on buildings and residential layouts" [19]. Friedrich [20] observed that most Yoruba traditional compounds were built around a central hall or corridor to accommodate a plurality of polygamous families linked together by agnatic relations of senior male members. This and other reasons made it possible for "motifs of the modern movement [to] become very popular by the late 1930s. One of these was the modern flat roof, which proved to be a bad solution for the tropics with heavy rainfall" [4].

Contemporary Yoruba architecture has departed from traditional architecture, which was defined by culture and family life. The shift is traceable to when returning slaves got to Lagos and started building houses that reflected status and aesthetics more than the traditional functionality of designed spaces. Since the urban space lost the traditional designs, the "contemporary modern Nigerian architecture is dominated by the International Style. This is architecture of the Nigerian urban landscape" [4].

With foreign influence and the need for housing, the government began intervention programmes and engaged Low-trop Architecture, which represents "architecture of the masses, architecture of draughtsmen, and low income estates reminiscent of monotonous housing estates in Eastern Europe. Here apart from lack of good aesthetics, the standard of living is very low" [21].

This change from the traditional architecture has brought about a departure from traditional designs, their meanings and morality. The vernacular architecture has "a traditional base in the socio-cultural organization of the Nigerian society and the interaction between it and the other influences have crystallized into the Nigerian Vernacular Architecture" [22]. Morality has been excluded from the elements that inform space and design, just as the communal life that defined the Yoruba has also been affected; children do not grow up sharing food, water and utensils. Architecture does not intend to promote virtues or strong family bond. Where it is possible, individuals have all they need without interacting with neighbours. According to Rikko and Gwatau, "Contemporary world civilization influences architecture in manner responsive to technological advancement, modern materials available, socio-economic status of individuals and lifestyles of people." [19].

5.2 Conclusion

The contemporary architecture of southwest Nigeria is a blend of Eurocentric ideas that do not reflect the cultural reality of the people. It also sometimes continues the master-servant relationship through the separation of the main building from what is called the boys' quarters. The reality shows that the pressure of housing, among other factors, has been eroding the African values of family life, communal interactions and morality that the traditional architecture protected or sustained. With international influence came a kind of architecture that blended the European (nuclear family) design with the aesthetics and economy of space. It is important to stress that contemporary designs should respect culture and tradition, and one way to do that is to account for the knowledge possessed by the older generation.

References

1. Adeosun A (2017) Nigeria: the evolution of traditional Yoruba architecture. https://allafrica.com/stories/201710110696.html. Accessed 5 Sept 2019
2. Adegoke AK (2015) Ornamentation in Yoruba domestic architecture in Osogbo: a study in form, content and meaning. A Thesis Presented in the Department of Architecture, School of Environmental Studies, College of Science and Technology, Covenant University, Ota. In Fulfillment of the requirements for the award of Ph. D in Architectural History

3. Martinet A (1966) Structure and language, in Yale French studies: structuralism. Yale University Press, New Haven
4. Prucnal-Ogunsote B (1993) Classification of Nigerian architecture. AARCHES J 1(6):48–56
5. Osasona CO (2007) From traditional residential architecture to the vernacular: the Nigerian experience
6. Vitruvius ten books on architecture, with regard to landscape and garden design. garden-visit.com. Archived from the original on 12 Oct 2007. Retrieved 14 November 2018
7. Ruskin J (1880) The seven lamps of architecture. In: Allen G (ed) Reprinted Dover, (1989)
8. Corbusier L (1985) Towards a new architecture. Dover Publications, United States
9. Vlach JM (1984) The Brazilian house in Nigeria: the emergence of a 20th-century vernacular house type. J Am Folklore 97(383):3–23
10. Adeokun CO (n.d.) The Orowa House: a typology of traditional Yoruba architecture in Ile-Ife, Nigeria. https://Pdfs.Semanticscholar.Org/C99a/54ddfcee2224ff3a9534c30adf4a27761a78.Pdf Accessed 27 July 2019
11. Ekhaese EN (2011) Domestic architecture in Benin: a study of continuity and change. Unpublished thesis, Covenant University, Ota, Nigeria
12. Isaac-Sodeye F (2012) The kitchen in domestic space: a comparative study of Kitchens cooking and culinary practice in Ile-Ife, Nigeria. In: Laryea S, Agyepong SA, Leiringer R, Hughes W (eds) Proceedings of 4th West Africa Built Environment Research (WABER) conference, Abuja Nigeria, pp 589–605
13. Muhammad-Oumar AA (1997) Gidaje: the socio-cultural morphology of Hausa living spaces. Unpublished thesis. University of London
14. Herskovits MJ (2013) The human factor in changing Africa. Routledge, USA
15. Crooke P (1966) A sample survey of Yoruba rural buildings. Odu 20:41–71
16. Marris P (1962) Family and social change in an African city: a study of rehousing in Lagos. North western University Press, Evanston. Accessed 25 July 2019
17. Adewumi J (2016) Nigerian traditional architecture: what are the building types? https://enititanblog.wordpress.com/2016/09/27/nigerian-traditional-architecture-what-are-the-building-types/. Accessed 11 Sept 2019
18. Architecture (n.d.) https://en.wikipedia.org/wiki/Architecture. Accessed 5 Sept 2019
19. Rikko LS, Gwatau D (2011) The Nigerian architecture: the trend in housing development. J Geogr Reg Plann 4(5):273–278
20. Friedrich WS (1982) Traditional housing in African cities. a comparative study of houses in Zaria, Ibadan and Marrakech. John Willey and Sons, New York
21. Izomoh SO (1997) Housing provision and management in Nigeria. Emiola Publishers Ltd, Ogbomoso
22. Adeyemi EA (1975–1976) Changing traditional culture and modern architecture. Archiforum Mag. 1:12–19

Urban Sprawl and Housing: A Case for Densification in Nigerian Cities

Saidat Damola Olanrewaju and Olumuyiwa Bayode Adegun

Abstract The aspiration for urban sustainability is captured within the Sustainable Development Goals (SDGs) and other global agendas. Urban sustainability cannot be achieved without significantly changing the way housing and other urban spaces are planned, designed and developed. Urban morphology plays a critical role in achieving sustainable housing and resilient communities. In Nigeria and elsewhere, housing development in the context of rapid urbanisation generally involves two types of residential development patterns: densification/compactness and urban sprawl. This chapter is based on a review of literature on studies conducted in relation to urban sprawl and densification in Nigeria and elsewhere. Through the review of relevant studies, the chapter shows the nature of urban sprawl in Nigeria and its economic, environmental and social impacts. These establish the fact that sprawls are an undesirable urban housing development pattern. Compact urban housing development, given its advantages, is proposed as the urban form suited for sustainability. Knowledge established through this review provides a basis for housing densification policy in Nigeria's main cities and the rapidly growing and, at times overlooked, secondary cities.

Keywords Compact city · Densification · Urban sprawl · Land use · Suburban housing · Peri-urban settlements

1 Introduction

Urban areas across the globe are growing rapidly in population and territorial coverage. According to estimates from the United Nations, half of the world's population were urban dwellers as at 2015. This proportion is expected to increase to 66% by year 2050, with more growth occurring in African cities. For many countries, urbanisation processes usually boost the urban economy, social development and industrial activities [68]. Unprecedented increases in the urban population also

S. D. Olanrewaju (✉) · O. B. Adegun
Department of Architecture, Federal University of Technology, Akure, Nigeria
e-mail: olanrewajuso@futa.edu.ng

results in housing deficit, overstretched infrastructural facilities and environmental pollution, etc. Notably, this growth in population also affects urban land-use patterns.

Globally, urban land management is seen as a challenge to sustainable development since land is a limited resource. The relationship between urban form and sustainability cannot be separated from patterns of land use for various purposes in the city. Sustainable urban form generally aims to mitigate climate change, create a livable environment, preserve natural ecosystems, encourage self-sufficiency in terms of energy and food production, as well as improve quality of life [19]. The pattern of growth for residential areas is either associated with urban sprawl (expansion) or compact development (intensification). Urban sprawl generally refers to the development of new residential areas on formerly vacant land and is characterised by low-density housing [23]. It is also linked to high carbon emissions due to automobile dependency and land-use segregation [38, 3]. According to Couch et al. [26], urban sprawl refers to low-density suburban development around the borders of a city. It can be characterised by informality in housing, a situation whereby dwellings are developed without relevant planning permits in areas lacking proper layouts [58]. It also involves opportunity for incremental housing, which emphasises the role of the self-built and owner-occupied housing process [8].

African cities have grown in terms of population and built-up areas by more than 5% annually and have less compact urban forms compared to many European cities with similar populations [78]. The continuous increase in population simultaneously increases the demand for land for housing. Majority of people migrating to urban areas in Nigeria reside in the suburban or peri-urban areas due to land availability and affordability on the fringes. The sprawling nature of the cities involves formation of peri-urban settlements characterised by haphazard (informal) housing development, disordered land-use patterns and low residential densities [13, 20]. Urban sprawl has also been recognised as a product of governance challenges and poor urban planning [58, 5]. While the Nigerian government made efforts to address the urban sprawl, this growth pattern continued due to the dependence on outdated city master plans coupled with implementation and enforcement negligence, among others.

As will be shown in this chapter, urban sprawl is generally understood to be unaligned to the goals of sustainable development, hence the need to engage it within the Sustainable Development Goals (SDG) framework and associated global development agendas. In relation to housing, urban sprawl is directly linked to SDG 11, which seeks to "make cities and human settlements inclusive, safe, resilient and sustainable." It is indirectly linked to other goals, especially to ensure sustainable consumption and production patterns (Goal 12), action to combat climate change and its impacts (Goal 13) and sustainable use of terrestrial ecosystems (Goal 15). The idea of appropriate densities and compactness, as well as of containing sprawl in cities, is clearly presented within Habitat III's New Urban Agenda, which was adopted in 2016.

The subject of sprawl has attracted considerable scholarly attention locally and internationally. Using various Geographic Information Systems and Remote Sensing techniques and indicators (e.g. population density, development density and land-use mix), many scholars have measured and examined the manifestations, trends and dynamics of urban sprawl across the country (e.g. [4–7, 17, 62]). In addition, international literature has established a relationship between sprawl and spatial inequality from economic, social and environmental perspectives.

This chapter presents a review of literature on urban sprawl in Nigeria and elsewhere. It aims to show why sprawl is undesirable (or otherwise) and makes a case for housing densification in urban areas. Specifically, the review seeks to answer the question, "What are the driving forces, disadvantages and advantages of urban sprawl versus compact development in relation to residential/housing development?" Based on the impacts identified, a case is made for housing densification as a means for sustainability and achieving the Sustainable Development Goals (SDGs) in the Nigerian context.

2 Method

This review documents literature and analyses the key threads of thought regarding densification and urban sprawl in order to identify the way forward in the Nigerian context. The literature comes from peer-reviewed outlets such as journals, conference proceedings and book chapters. The study was conducted using a four-stage process. The first stage was to identify and collect relevant studies through extensive search within the Google Scholar (also covering Science Direct) online database. The main keywords used for the search included *urban form, urban expansion, urban sprawl, housing densification, infill development, housing density*. Combination of the keywords yielded over 16,000 outcomes, out of which 124 potentially relevant articles were retrieved. After the initial collection, quality of the retrieved articles was assessed. A total of 67 articles were eventually reviewed. Of the 67 articles, 30 emanated from Nigeria while the rest are from Australia, Belgium, China, Sydney, Spain, Iran and USA. Literature from 2009 onwards was used in order to access the current state of knowledge in the research area.

3 Driving Forces of Urban Sprawl

Understanding the urban expansion process is beneficial to urban planning and management. The underlying driving forces of urban expansion in Nigerian cities are associated primarily with geo-physical, socioeconomic and institutional factors. The literature shows that simultaneous increase in population and income [55], ineffective land use [63], urban land price [34], land-use pattern in the city centre [58] and housing affordability [57] are the major driving forces of urban sprawl in Nigerian cities. For instance, Fasakin et al. [34] examined the residential density pattern in the suburban areas of Akure. They found a significant relationship between the price of land and residential density patterns, which shows the need for having an appropriate residential density mix. From the remote imagery perspective, Owoeye and Ibitoye

[63] analysed the pattern of land-use change and extent of urban expansion. Their study confirmed that the rate of urban expansion is influenced by discordant land-use changes.

Similarly, within the context of Iran, Bagheri and Tousi [16] established population growth as the main driving force of urban expansion due to employment opportunities, higher relative household income and affordable housing policies. The authors also identified transit-oriented developments as a driver of urban expansion towards rural and invaluable natural areas. The driving forces behind the urban sprawl in China are similar [79]. Furthermore, Li et al. [49] highlighted other drivers to include socioeconomic, physical, proximity, accessibility, and neighborhood factors on the national and regional scales in countries with rapid urbanisation.

Geological characteristics such as slope, elevation and availability of developable land have been established as one of the influences for sprawl formation in some countries [23, 37]. However, governance dispensation has not been confirmed to influence land use and emergence of urban sprawl in Nigeria. In their analysis, Taiwo et al. [71] found no significant difference in land-use patterns across different governance regimes, i.e. between military and civilian regimes in the country. All the same, spatial policy variables are responsible for sprawl residential patterns [5]. As the urban population grows, the demand for housing increases, usually leading to an increase in rental values in the inner cities. In this situation, low-income residents are driven to the fringes, hence the link between urban expansion, tenure insecurity and urban poverty.

4 Effects of Urban Sprawl

4.1 Environmental Impacts

Research links the urban sprawl with environmental degradation. As shown in this section, the urban sprawl is increasingly being associated with loss of biodiversity, increased land surface temperature, loss of agricultural land and reduction in water bodies. Atu et al. [14] found that sprawl development threatens biodiversity directly through habitat loss and indirectly through habitat fragmentation, degradation and homogenisation of the native biota. Species diversity in agricultural lands is declining due to the encroachment associated with sprawl. Suburban and peri-urban land that is valuable for farming has been significantly affected due to conversion of agricultural land to other uses [75, 59]. Corroborating this, Atu et al.'s [13] study shows significant variation in the sizes and densities of farms between agricultural lands within sprawl and agricultural lands outside sprawl. These impacts contribute to climate change.

The unplanned nature of residential areas resulting from the urban sprawl involves indiscriminate dumping of solid waste, unclean environment and contamination of water bodies [7, 42, 59]. Encroachment on water bodies is particularly notable [54]. About 70% of ecosystem functions—ranging from provisioning (e.g. food, water and fuel) to regulating (e.g. climate regulation and water purification) and cultural services (e.g. recreation and aesthetic values) to supporting services (e.g. soil formation and erosion control)—have been degraded due to improper urban land consumption in Nigeria [11]. Urban expansion and land-use conversions account for significant land surface temperature variations [41]. If these continue, it increases the risk of communicable diseases, pollution and natural disasters such as flooding.

In addition, sprawl is associated with environmental injustice, particularly in poor, minority and disadvantaged neighbourhoods [27]. For instance, Schweitzer and Zhou [67] examined air quality in compact and sprawl regions at neighbourhood levels. The study found that ozone concentrations are significantly higher in the sprawling regions. Sprawl is a major contributor to energy consumption, climate change and air pollution, consequently affecting air quality [29].

4.2 Social Impacts

In urban planning, quality of life is an important goal in policy development [74]; therefore, enriching a people's quality of life is viewed as an indicator of social sustainability. Studying the quality of life has become crucial because of its impacts on health and longevity [28]. Sprawl settlements epitomise urban fragmentation and this impacts quality-of-life outcomes. In sprawl development, residents' access to basic facilities such as water, electricity, health care and educational facilities is generally poor, thus fostering urban inequality [58]. Another social problem in suburban areas is insecurity, which manifests there partly due to increase in population not being matched by availability of employment opportunities [47, 59].

Some studies have documented the impact of sprawl on social inequality. Living in sprawl development promotes educational inequality [18], residential segregation [64], reduced social capital [53], as well as income and health inequality [30, 35]. Ewing et al. [32] noted that living in sprawling communities is associated with less walkability and increased obesity than in compact communities. Similarly, Alabi [7] assessed the socio-economic impacts of sprawl in Akure. The study revealed that fatigue/tiredness is a major health impact of sprawl. Talen et al. [72] examined the relationship between urban form and sociological factors such as race and household characteristics. The study found a significant relationship between urban sprawl and social inequalities because of the uneven distribution of public services and amenities.

The sprawling nature of major cities in Nigeria shows the complex interaction between traffic and land use. Commuting from peripheral settlements increases urban trip length and time, in addition to causing traffic congestion [56, 66]. Residents of sprawl areas are often faced with high cost of living due to transportation cost to the city center [7]. Zolnik's [82] study from the USA confirmed the relationship between sprawl and private-vehicle commuting distances and times. Thus, sprawl is associated with significantly higher fatal crash rates and reduced opportunities for physical activities [32]. In Nigeria, during traffic congestion, criminals at times have opportunity to rob commuters. In addition, congestion leads to loss of productive hours and causes environmental pollution [66].

4.3 Economic Impacts

In terms of its economics, urban sprawl is associated with increased travel costs and higher costs of infrastructural provision [59]. Continuous conversion of agricultural land can lead to shortage of fresh food supply and increase in cost of food [54]. Due to the concentration of poverty, sprawl settlements are likely to degenerate, thus perpetuating the cycle of disadvantage. Furthermore, the influx of people into peri-urban settlements can lead to increase in land value and rental costs [47]. Nevertheless, some urban residents have positively benefited from urban sprawl through access to affordable housing on cheaper land located on the urban periphery [34, 57]. Mueller et al. [52] have argued, however, that lower housing prices on the periphery are offset by higher transportation costs.

The development process of urban sprawl limits equality and increases spatial inequality. Low-income earners often reside on the periphery of urban regions while high-income earners reside at the centre due to availability of infrastructure and services [80]. Moreover, urban sprawl has significant impact on income inequalities [77]. International literature has established a significant relationship between household income and urban density. In the USA, Lee et al. [48] confirmed that lower levels of sprawl are associated with lower levels of income inequality and correspond to higher levels of financial well-being. Urban sprawl offers access to new job opportunities with relatively low pay and offers little upward mobility, thereby leading to higher income gaps and segregation [30]. Some studies found that urban sprawl is significantly associated with net job quality and upward mobility [24, 30].

5 Compact Urban Housing Development: A Path to Sustainability

Since the emergence of the sustainable development agenda, debate has been intensifying on urban form in terms of the physical characteristics, size and density to enhance the quality of life. A paradigm shift to urban sustainability suggests more sustainable urban forms. These include neo-traditional development, urban containment, the compact city and the eco-city concepts [44]. Advocacy has thus increased for adopting a compact urban form with a view to combating the menace of sprawl. According to Tang [73], the compact urban form enables equitable access to services and offers better quality of life. It also leads to decrease in social inequality because it enables social interactions and heterogeneity. Development of appropriate compact city policies plays a critical role in achieving urban sustainability goals (environmental quality, economic viability and social equality). Countries working towards sustainability have adopted urban compaction polices to counteract sprawl on the edge of the city and encroachment on terrestrial ecosystem services.

5.1 Urban Densification

Densification or intensification of housing development is a possible way to create compact cities [81]. Urban densification refers to a gradual progression in residential density and scholars have argued that densification is a more efficient and less-polluting strategy for compact city development. Intense development encourages redevelopment of existing built-up areas, development corridors and open areas within the urban edge [2]. The proponents of housing intensification have argued for its spatial, social, economic and environmental benefits to validate its relevance, although the adverse effects of highly dense urban forms are also evident.

Housing densification improves access to services, lowers the cost of infrastructural provision, encourages economic opportunities and promotes public transportation. The provision of public social goods such as water, transportation and healthcare becomes easier [8]. As a Hong Kong case study shows, living in dense neighbourhoods improves accessibility to facilities and services [46]. Lieske et al. [50] reported a decrease in the cost of infrastructure in higher-density development, while Glaeser and Joshi-Ghani [36] found that high density encourages more productive economies.

Compact and connected development has potential to provide a better quality of urban life through enhanced social interaction, walkability and reduced greenhouse emissions by minimising transport and travel [31]. High building density is economically beneficial because it enhances public transportation. Moreover, Ewing et al. [29] argued that dense urban form reduces traffic congestion when the width of street surfaces is adequate and highway user fees are high. Compact cities promote effective use of land, thereby preserving agricultural land and environmental resources [22]. In California, Smith [70] confirmed the potential benefit of urban densification as opposed to edge development by showing that densified projects can save about five kilometers of agricultural land or land that could be assigned to further uses.

Contrary to the above assertions, however, some scholars argue against densification, claiming that higher density leads to lower living quality, high crime rates, overcrowding, environmental pollution, traffic congestion, loss of open space and lack of green space [12]. Reiter [65] argues that densification creates urban heat islands and wind discomfort while also neglecting the urban heritage. According to Zhu [81], high density can foster crime and social disorganisation under certain socioeconomic characteristics. Local residents often have a negative perception of housing densification due to its association with overcrowding [10]. In addition, residents often see infill developments as negative externalities on their neighbourhood character and as additional strain on available infrastructure and services. One of the main constraints of the compact city development and densification process is provision of the urban green space that is the main supplier of ecosystem services supporting human well-being [39]. According to Lin et al. [51] greater dwelling density leads to a general loss of public parkland and residential trees. Similarly, Arnberger [10] confirms that densification around public green areas leads to loss of recreational value. Loss of urban green space can be mitigated by using vertical greening systems and provision of quality public green spaces to augment loss of private green spaces.

When properly planned, housing densification may be regarded as a solution since the negative consequences of sprawl continues to prevail. Densification strategies include infill development, backyard filling and replacement of existing low-density buildings with high-rise buildings. Other strategies include transforming and renovating saddle roofs on the top of buildings into wider and liveable spaces and roof stacking [9]. The combination of residential densification and upgrading of streetscapes through provision of open space amenities could result in positive synergies in compact development [21]. However, densification planning must be contextually sensitive to achieve its primary objectives [45].

5.2   Densification in Nigeria

Most of Nigeria's urban centres are currently undergoing densification processes. The most notable strategies employed in Nigeria are infill development, backyard filling and replacement of existing low-density buildings with high-rise buildings [25, 33]. Infill development occurs on vacant land existing between buildings in built-up areas, thus increasing the density of the residential neighborhood. Specifically, Government Reserved Areas (GRA) present opportunities for infill development due to hitherto large plot sizes, open spaces and infrastructural facilities [61, 15, 69]. Infill development also occurs on available green spaces, new plots and leftover space of previously allocated plots that have been sold out [43]. However, most of the housing densification occurs without the approval of relevant planning authorities. For example, floors are added to existing buildings without adequately considering structural stability, which can lead to building collapse [60]. These forms of unlawful densification result in poor environmental conditions, thus affecting quality of life [61, 15].

Since it is imperative to monitor urban densification processes, existing patterns of urban densification can be studied and information elicited to predict future densification trends using land-use change models such as Markov chain, economic-based systems, agent-based systems, cellular automata and artificial neural networks (ANN) [1, 76]. In predicting urban densification, the categories of urban density and relevant predictor variables are indispensable. It is noteworthy that densification processes should be guided to prevent gentrification or unreasonable increases in land prices. In carrying out densification, residents' attitudes and perceptions should be elicited for relevant input and feedback [40].

6 Conclusion

As the population of urban dwellers in Nigeria and the accompanying housing demand continue to increase, it is important to pay attention to sustainable urban development patterns, in line with the Sustainable Development Goals (SDGs) and New Urban Agenda. This chapter has shown the nature, drivers, characteristics and impacts of urban sprawl generally and in relation to the housing sector in Nigerian cities. As the reviewed studies show, urban sprawl is a problem due to its detrimental impacts and unsustainable nature, although a few benefits of sprawl were also identified. The compact urban form for housing development, though not without its own disadvantages, is better than sprawl.

Densifying Nigeria's major cities is an imperative for a more sustainable urban future. As such, densification policies that catalyse compact housing development are needed. This should be context-dependent, as it must consider the local conditions of the cities to be densified. Compaction policies should focus on land-use restriction, development ratio and mixed land uses. Since densification can exert pressure on existing infrastructural facilities and services, the densification process needs to be monitored and commensurate infrastructure and services provided. Moreover, densification policies that restrict land supply should be done carefully to avoid unnecessary increase in land price that reduce access and affordability. Furthermore, residents' perception and sociocultural disposition should be acknowledged in formulating and implementing the densification process. Housing mortgage can also be provided to landowners who meet the developmental guidelines in densification planning. In addition, principles of densification such as effective transportation, appropriate densities, social cohesion and environmental consciousness need to be considered holistically.

References

1. Abdullahi S, Pradhan B, Mojaddadi H (2017) City compactness: assessing the influence of the growth of residential land use. Urban Technol. https://doi.org/10.1016/j.scs.2017.09.021
2. Abdullahi S, Pradhan B (2016) Sustainable brownfields land use change modeling using GIS based weights-of-evidence approach. Appl. Spat. Anal Policy 9:21–38
3. Abdullahi S, Pradhan B, Jebur MN (2015) GIS-BASED sustainable city compactness assessment using integration of MCDM, Bayes Theorem and RADAR Technology. Geocarto Int 30(4):365–387
4. Ade MA, Afolabi YD (2013) Monitoring urban sprawl in the federal capital territory of Nigeria using remote sensing and GIS techniques. Ethiop J Environ Stud Manag 6(1):82–95. https://doi.org/10.4314/ejesm.v6i1.10zx
5. Aguda AS, Adegboyega SA (2013) Evaluation of spatio-temporal dynamics of urban sprawl in Osogbo, Nigeria using satellite imagery & GIS techniques. Int J Multi Curr Res 60–73. ISSN: 2321-3124
6. Akintunde JA, Adzandeh EA, Fabiyi OO (2016) Spatio-temporal pattern of urban growth in Jos Metropolis, Nigeria. Remote Sens Appl Soc Environ. https://doi.org/10.1016/j.rsase.2016.04.003

7. Alabi MO (2019) Encroachment on green open space, its implications on health and socio-economy in Akure, Nigeria. Cities Health. https://doi.org/10.1080/23748834.2019.1639421
8. Andreasen MH, Agergaard J, Møller-Jensen L (2016) Sub-urbanisation, homeownership aspirations and urban housing: exploring urban expansion in Dar es Salaam. Urban Stud 1–17
9. Amer M, Attia S (2017) Roof stacking: learned lessons from architects. SBD Lab, Liege University, Belgium. Retrieved from http://orbi.ulg.ac.be/handle/2268/210472
10. Arnberger A (2012) Urban densification and recreational quality of public urban green space: a Viennese case study. Sustainability 4:703–720
11. Arowolo OA, Deng X, Olatunji OA, Obayelu AE (2018) Assessing changes in the value of ecosystem services in response to land-use/land-cover dynamics in Nigeria. Sci Total Environ 636:597–609. https://doi.org/10.1016/j.scitotenv.2018.04.277
12. Arvola A, Pennanen K (2014) Understanding residents' attitudes towards infill development at Finnish Urban suburbs. In: Conference proceedings for world SB14 Barcelona 3, pp 19–28
13. Atu JE, Offiong RA, Eni DI, Eja EI, Essien OE (2012) The effects of urban sprawl on peripheral agricultural lands in Calabar, Nigeria. Int Rev Soc Sci Humanit 2(2):68–76
14. Atu JE, Offiong RA, Eja EI (2013) Urban sprawl effects on biodiversity in peripheral agricultural lands in Calabar, Nigeria. J Environ Earth Sci 3(7):219–231
15. Ayotamuno A, Gobo AE, Owei OB (2010) The impact of land use conversion on a residential district in Port Harcourt, Nigeria. Environ Urbanization 22(1):259–265
16. Bagheri B, Tousi SN (2017) An explanation of urban sprawl phenomenon in Shiraz Metropolitan Area (SMA). Cities. https://doi.org/10.1016/j.cities.2017.10.011
17. Balogun IA, Adeyewa DZ, Balogun AA, Morakinyo TE (2011) Analysis of urban expansion and land use changes in Akure, Nigeria, using remote sensing and geographic information system (GIS) techniques. J Geogr Reg Plann 4(9):533–541
18. Batchis W (2010) Urban sprawl and the constitution: educational inequality as an impetus to low density living. Urban Lawyer 42(1):95–133
19. Beatley T, Newman P (2009) Green urbanism down under: learning from sustainable communities in Australia. Island Press, New York
20. Bijimi CK (2013) The relevance of a good urban design in managing urban sprawl in Nigeria. Int J Technol Enhancements Emerg Eng Res 1(4):123–125
21. Bolleter J (2016) On the verge: re-thinking street reserves in relation to suburban densification. J Urban Des. https://doi.org/10.1080/13574809.2015.1133229
22. Boyko C, Cooper R (2011) Clarifying and re-conceptualising density. Prog Plann 76:1–61
23. Broitman D, Koomen E (2015) Residential density change: densification and urban expansion. Comput Environ Urban Syst 54:32–46
24. Chapple K (2018) The fiscal trade off: sprawl, the conversion of land use, and wage decline in California's metropolitan regions. Landscape Urban Plann 177:294–302. https://doi.org/10.1016/j.landurbplan.2018.01.002
25. Chiroma MA, Shah MZ, Isa AH, Usman AS, Kagu A, Ijafiya I (2018) Impacts of infill development on land use in Ibrahim Taiwo Housing Estate, Maiduguri, Nigeria. Adv Sci Lett 24(5):3758–3764
26. Couch C, Leontidou L, Petschel-Held G (2007) Urban sprawl in Europe, landscapes. Land use changes and policy. Blackwell publishing, Oxford. ISBN 978 1405139175
27. Darby KJ, Atchison CL (2014) Environmental justice: insights from an interdisciplinary instructional workshop. J Environ Stud Sci 4:288–293
28. Diener E, Chan M (2011) Happy people live longer: subjective well-being contributes to health and longevity. Appl Psychol Health Well-Being 3(1):1–43
29. Ewing R, Tian G, Lyons T (2018) Does compact development increase or reduce traffic congestion? Cities 72:94–101
30. Ewing R, Hamidi S, Grace JB (2016) Urban sprawl as a risk factor in motor vehicle crashes. Urban Stud 53(2):247–266
31. Ewing R, Hamidi S (2014) Measuring urban sprawl and validating sprawl measures. National Institutes of Health and Smart Growth America, Washington, DC

32. Ewing R, Meakins G, Hamidi S, Nelson AC (2014) Relationship between urban sprawl and physical activity, obesity, and morbidity–update and refinement. Health Place 26:118–126
33. Ezema I, Oluwatayo A (2014) Densification as sustainable urban policy: the case of Ikoyi, Lagos, Nigeria. In: Proceedings of the CIB W107 2014 international conference, Lagos, Nigeria, 28th–30th January 2014, pp 695–704
34. Fasakin JO, Basorun JO, Bello MO, Enisan OF, Ojo B, Popoola OO (2018) Effect of land pricing on residential density pattern in Akure, Nigeria. Adv Soc Sci Res J 5(1):31–43
35. Frenkel A, Israel E (2018) Spatial inequality in the context of city-suburb cleavages enlarging the framework of well-being and social inequality. Landscape Urban Plann 177:328–339. https://doi.org/10.1016/j.landurbplan.2017.02.018
36. Glaeser E, Joshi-Ghani A (2013) The urban imperative: towards shared prosperity. World Bank, Washington
37. Gouda AA, Hosseini M, Masoumi HE (2016) The status of urban and suburban sprawl in Egypt and Iran. GeoScape 10(1):1–15
38. Gu Z, Sun Q, Wennersten R (2013) Impact of urban residences on energy consumption and carbon emissions: an investigation in Nanjing, china. Sustain Cities Soc 7:52–61
39. Haaland C, Bosch CK (2015) Challenges and strategies for urban green-space planning in cities undergoing densification: a review. Urban For Urban Greening. https://doi.org/10.1016/j.ufug.2015.07.009
40. Holden M (2019) Bringing the neighbourhood into urban infill development in the interest of well-being. Int J Commun Well-Being 1:137–155
41. Ibitoye MO, Aderibigbe OG, Adegboyega SA, Adebola AO (2017) Spatio-temporal analysis of land surface temperature variations in the rapidly developing Akure and its environs, southwestern Nigeria using Landsat data. Ethiop J Environ Stud Manag 10(3):389–403
42. Ifatimehin OO, Musa SD, Adeyemi JO (2009) An analysis of the changing land use and its impact on the environment of Anyigba Town, Nigeria. J Sustain Dev Afr 10(4):357–367
43. Imam MZ, Rostam K (2011) The impacts of unauthorized subdivisions of residential plots in Gadon Kaya, Kano City, Nigeria. Geogr: Malays J Soc Space 7(2):1–10
44. Jabareen YR (2006) Sustainable urban forms: their typologies, models, and concepts. J Plann Educ Res 26:38–52
45. Kyttä M, Broberg A, Tzoulas T, Snabb K (2013) Towards contextually sensitive urban densification: location-based soft GIS knowledge revealing perceived residential environmental quality. Landscape Urban Plann 113:30–46
46. Lang W, Chen TT, Chan EHW, Yung EHK, Lee TCF (2019) Understanding livable dense urban form for shaping the landscape of community facilities in Hong Kong using fine-scale measurements. Cities 84:34–45
47. Lawanson T, Yadua O, Salako I (2012) An investigation of rural-urban linkages of the Lagos megacity, Nigeria. J Constr Proj Manag Innov 2(2):464–481
48. Lee WH, Ambrey C, Pojani D (2018) How do sprawl and inequality affect wellbeing in American cities? Cities 79:70–77
49. Li G, Sun S, Fang C (2018) The varying driving forces of urban expansion in China: insights from aspatial-temporal analysis. Landscape Urban Plann 174:63–77
50. Lieske SN, McLeod DM, Coupal RH, Srivastava SK (2012) Determining the relationship between urban form and the costs of public services. Environ Plann Part B 39(1):155
51. Lin B, Meyers J, Barnett G (2015) Understanding the potential loss and inequities of green space distribution with urban densification. Urban For Urban Greening 14:952–958
52. Mueller EJ, Hilde TW, Torrado MJ (2018) Methods for countering spatial inequality: incorporating strategic opportunities for housing preservation into transit-oriented development planning. Landscape Urban Plann 177:317–327
53. Nguyen D (2010) Evidence of the impacts of urban sprawl on social capital. Environ Plan 37(4):610–627
54. Oladele BM, Oladimeji BH (2011) Dynamics of urban land use changes with remote sensing: case of Ibadan, Nigeria. J Geogr Reg Plann 4(11):632–643

55. Oduwaye L (2015) Urban land use planning and reconciliation. Inaugural Lecture Series 2015, University of Lagos, Nigeria
56. Oduwaye L, Alade W, Adekunle S (2011) Land use and traffic pattern along Lagos—Badagry corridor, Lagos, Nigeria. In: Real Corp 2011: change for stability: lifecycles of cities and regions, pp 525–532
57. Olujimi J, Gbadamosi K (2007) Urbanisation of peri-urban settlements: a case study of Aba-Oyo in Akure, Nigeria. Soc Sci 2(1):60–69
58. Olujimi J (2009) Evolving a planning strategy for managing urban sprawl in Nigeria. J Hum Ecol 25(3):201–208. https://doi.org/10.1080/09709274.2009.11906156
59. Onaiwu DN, Onaiwu FO (2019) Impact of sprawl on the peri-urban areas. Malays J Soc Space 15(20):1–14. https://doi.org/10.17576/geo-2019-1502-01
60. Oni AO (2010) Analysis of incidences of collapsed buildings in Lagos Metropolis, Nigeria. Int J Strateg Property Manag 14(4):332–346
61. Orekan AA (2014) An assessment of the impact of plot standard on physical development: the case study of Kano Metropolis, Nigeria. Int J Eng Sci 2(2):46–52
62. Owoeye JO, Popoola OO (2017) Predicting urban sprawl and land use changes in Akure region using Markov chains modeling. Geogr Reg Plann 10(7):197–207
63. Owoeye JO, Ibitoye OA (2016) Analysis of Akure urban land use change detection from remote imagery perspective. Urban Stud Res 1–9
64. Ragusett JM (2016) Black residential segregation in the era of urban sprawl. Rev Black Polit Econ 3–4:253–272
65. Reiter S (2010) Assessing wind comfort in urban planning. Environ Plann B: Urban Anal City Sci 37(5):857–873. https://doi.org/10.1068/b35154
66. Salau T, Lawanson T, Odumbaku O (2013) Amoebic urbanization in Nigerian cities (the case of Lagos and Ota). Int J Archit Urban Dev 3(4):19–26
67. Schweitzer L, Zhou J (2010) Neighborhood air quality outcomes in compact and sprawled regions. J Am Plann Assoc 76(3):363–371
68. Seto KC, Sánchez-Rodríguez R, Fragkias M (2010) The new geography of contemporary urbanization and the environment. Annu Rev Environ Resour 35:167–194
69. Shittu AO (2010) The need for inter-agency collaboration in urban infilling. Maiduguri J Arts Soc Sci (MAJASS, UniMaid) 8(2):234–245
70. Smith M (2013) Project densification saves how much farmland? Planned densification and urban betterment [online]. Available from: http://www.planneddensification.com/2013/06/04/project-densification-saves-how-much-farm-land/. Accessed 13 Sept 2019
71. Taiwo OJ, Abu-Taleb KA, Ngie A, Ahmed F (2014) Effects of political dispensations on the pattern of urban expansion in the Osogbo Metropolis, Osun state, Nigeria. In: Proceedings of the 10th international conference of AARSE, pp 242–251
72. Talen E, Wheeler SM, Anselin L (2018) The social context of U.S. built landscapes. Landscape Urban Plann 177:266–280
73. Tang B (2017) Is the distribution of public open space in Hong Kong equitable, why not? Landscape Urban Plann 161:80–89
74. Thin N (2012) Social happiness: theory into policy and practice. Policy Press, Bristol
75. Wahab B, Abiodun O (2018) Strengthening food security through peri-urban agriculture in Ibadan, Nigeria. Ghana J Geogr 10(2):50–60. https://doi.org/10.4314/gig.v10i2.4
76. Wang L, Omrani H, Zhao Z, Francomano D, Li K, Pijanowski B (2019) Analysis on urban densification dynamics and future modes in southeastern Wisconsin, USA. PLoS One 14(3). https://doi.org/10.1371/journal.pone.0211964
77. Wei YHD (2012) Restructuring for growth in urban China: transitional institutions, urban development, and spatial transformation. Habitat Int 36:396–405
78. Xu G, Dong T, Cobbinah PB, Jiao L, Sumari NS, Chai B, Liu Y (2019) Urban expansion and form changes across African cities with a global outlook: spatiotemporal analysis of urban land densities. J Clean Prod. https://doi.org/10.1016/j.jclepro.2019.03.276
79. Zeng C, Liu Y, Stein A, Jiao L (2015) Characterization and spatial modeling of urban sprawl in the Wuhan Metropolitan Area, China. Int J Appl Earth Obs Geoinf 34:10–24

80. Zhao P (2013) The impact of urban sprawl on social segregation in Beijing and a limited role for spatial planning. Tijdschrift voor Economische Sociale Geografie 104(5):571–587
81. Zhu J (2012) Development of sustainable urban forms for high-density low-income Asian countries: The case of Vietnam. Cities 29:77–87
82. Zolnik EJ (2011) The effects of sprawl on private-vehicle commuting distances and times. Environ Plan 38:1071–1084

Environmental Planning in Mass Housing Schemes: Strategies for Achieving Inclusive and Safe Urban Communities

Foluke O. Jegede, Eziyi O. Ibem, and Adedapo A. Oluwatayo

Abstract The layout of plots, arrangement of buildings and management of spaces between buildings in residential neighbourhoods are vital components of the planning and designing of housing schemes. However, there is insufficient empirical evidence on how these environmental planning and architectural design strategies can contribute to security of lives and property in mass housing schemes, especially in a developing country like Nigeria that is confronted with different kinds of security challenges. This chapter presents and discusses the findings of a study conducted to examine the influence of residential neighbourhood planning and design of housing units on the security of lives and property in 12 selected public mass housing estates developed by the Lagos State Development and Property Corporation (LSDPC) in Lagos Metropolis, Nigeria. The data were derived from a household survey involving 1036 residents in different LSDPC housing estates in the Lagos metropolis. The results of the descriptive statistics and content analysis of the data reveal the predominant layout patterns of the estates, the design and construction features of the housing units and spaces, and how these have influenced residents' perception of security of lives and property in the residential estates. The results also show aspects of neighbourhood planning and design with the most significant influence on security of lives and property in the estates investigated. Moreover, the study identifies areas that need to be strengthened by housing experts, developers, urban designers and managers to ensure that mass housing schemes in rapidly growing cities are secure for residents, thereby contributing to the attainment of goal 11 of the Sustainable Development Goals (SDGs) in Nigeria.

Keywords Lagos · Mass housing · Neighbourhood design · Sustainable development goals · Security · Urban neighbourhood

F. O. Jegede (✉) · E. O. Ibem · A. A. Oluwatayo
Department of Architecture, Covenant University, Ota, Nigeria
e-mail: foluke.jegede@covenantuniversity.edu.ng

© The Author(s), under exclusive license to Springer Nature Singapore Pte Ltd. 2021
T. G. Nubi et al. (eds.), *Housing and SDGs in Urban Africa*, Advances in 21st Century Human Settlements, https://doi.org/10.1007/978-981-33-4424-2_17

1 Introduction

Goal 11 of the United Nations Sustainable Development Goals (SDGs) seeks to make cities and human settlements inclusive, safe, resilient and sustainable for all categories of people by the year 2030 irrespective of gender, age, socioeconomic status and political, as well as religious affiliations. Following from this, different professions have been seeking ways to contribute their quota to the realisation of this goal. Not left out in the quest to achieving a better and more sustainable future for all are the built environment planners, as well as architecture and design professions. In line with this, housing designers and developers have come to realise that the manner and ways in which residential environments are planned, designed, constructed and managed have influence on the level of security and safety of people who live in or use such environments. Indeed, there is no gainsaying the fact that in planning residential neighbourhoods, especially mass housing schemes or estates, adequate attention must be paid to layout in order to ensure the reasonable creation of an ambience of sanity, beauty, safety and ease of mobility.

In many cities, residential developments take a large portion of land use, providing different categories of housing for the teeming population. Therefore, in planning residential neighbourhoods in contemporary cities, attention is given not only to the placement of houses but also the effectiveness of such placements in relation to access to services and security of residents. Residents' activities and lifestyles within a housing environment sometimes give rise to crime and other antisocial behaviours. To achieve a reasonable level of security, the layout of a residential environment comprises all the entities that make up the surroundings, including infrastructural facilities such as roads, walkways, landscape, car parks, footpaths, gates, open spaces and gardens must be given adequate consideration. The arrangement of buildings, spaces and other components within residential neighbourhoods can be achieved through site layout designing and planning that give careful consideration to building scale and forms, as well as movement patterns, the external spaces available and the interrelationship between them.

One of the factors that affect the functionality and conduciveness of any residential neighbourhood is the extent to which residents feel secured within the environment. In fact, UN-HABITAT [28] identifies security as a vital aspect of adequate housing provision and describes its absence in any residential development as housing inadequacy that might cause social problems and political instability. This is of course understandable, considering that security is one of the most fundamental needs of man in Maslow's hierarchy of needs[12, 18]. Therefore, security is a major factor to be considered in the design of residential neighbourhoods if goal 11 of the SDGs must be achieved in sub-Saharan Africa.

Insecurity is considered as a social problem that is daily affecting millions of people across the world. According to Roh et al. [25], the fear of crime, which is a manifestation of insecurity, has several negative effects in the community. Among others, it can cause restriction to people's freedom of movement and can also prevent them from participating fully in several community activities, which will ultimately constitute a barrier to social equity. This implies that if crime is not curtailed within neighbourhoods it may hinder social and economic activities [13, 14]. In view of this, there is a need to explore various design and planning and strategies that can be used to enhance the security of lives and property in urban communities. This is particularly important because the review of extant literature shows that this aspect of housing research has not received adequate attention, especially in developing countries.

Consequently, this chapter discusses and presents the findings of a study on how residential neighbourhood design and planning, as well as the layout of housing units, tend to influence security of lives and property in twelve (12) selected public mass housing developments by the Lagos State Development and Property Corporation (LSDPC) in the Lagos metropolis, Nigeria. Assessment of the extent to which the security needs of residents in the existing public housing schemes has been met will inform design and management practices on the specific spatial planning and architectural design strategies that can enhance overall security of lives and property in urban neighbourhoods. It will also create awareness and re-awaken the consciousness of built environment and construction professionals on the need to incorporate security design measures in mass housing schemes, with a view to contributing to the attainment of inclusive, safe, resilient urban communities in Nigeria and other countries in the global South.

2   Theoretical Framework

There is increasing emphasis on sustainable architecture, especially in the adoption of architectural design principles, practices and design measures that ensure the safety of people and security of property while also minimising the negative environmental impact of buildings. This suggests that there is a relationship between the design of houses and security, particularly in the arrangement of spaces within dwelling units, location and sizes of fenestrations and the relationships between internal and external spaces. All these aspects of housing design are very important in the attainment of goal 11 of the SDGs and must be carefully examined in the planning and design of residential neighbourhoods. Moreover, how these aspects of housing design are handled is determined by the need to ensure space optimisation, energy efficiency and functionality of spaces within housing units. Therefore, in addition to ensuring security of lives and property, contemporary housing design practice is concerned with measures that relate form to site and climatic conditions while also achieving a harmonious and having a lasting relationship between inhabitants and their surrounding areas [15].

In line with such thinking, the idea of defensible space emerged. Defensible space is a model type for residential environment design that inhibits crime by creating a physical expression and social fabrics that defend it [19]. The different elements specified in defensible space model, have a common goal of creating an environment in which latent territoriality is created making sure there is a safe, productive, pleasing and well-maintained living space. A potential criminal minded person, perceives the living spaces controlled by residents, as areas that are not favourable to crime because unauthorised visitors or users of such places are viewed as intruders to be dealt with [19]. Covington and Taylor [7] as well as Cozens et al. [8] have argued that the defensible space phenomenon is a strategy for crime prevention and a means of controlling crime when adequately implemented. Based on these, architectural design and physical planning principles are gaining recognition as tools for achieving secured built environments that allow people to live peacefully and participate effectively in socioeconomic, political and religious activities. In support of this assertion, Newman [19, 20] explained that a good design might help people to develop a sense of ownership and responsibility within their places of residence, thus encouraging them to defend such places and make them safe for habitation human habitation.

3   Crime in Residential Areas

Especially in today's rapidly growing cities, crime has become an alarming reality of daily life, hence the need for development experts and governments across the world to pay more attention to the security needs of city residents. Discussions on this subject have traditionally focused more on the arrest and punishment of criminals, rather than on crime prevention strategies. In Europe, for example, the government had noted the failure and inability of the British Criminal Justice System (CJS) existing throughout the twentieth century to contain or reduce crime [8]. This issue continues to be highly controversial and this area of research hotly debated. In Britain, since 1918, there has been increase in crime, with an average of 5.1 percent increase per year and it's reaching 4.5 million cases in 1997 [8]. The situation is the same in America and other countries, including those in sub-Saharan Africa, where criminality and terrorism have combined to worsen security challenges.

Notably, the most obvious and visible consequences of urbanisation waves experienced in most developing countries, including Nigeria, is the rapidly deterioration in urban housing provisions and living conditions of the citizens and residents [17, 22]. Residents of most of Nigeria's urban areas worry about insecurity and criminal activities. In fact, violent crimes and property crimes within residential areas in towns and cities include burglary, robbery, larceny, assault, auto theft, drug trafficking, vandalism as well as abuse and rape. According to Olotuah [22], one of the reasons for this state of insecurity is that urban centres in Nigeria have large numbers of people who earn very low wages and experience irregular employment, leading to criminal thoughts and actions.

Crime in residential neighbourhoods in urban areas does not necessarily depend on the type of urban area but the quality of housing there. This suggests that the prevalence of crime may be dependent on factors of housing quality such as building design, environment and space layouts. Crime control and crime prevention strategies in urban environments are often hampered by poor building layout and design, poor

Table 1 Fear of crime in Lagos 2013–2015

Level of crime	82.69	Very high
There is increasing crime in the past 3 years	70.69	High rate
Fear of home broken and things stolen	74.14	High rate
Fear of being mugged or robbed	78.70	High rate
Fear of car stolen	72.41	High rate
Fear of my belongings taken from car	75.00	High rate
Fear being attacked by hoodlums	70.83	High rate
Fear of being insulted	68.75	High rate
Fear of being subject to a physical attack because of my skin colour, ethnic origin or religion	62.50	High rate
Fear of drug addicts or abusers	73.21	High rate
Fear of property crimes such as vandalism and theft	75.89	High rate
Fear violent crimes such as assault and armed robbery	81.03	Very high
Fear of problems of corruption and bribery	90.52	Very high

Source Numbeo [21]

circulation planning and undesignated utility space allocations such as drainages and refuse dumps. Recent increases in criminal activities also point to the urgent need for information on how best to ameliorate people's anxiety and the threats posed by the increasing rate of crime in Nigeria, especially within the residential areas [3].

Numbeo [21] provides some data showing Lagos residents' perception of some common criminal activities within their localities from 2013 to 2015. As Table 1 shows, a large number of residents live in fear of crime. In fact, most of the survey participants reported high rates of all the 13 criminal activities investigated. Despite the passage of time since the study was conducted, its data offer insight into the type of criminal activities that are prevalent in the residential areas of Lagos. This suggests that insecurity of lives and property constitutes a major threat to sustainable development in the city of Lagos and that this might be the case in several other cities and urban areas in Nigeria.

4   Design Strategies for Security in Residential Environments

The housing environment comprises housing units, housing unit support services and the neighbourhood environment. The housing units are the physical structures (buildings), which include design and structural features such as the house type and the spaces within the house. Housing support services include basic services such as electricity, water supply, domestic waste management facilities and gas supply, while the neighbourhood facilities include educational, healthcare, recreational and shopping facilities, as well as other housing infrastructures such as roads, drainages, waste disposal facilities and others. All these constitute the bundle of housing that guarantee security, a control, a sense of attachment, a permanence and having continuity [26].

In the mass housing environment of most countries, security challenge has existed for many years [30]. In the context of the USA, Newman [20] suggests that there is a link between crime and mass housing schemes developed by government. Based on this, the author recommends extensive measures to eliminate crime from such environments through rebranding of the physical appearances of such housing developments. On their part, Ready et al. [23] suggest starting social programmes for residents in such environments. Their suggestion emanated from evidence in selected housing developments in Philadelphia and Jersey City in the USA, where health codes were aggressively enforced and city ordinances were used to check drug addiction and incivilities, leading to a reduction in crime levels.

These developments underscore the importance of security of lives and property in planned mass housing environments. Ready et al. [23] presented a model of a problem-solving approach to checking crime in planned mass housing environments. The authors explained that the traditional policing approach, which involve arrests and prosecution, have not been effective in checking crime. This view was corroborated by Vilkinyte [29], who also argues that traditional policing methods alone cannot effectively tackle the issue of security in planned mass housing environments.

Key features in the design of a residential housing environment, especially in public housing schemes, include car parks, greenery, footpaths, walkways, open spaces and gates. These features, as highlighted in Lancashire County Council [16], should be designed to ensure security within the neighbourhood. For instance, parking spaces should be visible from some of the dwellings, while communal parking areas should be in small groups and open to views in addition to being well-lit. Where there is a gate, it should be positioned as close to the front elevation as much as possible in order to deny would-be intruders a hiding place for perpetrating crime. In addition, landscape elements should not obstruct opportunities for natural surveillance, while the frontage of houses should be in open view with walls and hedges not higher than one meter. Trees and plants should not be those that may grow to obscure street lighting or serve as ladders for criminally minded persons. In the same vein, footpaths should be well-lit and serve the circulation needs of residents without providing unnecessary access to non-residents.

5 Spatial Layout of Residential Neighborhoods and Security

Notably, one of the objectives of a good mass housing design is to create a secure neighbourhood environment, which will present a less attractive target for criminals.

Goodchild [11] argues that one of the key imperatives of sustainable residential development is the use of spatial planning layouts in the planning of residential areas. Having a good design and layout for residential estates requires proper design of features that contribute to the security requirements of the dwelling units and support infrastructure [11]. This means that adequate attention should be given to the planning of residential neighbourhoods if security of lives and property must be achieved.

There are different types of spatial layouts for the planning of residential neighbourhoods [24]. Each layout has its own unique features and security implications. Without a doubt, the layout of individual housing units in residential neighbourhoods or estates contributes to security [4], hence there is a need to consider the various layout options available when developing mass housing schemes. Figures 1, 2, 3 and 4 are graphical representations of the different types of residential street layout patterns, such as the grid pattern (gridiron patterns), curvilinear loop pattern and the conventional cul-de-sac pattern. These three patterns also come with their variations and include the square grid, oblong grid, radial grid, irregular grid, loops 1 and 2, as well as the cul-de-sac redburn patterns (see Figs. 2 and 4). Figure 4 shows a typical example of estate planning using the cul-de-sac layout pattern.

Furthermore, development of sites should contribute to the creation of public and private open spaces that are accessible, safe and attractive and is comfortable for its users [11, 27]. In a study of different classes of residential neighbourhoods in China, Shu [27] found links between the urban areas, spatial layouts of housing estates and the spatial distribution of property crimes. The author emphasised the importance of territoriality and defensible space and suggested the need for the inclusion of other parameters. Shu's finding revealed that property crimes are more in areas that are locally segregated such as footpaths, cul-de-sacs and rear dead-end alleys, thus suggesting that designs incorporating lonelier roads or dead ends encourage crime, as there is less surveillance and territoriality there.

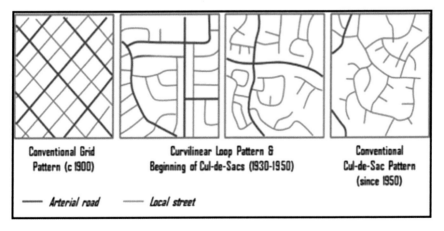

Fig. 1 The main layout patterns of residential estates. *Source* Research Highlights [24]

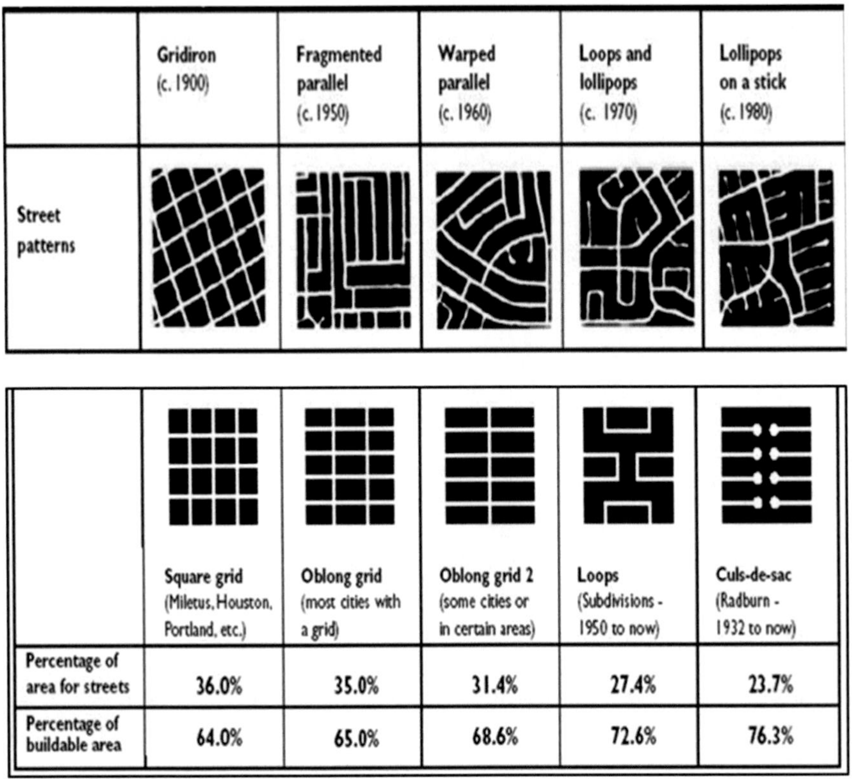

	Gridiron (c. 1900)	Fragmented parallel (c. 1950)	Warped parallel (c. 1960)	Loops and lollipops (c. 1970)	Lollipops on a stick (c. 1980)
Street patterns					
	Square grid (Miletus, Houston, Portland, etc.)	Oblong grid (most cities with a grid)	Oblong grid 2 (some cities or in certain areas)	Loops (Subdivisions - 1950 to now)	Culs-de-sac (Radburn - 1932 to now)
Percentage of area for streets	36.0%	35.0%	31.4%	27.4%	23.7%
Percentage of buildable area	64.0%	65.0%	68.6%	72.6%	76.3%

Fig. 2 The variations to the grid layout patterns of residential streets. *Source* Research Highlights [24]

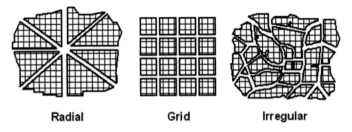

Radial   Grid   Irregular

Fig. 3 The variations to the grid layout patterns of residential streets. *Source* Research Highlights [24]

Based on this finding on the surveillance issues with cul-de-sacs (i.e. streets that are closed at one end), some scholars are strongly in support of the traditional grid design of layouts as against the curvilinear loop design and cul-de-sac design. Badger [5], for examples, explained that most of the oldest cities in America, Europe and Roman Empire were laid out in a neat, densely interconnected grids that enabled

Fig. 4 The variations to the grid layout patterns of residential streets. *Source* Research Highlights [24]

people to get around easily and this was before cars came to be a hindrance to movements. Badger [5] argued that newer residential layouts and designs using cul-de-sacs were merely building communities for cars, not people. This makes it clear that street design is for humans, not for objects or the movement of these objects. In the USA, a study of City of David which is laid out in the grid pattern, found a city free from road crashes, as residents rely more on motorbikes to move around the city [5]. This is possible because of the traditional monotonous grid city design of the city.

6 Research Methods

Conducted in July 2017, this study adopted the survey method, which involved the use of a questionnaire to extract data from residents of selected public housing estates developed by the LSPDC in the Lagos metropolis. In addition to the use of structured questions as the principal data collection instrument, the researcher also deployed observation and photographic materials in generating data.

The research population consisted of selected LSDPC housing estates from both the Island and the Mainland areas of Lagos in the 16 Local Government Areas of Metropolitan Lagos. The housing estates studies were constructed between 1979 and 1992 and the dwelling units included both the low-cost housing (comprising 13,644 housing units) and the medium-income housing (with 916 units). In total, the study covered 14,560 housing units in 12 residential estates (see Table 2).

The research adopted the Yamane [31] formula for finite population to estimate the minimum sample size for the survey. This formula has been used by previous

Table 2 The percentages of housing units sampled relative to the total population in each housing estate

S. No.	Location	No. of housing units and households ($N =$ 14,560)	The percentage of the total population (%)	Sample size for each estate (housing units)
1	Abesan	4272	29.3	299
2	Amuwo-Odofin	2068	14.2	144
3	Dairy farm/ijaiye	708	4.8	52
4	Dolphin/Ikoyi	576	3.9	40
5	Iponri	1026	7.0	72
6	Isolo	3664	25.1	257
7	Ojokoro	534	3.6	41
8	Ijaiye/Ogba Phase II	824	5.6	65
9	Opebi/Maryland	120	0.8	8
10	Ebute-Meta	528	3.6	41
11	Alapere	140	0.9	10
12	Omole	100	0.6	7
	Total	14,560	100	1036

Source Jegede et al. [13, 14]

researchers [1, 2, 6, 12]. The level of precision is 3%, with the level of confidence or risk put at 97%, using the formula:

$$n = \frac{N}{1 + N(e)^2}$$

where n = sample size for finite population, N = population size = 14,560, e = level of precision expressed as a proportion of (0.03). The minimum sample size obtained was 1036 housing units, from which the survey respondents were selected.

The questionnaire was designed to elicit responses from estate residents on their perception of general level of security of lives and property in their residences. Questions were also asked on their perception of 12 planning features of the housing estates and how these influence general security, using a 5-point Likert-type scale of '1' for *Strongly Disagree*, '2' for *Disagree*, '3' for *Not Sure*, '4' for *Agree* and '5' for *Strongly Agree*. Similarly, the residents were asked to indicate their level of satisfaction with security in the housing estates using a 5-point Likert-type scale ranging from '1' for *Very Dissatisfied* to '5' for *Very Satisfied*. For questionnaire administration, the random sampling technique was used to select the household head or one adult member in the selected 1036 housing units in the 12 residential estates. One copy of the questionnaire was administered per housing unit selected for the study. Observations were also made on the design and layouts of each of the 12 housing estates studied and findings were documented using observation schedules and photographic materials. The quantitative data collected via the questionnaire were analysed using descriptive statistics such as percentages and frequencies, as well as categorical regression analysis. The qualitative data were subjected to content analysis based on the site layout design of each of the 12 estates.

7 Spatial Planning Features of the Residential Estates

Table 3 shows a summary of the features of the 12 estates investigated. The significant features with security implications are zoning, layout pattern, availability of open spaces, road network/walkways, availability of blind spots, blockage of streets and open spaces around the buildings. The data on these features were collected using observation schedules and photographic materials. From Table 3 it can be seen that eight of the 12 estates have some level of zoning of land use, meaning that the estates benefitted from proper zoning, while the layout pattern is predominantly gridiron. It was also observed that in spite of the zoning of facilities in the estates, the layouts appear to be crowded with few showing evidence of well-planned residential neighbourhoods. Moreover, in most of the estates, open spaces were found but were not properly taken care of, while some of the existing roads and walkways were either in good condition or in disrepair. Although several blind spots were found in most of the estates, cases of blockage of streets, roads and alleys were minimal and the spaces around the buildings in most of the estates appeared to be inadequate when compared to the sizes of the buildings.

Considering that all these features have diverse security implications, it was important to understand the views of estate residents on this. The study findings show that around 52.52% of the respondents were satisfied with the general level of security of lives and property in the estates. A total of 16.12% were dissatisfied and 39.94% indicated that they were not sure of their level of satisfaction with estate security. Table 4 contains a summary of the results on residents' perception of the planning attributes of estates and how they influence security of lives and property. The results (Table 4) show that most of the participants in the survey indicated that the planning of the estates was largely not a source of security concern to them.

Furthermore, the layout and arrangement of buildings within the estates seem to promote surveillance, as the buildings are arranged in rows facing each other (see Figs. 5 and 6). Although some of the estates have very large numbers of buildings, housing units and blocks, the large number appears not to have any effect on residents' security. This is probably because the estates are gated communities, since they have perimeter fences and mini gates at the entrance of each building (see Figs. 7 and 8).

Table 3 Summary of the characteristics of the housing estates

Name of Estates	Investigated attributes of the planning, design and layout of the estates						
	Zoning	Layout pattern	Open space/greens	Road type/demarcation of walkways	Blind spots	Blocking of street/alleys	Spaces around building
Abesan	Present	Grid/crowded	Available	Disrepair/present	Yes	No	Inadequate
Dairy	Present	Grid/spacious	Available	Good/present	Yes	Yes	Adequate
Dophin/Ikoyi	Absent	Grid/well planned	None	Good/present	Yes	No	Inadequate
Ebute-meta	Absent	Grid/well planned	Available	Good/present	No	No	Inadequate
Iponri	Present	Cul-de-sac/Haphazard	None	Disrepair/present	Yes	Yes	Inadequate
Ojokoro	Present	Grid/well planned	Available	Disrepair/present	Yes	No	Inadequate
Alapere	Absent	Grid/spacious	Available	Good/present	No	No	Adequate
Omole	present	Cul-de-sac/spacious	Available	Disrepair/present		Yes	Adequate
Opebi	Absent	Grid/well planned	Available	Disrepair/present	Yes	No	Inadequate
Isolo	Present	Grid/crowded	Available	Disrepair/present	Yes	No	Inadequate
Amuwo-Odofin	Present	Grid/crowded	None	Disrepair/present	Yes	Yes	Inadequate
Ijaiye/Ogba	Present	Grid/well planned	Available	Good/present	Yes	No	Adequate

Source Authors' fieldwork (2017)

8 Influence of Planning on Security in the Housing Estates

To investigate the influence of the planning attributes of estates on residents' perception of security of lives and property in the estates, Categorical Regression Analysis was conducted using the overall security of lives and property as the dependent variable and the 12 planning attributes as the independent variables. The result shows $R^2 = 0.146$ (F 37, 998 = 7.545, $p = 0.000$), indicating that the regression model explained about 14.6% of the variance in perceived level of security in the estates.

Table 4 Residents' perception of planning features and security in the housing estates

Variables	Strongly disagree	Disagree	Not sure	Agree	Strongly agree
Number of housing units in a plot estate makes me feel unsafe	124 (23.8)	367 (35.4)	147 (14.2)	124 (12.0)	21 (2.0)
Planning of the estate is a source of insecurity	247 (26.2)	337 (35.8)	225 (23.9)	107 (11.4)	25 (2.7)
Planning of estate does not encourage knowing people easily	215 (22.8)	335 (37.9)	195 (20.7)	142 (15.1)	34 (3.6)
Size of estate does not encourage communal living	190 (20.5)	371 (40.0)	213 (23.0)	120 (12.9)	34 (3.7)
Road network discourages free movement day and night	225 (24.1)	406 (43.4)	148 (15.8)	121 (12.9)	35 (3.7)
Quality of road and walkways constitute security risk	190 (20.0)	357 (37.6)	233 (24.6)	135 (14.2)	33 (3.5)
Location of building makes it easy committing crime unnoticed	199(20.9)	323 (33.9)	243 (25.5)	147 (15.4)	41(4.3)
Location of open space/recreational facilities instils fear in intended users	168 (18.0)	345 (36.9)	265 (28.3)	130 (13.9)	26 (2.8)
Location of house brings feeling of not being safe	262 (28.1)	399 (42.7)	140 (15.0)	109 (11.7)	24 (2.6)

(continued)

Table 4 (continued)

Variables	Strongly disagree	Disagree	Not sure	Agree	Strongly agree
Number of housing units makes it unsafe for residents	248 (27.3)	366 (40.4)	146 (16.1)	125 (13.80	21 (2.3)
Adequate open space and green areas in this housing estate	110 (10.6)	160 (15.4)	222 (21.4)	385 (37.2)	90 (8.7)
Estate is zoned into different section	64 (6.2)	62 (6.0)	134 (12.9)	459 (44.3)	222(21.4)

Note The figures outside the brackets are frequencies; those in brackets are percentages

Source Authors' fieldwork (2017)

Fig. 5 Buildings arranged in rows facing the roads at Ojokoro and Ebute-Meta Estates for easy viewing of persons walking on the road and the use of kerbs to demarcate the car parks from green areas

Fig. 6 Buildings arranged in rows facing the roads at Ojokoro and Ebute-Meta Estates for easy viewing of persons walking on the road and the use of kerbs to demarcate the car parks from green areas

Fig. 7 Design layout to include a manned entrance gate at Ijaiye/Ogba Low-cost housing estate and mini gate at each car park for security purpose at Ijaiye/Ogba estate respectively

Figure 8 Design layout to include a manned entrance gate at Ijaiye/Ogba low-cost housing estate and mini gate at each car park for security purpose at Ijaiye/Ogba estate respectively

Table 5 shows that four out of the 12 variables investigated are significant predictors of the general level of security as perceived by the residents. Based on the β values, the planning attribute with the most significant predictor of security in the residential estate is the number of housing units built in a plot (β = 0.208, F = 15.079, p = 0.000), followed by adequacy of open space and green areas (β = 0.154, F = 15.888, p = 0.000), location of open spaces/recreational areas instilling fear in intended users (β = 0.152, F = 11.246, p = 0.001) and planning of the estate (β = 0.119, F = 7.154, p = 0.001).

Residents in the estates described planning as one of the factors contributing to the level of security, thus suggesting that respondents were more aware of the role of planning in achieving security in the estates. This seems to agree with the suggestion of Goodchild [11] that good layouts and planning are sacrosanct in achieving security within the built environment. The gridiron type of estate layout is identified as a factor that affects the level of security in the estates investigated. This type of layout encourages the effectiveness of territoriality, accessibility and natural surveillance, which are elements of the defensible space phenomenon as identified by Newman [19]. It gives residents an easy view of exterior spaces, roads and streets from within their houses. Long stretches of road are easily visible in most of the estates studied. As Shu [27] observes, property crime is more common in locally segregated areas such as cul-de-sacs, footpaths, rear dead-end alleys and lonely roads or dead ends, since such designs encourage crime owing to less surveillance and territoriality. Prevalence of the grid layout design, as against the cul-de-sac or curvilinear loop layout design in estate planning, enables easy movement of people and cars within neighbourhoods, as confirmed by Badger [5] with regard to the US neighbourhood of the City of David.

Also important is the finding that the adequacy of open spaces in the estates is a significant factor that influences security in the estates. Table 3 reveals that the open spaces around the buildings were inadequate, a fact that was also confirmed by the residents in Table 4, with about 46% of them agreeing that open spaces in the housing estates were inadequate.

Considering that having well-maintained open spaces in residential areas promotes territoriality, surveillance and communal activities, thereby improving security and inclusiveness through provision of recreational areas for all categories of people, it can be inferred that adequate provision of open spaces contributes to the creation of sustainable residential neighbourhoods in cities. Related to this is the location of open spaces and recreational facilities. The role of open spaces and recreational facilities in promoting good health and well-being, as well as social interaction and security, is undeniable. However, to ensure that this is achieved, open spaces that are meant for recreational purposes should be properly located in areas that are not isolated and therefore do not instil fear in users. On the contrary, the location, planning and design of such spaces should instil fear in criminally minded persons and thus deter them from perpetrating crime.

The number of houses in a plot also influences level of satisfaction with security in the estates, since the number and size of buildings on a plot of land determines the level of crowding in residential neighbourhoods. The more the number of buildings and the larger the sizes of the buildings, the less the amount of open spaces that will be available around the buildings. Consequently, this will influence residents' perception on the link between rowdiness and insecurity within and around their homes. The results show that residents believed that number of buildings on plots did not make them feel insecure; therefore, this may explain why many of them expressed satisfaction with security in the housing estates investigated.

9   Implications of Findings and Conclusion

This study investigated the influence of neighbourhood planning and design features on the security of lives and property in mass housing environments such as public

Table 5 Coefficients of the regression analysis to identify factors that predict the level of general security in the housing estates

Coefficients

Variables	Standardized coefficients		df	f	p
	Beta	Bootstrap (1000) estimate of std. error			
Number of housing units and feeling safe	−0.023	0.071	4	0.110	0.979
Planning of the estate is a source of insecurity	−0.119	0.045	2	7.154	0.001*
Planning of estate does not encourage knowing people easily	−0.045	0.072	2	0.388	0.678
Size of estate does not encourage communal living	−0.041	0.071	2	0.336	0.715
Road network discourages free movements day and night	−0.043	0.078	1	0.303	0.582
Quality of road and walkways as security risk in the estates	−0.056	0.058	2	0.936	0.393
Location of building and committing crime unnoticed	−0.048	0.079	2	0.368	0.692
Location of open space /recreational facilities instil fear in intended users	0.152	0.045	1	11.246	0.001*
Location of house and feeling of not being safe	−0.081	0.080	2	1.021	0.360
Number of housing units in a plot and security	0.208	0.054	2	15.079	0.000*
Adequacy of open space and green areas	0.154	0.039	2	15.888	0.000*
Zoning of the estates	−0.036	0.063	1	0.324	0.569

*Significant predictors
Dependent Variable Satisfaction with general security

housing estates. The essence is to highlight the role that environmental design strategies can play in achieving safe and inclusive urban communities in rapidly growing cities. The findings of the study suggest that many residents were satisfied with the level of security of lives and property in the study areas. The study also established a relationship between the architectural and spatial planning of the estates and residents' satisfaction with level of security there. Among the key aspects of the estates with significant influence on the level of satisfaction were number of housing units on a plot, adequacy of open spaces and green areas, spatial layout of the estates, and whether the location of open spaces/recreational areas instils fear in intended users.

These findings have implications for practice, especially for architectural design and planning strategies for attainment of goal 11 of the Sustainable Development Goals (SDGs) in Nigeria. Specifically, the findings suggest that environmental planning and design strategies, when adequately engaged, can promote sustainable urban communities by ensuring that residents are fully engaged in securing their environments against the activities of criminals. As the literature suggests, the fear of crime is inimical to inclusive and safe neighbourhoods because it prevents people from participating effectively in socioeconomic and political activities. Therefore, to ensure inclusive and safe residential neighbourhood environments, architects, urban designers and planners, as well as housing experts, should pay attention to the overall layout of mass housing schemes. As much as possible, the gridiron layout should be preferred to other patterns, since it ensures visual surveillance in housing schemes as this would have effect on the overall urban planning, as seen in most cities designed since the renaissance era in places such as India, Egypt, Asia Minor, Hippodamic Greece, Rome and Central American civilization; San Francisco and Manhattan, as described by Crawford [9]. The suggestion in the adoption of gridiron layout pattern is based on the evidence that it have no chaotic and unplanned development features. This may be explained as mankind's generic urge for order and regularity in contrast to having a chaotic growth that nature will provide.

The positive impact of grid pattern cannot be over emphasised, as it is characterised by easy linkages of buildings, streets and roads, which supports social interactions and exchange within neighbourhoods, helping in the security of neighbourhoods and its creates safe, quiet and healthy environment. An example of such is the Land Ordinance of 1785 in the downtown grids of Chicago and many U.S cities that are oriented on a true north–south axis [10]. Another suggestion from this study is that the minimum built-up area within a plot should be maintained to prevent overcrowding of neighbourhoods so that adequate open spaces are provided for green infrastructure and recreational facilities. Finally, it is also suggested that recreational areas should be located in areas that are easily accessible and do not instil fear in any category of intended users. The effects of green infrastructures and recreational facilities can soften and civilize public urban spaces, and as a result the thought of crime can vanish from intending criminals within this in place.

References

1. Adebayo MA (2006) The state of urban infrastructure and its effects on property values in Lagos, Nigeria. J Land Use Develop Stud 2:50–59
2. Adewale BA (2014). Assessment of core area housing in Ibadan, Oyo State, Nigeria. Unpublished doctoral dissertation submitted to the School of Postgraduate Studies, Covenant University, Ota, Nigeria

3. Agbola AG (2013) Nigeria crime index assessment and classification. J Appl Secur Res 8:404–418
4. Akinjiyan A (2014) Six types of residential estate housing schemes in Nigeria. Real estate tutorial. https://www.Realestatesurveyor.blogspot.com/2014/02/6-types-of-residential estate-housing.html. Accessed 22 Oct 2015
5. Badger E (2015) Debunking the Cul-de-sac; The design of American's suburb has actually made our streets more dangerous. https://www.printerest.com.au. Accessed 22 Oct 2015
6. Bartlett JA, DeMasi R, Quinn J, Moxham C, Rousseau F (2001) Overview of the effectiveness of triple combination therapy in antiretroviral-naive HIV-1 infected adults. Aids 15(11):1369–1377
7. Covington J, Taylor BR (2013) Fear of crime in urban residential neighbourhoods: Implications of between–and within–neighbourhood sources for current models. Sociol Q Wiley Online Libr 32(2):231–249
8. Cozens P, Hillie D, Prescott G (2001) Crime and the design of residential property exploring the theoretical background. Property Manage 19(2):136–164
9. Crawford JH (2005) A brief history of urban form. Street layout through the ages. https://www.carfree.Com. Accessed 12 Dec 20 16
10. Ellickson RC (2013) The law and economics of street layouts: how a grid pattern benefits a downtown. Ala L Rev 64(3):463–510
11. Goodchild B (1994) Housing design, urban form and sustainable development: reflections on the future residential landscape. Town Plann Rev65(2):143–158
12. Jegede FO, Ibem EO, Oluwatayo AA (2018) Manifestation of defensible space in Lagos state development and property cooperation housing estate Lagos, Nigeria. Int J Civil Eng Technol 9(12):491–505
13. Jegede FO, Ibem EO, Oluwatayo AA (2019a) Resident satisfaction with security in public housing in Lagos Nigeria: the gender perspective. Int J Innov Technol Explor Eng 8(6):375–382
14. Jegede FO, Ibem EO, Oluwatayo AA (2019b) Influence of building maintenance practices on the security of lives and property in public housing in Lagos State, Nigeria. Int J Innov Technol Explor Eng (IJITEE) 9(2):2745–2751
15. Keitsch M (2012) Sustainable architecture, design and housing. Sustain Develop20(3):141–145
16. Lancashire County Council (2015) Reports to the leader of the county council, and cabinet members. https://council.lancashire.gov.uk/. Accessed 22 Oct 2015
17. Lewin AC (1981) Housing co-operatives in developing countries: a manual for self-help in low cost housing schemes. Wiley, London
18. Naghibi MS, Faizi M, Khakzand M, Fattahi M (2015) Achievement to physical characteristics of security in residential neighbourhoods. Proc Soc Behav Sci201:265–274
19. Newman O (1972) Creating defensible space. Washington D.C. U.S Department of Housing and Urban Development. Office of Policy Development and Research. USA
20. Newman O (1973) Defensible space: crime prevention through urban design. Am Polit Sci Rev 69(1):279–280
21. Numbeo (2015) Crime in Lagos, Nigeria 2009–2015. https://www.numbeo.com. Accessed 12 Dec 2017
22. Olotuah AO (2010) Housing development and environmental degeneration in Nigeria. Built Human Environ Rev 3:42–48
23. Ready J, Mazerolle GL, Revere E (1998) Getting evicted from public housing: an analysis of the factors influencing eviction decision in six public housing sites. Crime Prevent Stud 9:307–327
24. Research Highlights (2002) Residential street pattern design. Socio-economic series 75. https://www.realestate.wharton.upenn.edu.ca. Accessed 12 Dec 2015
25. Roh S, Kwak DH, Kim E (2013) Community policing and fear of crime in Seoul: a test of competing models. Polic Int J Police Strat Manage36(1):199–222
26. Sheuya S, Howden-Chapman P, Patel S (2007) The design of housing and shelter programs: the social and environmental determinants of inequalities. J Urban Health 84(1):98–108
27. Shu S (2000) Housing layout and crime vulnerability. Urban Des Int 5(3–4):177–188

28. UN-HABITAT (2006) State of the world's cities. The Millennium Development Goals and Urban Sustainability. Earthscan, London
29. Vilkinyte E (2015) Managing urban development: a case study of urban forest's sense of place. https://www.diva-portal.org/smash/record.jsf?pid=diva2%3A801398&dswid=-3822. Accessed 18 Dec 2017
30. Witte DA, Witte R (2001) What we spend and what we get: public and private of crime prevention and criminal justice. Fiscal Stud 22(1):1–40
31. Yamane T (1967) Statistics: an introductory analysis. Am J Med Biol Res 4(2):20–25

Housing and the SDGs in African Cities: Towards a Sustainable Future

Basirat Oyalowo and Taibat Lawanson

Keywords Keywords: African cities · Housing policy · Sustainable cities · SDGs · Urban Africa

1 Introduction: Towards Urgent Housing Interventions in Urban Africa

Despite the emergence of the Sustainable Development Goals (SDGs) and the pledge by African nations to commit to achieving the set goals and targets, the continent continues to experience housing shortage and slum proliferation, a problem that is compounded by inadequate infrastructure and basic services. These challenges are further exacerbated by poverty, inequality and conflict, especially on the city scale. Governance gaps and the adoption of models that do not reflect the peculiarities of African cultural systems have also impeded the implementation of fit-for-purpose urban processes and plans.

The population growth rates and the urban youth bulge across the continent are ticking time bombs that must be defused urgently. Statistics from the United Nations Department for Economic and Social Affairs [8] indicate that Ouagadougou, the Burkina Faso capital, is expected to experience a 126% urban growth by 2030, while Dar es Salaam (Tanzania), Nairobi (Kenya), Kinshasa (DR Congo) and Luanda (Angola) come behind at 70%. According to the United Nations, 226 million youth

B. Oyalowo (✉)
Department of Estate Management/Centre for Housing and Sustainable Development, University of Lagos, Lagos, Nigeria
e-mail: boyalowo@unilag.edu.ng

T. Lawanson
Department of Urban and Regional Planning/Centre for Housing and Sustainable Development, University of Lagos, Lagos, Nigeria
e-mail: tlawanson@unilag.edu.ng

© The Author(s), under exclusive license to Springer Nature Singapore Pte Ltd. 2021   321
T. G. Nubi et al. (eds.), *Housing and SDGs in Urban Africa*, Advances in 21st Century Human Settlements, https://doi.org/10.1007/978-981-33-4424-2_18

aged 15–24 lived in Africa in 2015, representing nearly 20% of Africa's population and making up one-fifth of the world's youth population. If all people aged below 35 are included, this number increases to a staggering three quarters of Africa's population. Moreover, the share of Africa's youth in the world is forecast to increase to 42% by 2030 and is expected to continue to grow throughout the twentyfirst Century, more than doubling from current levels by 2055 [11]. This has some consequences for both current and future access to urban resources and basic services, including housing.

Across Africa, the backlog of housing units is estimated to be over 51 million, with 17 countries having backlogs of over 1 million units [1] and the prevailing shortage being both qualitative and quantitative. Qualitative shortage manifests in the 50% of the population who live in slummy conditions, while quantitative deficits reflect in the low quality of the few available housing units. It should be noted that the failure to incorporate housing policy considerations in the development of national economic policies has resulted in the poor performance of the sector on the macro scale and as a contributor to national GDP.

2 Housing in Urban Africa: Transiting to a Sustainable Future

In Africa's bid to meet the SDGs, it will be necessary to examine housing from different perspectives. As such, the various chapters in this volume have attempted to address the issue by exploring the continent's housing challenge from the dimensions of governance, finance, urban management, urban health and community development. These chapters have also made various policy recommendations and strategic approaches to addressing these challenges and simultaneously meeting the SDG targets and indicators. Across the various cities and sectors interrogated, the following issues are reiterated:

1. Since housing straddles economic and social aspirations, housing policy should reflect this.
2. Global economic dynamics affect housing sector outcomes.
3. If the housing deficit will be bridged, housing finance needs to be restructured.
4. Housing is not a one-size-fits-all market.
5. Housing and neighbourhood quality are key to addressing environment and health challenges.
6. Local agency is required for sustainable slum upgrading and renewal.
7. Housing needs to be culturally contextual and coherent.

The above points are discussed more extensively below.

3   Need for Housing Policy to Reflect Economic and Social Aspirations

All SDGs are connected to housing, even as the question of whether housing is a public or economic good remains a long-standing one [2]. Consequently, the level of dependency of the state and the private sector in providing affordable and accessible housing needs to be reconciled.

In Chapter "Learning from Experience: An Exposition of Singapore's Home Ownership Scheme and Imperatives for Nigeria", Koleoso and Oyalowo illustrate how housing policy that straddles both macro-economic and social concerns has been successfully adopted. Using a case study of Singapore, which has achieved 90% homeownership in about 50 years, they highlight lessons for African cities, noting that in adopting strategies from other nations there is need to adapt them to local circumstances. They identify this gap as the reason why mortgages have not worked in Africa. According to them, mortgages as currently framed are not practicable due to the inherent exclusion of large segments of the economy who do not satisfy mortgage requirements, as well as the adoption of mortgage models from countries with vastly different socioeconomic and political realities. They further observe the importance of consistency and stability in policy formation and implementation, citing the case of Singapore, where a single agency has been responsible for housing provision for over 50 years, leading to the consolidation of market knowledge and policy consistency that has matured the country's housing sector. Juxtaposing this with the Nigerian experience, where the housing sector has been perpetually subjected to leadership, policy and strategy changes, it is easy to deduce the reason for the sectors' abysmal performance.

In that regard, Olowoyeye's Chapter "The Road not Taken: Policy and Politics of Housing Management in Africa" succinctly captures the situation in recounting how African governments missed the chance to embed housing provision in the continent's economic development priorities. Drawing from history, Olowoyeye begins with the 1963 meeting of the Economic Commission for Africa (ECA) and suggests that the ideological nuances that shape the political economy of housing have failed to identify housing as a veritable pathway to African development. The reader is thus able to deduce that government interventions in the housing sector have repeatedly failed due to poor systemic linkages between housing and economic development at national and sub-national levels.

Providing a wider perspective along the same lines, Chapter "Global Goal, Local Context: Pathways to Sustainable Urban Development in Lagos, Nigeria", by Lawanson, Oyalowo and Nubi, queries the local application of international development agenda and, via documentary analysis, seeks alignment between the goals and targets of SDG 11 and the strategic development policy of Lagos, Nigeria. They argue that where there is a misalignment between national priorities and local needs as is often the case, local governance is ignored and subsequently financing and achievement of goals at that level are neglected. Similarly, in Chapter "Housing and Possible Health Implications in Upgraded Informal Settlements: Evidence from Mangaung

Township, South Africa", Ntema, Anderson and Marais show that the South African government's attempts to align the National Development Plan (NDP) with Agenda 2030 have resulted in actions seeking to deploy health and social amenities as tools to achieve socially and economically integrated communities. However, there are significant gaps in localisation and mainstreaming across sub-national levels and portfolios. There is therefore a need for proper priority setting and systematic integration of the international agenda into local development plans. In the process of achieving this, there is the likelihood of creating strong institutions in tandem with SDG Goal 16—Peace, Justice and Strong Institutions.

4 Global Economic Dynamics and Housing Sector Outcomes

The housing sector is influenced by overarching economic indicators. Whether housing provision is targeted at affordable housing or not, macroeconomic variables weigh significantly on performance, especially in a globalised world where trade liberalisation has facilitated exchange and hence import shocks and stress from one country to others. Thus, in Chapter "Exchange Rate and Housing Deficit Trends in Nigeria: Descriptive and Inferential Analyses", John Agwu urges a recognition of the macroeconomy as a crucial underlying factor in housing deficits in Nigeria and possibly in many African countries. He points out that the housing deficit in the country has historically been associated with population increase and the mismanagement of the oil boom, with inefficiencies in the land tenure system, housing finance, legal procedures surrounding property and land procurement, further compunding the problem. With literature evidence he shows that foreign direct investment, migration, exchange rate, the technology gap, insecurity and climate change have recently begun to feature in the real-estate performance debate. He also identifies the exchange rate as a critical factor directly affecting consumer spending, capital investment, government spending and net export. Moreover, he highlighted the indirect impacts of the exchange rate on the interest rate, mortgage ecosystem, disposable income and government policy, as well as on inflation and rate of capital flows. He then recommends that more direct focus be given to the macroeconomic variables that impact on housing development activities on various scales. Agwu's analysis leads to the broad conclusion that a political-economic environment that supports the fledging housing sector is critical to development agenda such as the SDGs, which advocate social justice, economic development and environmental protection.

While there are wider debates on the appropriateness of neoliberalism as an economic tool in Africa, there are also concerns about the veracity of both micro- and macroeconomic variables, as well as the appropriateness of the policy and governance framework, as explained in the previous chapters. In Chapter "Analysing Hernando de Soto's The Mystery of Capital in the Nigerian Poverty Equation", Akeem Akinwale argues that capitalism as an economic system for alleviating poverty in Africa (with particular reference to Nigeria) is quite askance with the prevailing realities of these nations. Akinwale questions the common citing of de Soto's advocacy for land titling to Nigeria, pointing out the mismatch between policy recommendations meant for Asia, the Middle East and Latin America and those for Africa. This reflects back to discussions in Chapters "Africa's Housing Sector as a Pathway to Achieving the SDGs", Global Goal, Local Context: Pathways to Sustainable Urban Development in Lagos, Nigeria and Learning from Experience: An Exposition of Singapore's Home Ownership Scheme and Imperatives for Nigeria of this volume, all of which caution against policy 'copying and pasting'. Especially in the face of lack of data on the efficacy of de Soto's recommendations (and by extension, other transferred policies), Akinwale insists that there need to be more locally grown policies for addressing the crisis of extreme poverty as evidenced by proliferation of slum and informal settlements in Africa.

5 Bridging the Housing Deficit Through Housing Finance Restructuring

Funding the housing sector (from demand and supply angles) is a key issue in the twenty-first Century human settlements discussion. This is chiefly because, as far as human activities go, there are few other individual asset classes apart from housing that require one form of resource input or the other throughout its life cycle. On this score, Oluwaseun Oguntuase and Abimbola Windapo, in Chapter "Green Bonds and Green Buildings: New Options for Achieving Sustainable Development in Nigeria", establish that green bonds have a strong potential as financial instruments in the actualisation of green structures in Nigeria and possibly the rest of Africa. While directly linked to SDG 11 (Sustainable Cities and Communities), pursuing green finance for housing projects also puts Africa on the pathway to achieving the targets of SDG 12 (Responsible Consumption and Production) and SDG 13 (Climate Action). Citing the International Capital Market Association, Oguntuase and Windapo note that green bonds are debt instruments whose proceeds are specifically applied to finance or re-finance partially or entirely new and existing eligible green projects, thus resulting in environmental benefits.

Thus, with this equitable financing model, services associated with adequate housing and sustainable communities can be better provided at a neighbourhood level. Other projects such as housing regeneration, housing construction, slum upgrades and neighbourhood facilities can also be embarked upon. This will therefore help in addressing SDG 3 (Good Health and Well-being), 6 (Clean Water and Sanitation), 7 (Affordable and Clean Energy) as promoted by Ntema et al. in Chapter "Housing and Possible Health Implications in Upgraded Informal Settlements: Evidence from Mangaung Township, South Africa" and Johnson Falade in Chapter "Housing, Health and Well-Being of Slum Dwellers in Nigeria: Case Studies of Six Cities". Funds can also be directed towards mitigating the consequences of environmental pollution as suggested by Sogbamu et al. in Chapter "Adverse Impact of Human Activities on Aquatic Ecosystems: Investigating the Environmental Sustainability Perception of Stakeholders in Lagos and Ogun States, Nigeria", thus helping to protect life under water (SDG 14).

6 Peculiarities of Housing Market Goods

The quest for housing affordability is a normative one whose success rests on the ability of governments to develop equitable and accessible systems. In welfare-oriented economies, a clear bifurcation between social housing and housing markets has emerged, wherein social housing systems are developed for people who are unable to afford what is offered in the housing market and others, who are able to satisfy their needs from the market, have a recourse to do so. Although not perfect, this bifurcation has led to the emergence of special financial products (see subprime lending), structures (e.g., Registered Social Landlords, Armslength Management Organizations (ALMOS) and social housing in the pro-welfare administrative regimes of the UK, cooperative housing in several countries) and strategies (e.g., rental housing provision by private landlords in Germany), all of which have further grown the housing market and embedded it within the national economy. It would appear that a lack of clear policy segmentation that suits the generality of Africans has been the bane of housing policy in many countries.

The recognition of multiple alternatives in housing provision is important, as this helps in framing the strategies for providing for each segment. Thus, in Chapter "Homeownership in a Sub-Saharan Africa City: Exploring Self-Help Via Qualitative Insight To Achieve Sustainable Housing", Ebekozien identifies 'organised self-help' as a critical alternative through which informal communities have emerged in peri-urban areas. However, self-help is quite limiting, given that it leaves the individual builder navigating the complex housing development value chain as a sole traveler in a maze. Studies by Nubi [3], Oyalowo [4] have suggested that cooperative society activities need to be mainstreamed in order to cater for this category of builders by organising them into corporate entities. In Nigeria the instrumentality of cooperative society activities in the land market of peri-urban areas has gained some traction. Because they operate as workers' cooperatives, as well as community cooperatives and other forms, there are possibilities that this might provide a structure for individuals to actualise their housing aspirations in a more structured way.

Ebekozien also points out that housing mobility from city centre to peri-urban areas in recent times has been triggered by the homeownership aspirations of middle-income earners coupled with cost and bureaucratic bottlenecks in land administration in central city areas. This has resulted in urban sprawl and even informal settlements in the city's peri-urban interface, while high quality rental housing in city centres remain unoccupied. Thus, there is need for a shift from the blanket generalist study of the housing market to a more holistic understanding of the functions and externalities of the housing markets by disaggregating various sub-sets and interrogating housing aspirations and patterns of specific segments of the society. This will result in more fit-for-purpose interventions.

7 Addressing Environmental and Health Challenges Through Housing and Neighbourhood Quality

In Chapter "Housing and Possible Health Implications in Upgraded Informal Settlements: Evidence from Mangaung Township, South Africa", Ntema, Anderson and Marais suggest a redirection of policy focus from new build and the physical housing structure (the brick and mortar) to the relationship between housing, health and social amenities and infrastructure. In Chapter "Housing, Health and Well-Being of Slum Dwellers in Nigeria: Case Studies of Six Cities", Falade states that health outcomes, especially in informal communities, are often dependent on access to safe water, waste disposal facilities and health services. Therefore, the interconnections between health and well-being (SDG 3), water and sanitation (SDG 6) and decent housing (SDG 11) can be strengthened by the provision of adequate basic services, including health centres, and by integrating health considerations (water and sanitation) into the planning and housing process. By this, housing provision becomes integrated with the actualisation of several sustainable development goals and targets aimed at improving the well-being of the people.

Beyond the household, Temitope Sogbanmu, Opeyemi Ogunkoya, Esther Olaniran, Adedoyin Lasisi and Thomas Seiler offer empirical evidence in Chapter "Adverse Impact of Human Activities on Aquatic Ecosystems: Investigating the Environmental Sustainability Perception of Stakeholders in Lagos and Ogun States, Nigeria" to show how environmental pollution is aggravated by anthropogenic activities such as inappropriate space allocation, poor waste management and weak physical planning of communities and cities. Focusing on SDG 14 (Life Under Water), Sogbanmu et al. studied the process and impact of pollution of the Ogun River and the Lagos Lagoons by human activities (inefficient discharge of abattoir wastes) and construction activities (logging and sawmills) respectively. Their findings reveal a clear increase in the vulnerability of marine species, waterborne diseases and other negative environmental health outcomes, thus providing impetus for the implementation of evidence-based policies for the management of these ecosystems. The centrality of housing (and its macro environment) is also important to note. Where housing is placed in a societal context, human settlement forms can be broadly explored and the impact of activities can also be discerned.

Beyond outlining these risks, Sogbanmu et al. advance strategies for mitigating them within the ambit of SDG 12: responsible production and production systems. They propose leveraging the circular economy in which harmful by-products of industrial activities (such as effluents and wastes) are valued as inputs in the value chain of other products. For example, rather than discharging effluents of sawmills into the Lagos lagoons, operators of the Oko-baba Sawmill can turn these wastes into products of economic value such as pellets, carbon sequestering biofuels and wood blocks. Similarly, sludge and livestock wastes emanating from abattoirs can be processed for manure and green fuels such as biogas.

In Chapter "Green Bonds and Green Buildings: New Options for Achieving Sustainable Development in Nigeria" Oguntuase and Windapo, as well as Addo and Olajide in Chapter "Meeting the Sustainable Development Goals: Considerations for Household and Indoor Air Pollution in Nigeria and Ghana", draw attention to SDG 7 – the promotion of affordable and sustainable energy for human settlements. While Oguntuase and Windapo focus on the relevance of green and energy-efficient buildings, Addo and Olajide reiterate that where use of clean fuels is lacking at the domestic level, sustainability is undermined. According to them, the predominance of home-based enterprises in many informal communities and economic activities requiring widespread usage of solid fuels (where energy from the grid is unavailable) results in poor indoor air quality and consequent ill-health. The authors observe that indoor air quality as a factor that impacts human life and safety and the containment of infectious diseases has not been receiving the attention that ambient air quality is receiving due to climate change campaigns.

Therefore, it is necessary to develop policy actions that recognise housing as a system that links every facet of life. It is also necessary that the complex interactions of housing with land, life on water, energy usage and clean production are further interrogated for beneficial health outcomes. These recommendations align with the One-Health model being championed by the World Health Organisation [14].

8 Deploying Local Agency for Sustainable Slum Upgrading and Renewal

With this theme, the notion of housing as a human right is reasserted. This aligns with the opening statement of the New Urban Agenda averring the 'Right to the City'. Thus, in Chapter "A Study of Housing, Good health and Well-being in Kampala, Uganda", Mutyaba Musoke argues that recognising housing as a human right is unequivocally linked to promotion of the health and well-being of residents and is therefore necessary for implementation of SDG3. Musoke, writing about the Ugandan capital Kampala, points out that only decent housing can be considered as adequate housing and that it must be accessible to all residents.

In championing the right to housing, it is important that social injustices be condemned, especially those associated with violent forced evictions. Instead of forced evictions, which are fundamental breaches of the right to housing, the United Nations General Assembly [9] implores governments to take more proactive approaches towards protecting the housing rights of citizens. In Chapter "Housing, Health and Well-Being of Slum Dwellers in Nigeria: Case Studies of Six Cities" Johnson Falade suggests that comprehensive slum upgrading programmes are a fundamental step in this direction.

Even though there are numerous international conventions and frameworks to support upgrading of informal settlements [10, 12, 13], an often advocated one, especially in the World Bank ESS framework, is incorporating relocation strategies particularly in cases of slum clearance. However, in Chapter "Relocation and Informal Settlements Upgrading in South Africa: The Case Study of Mangaung Township, Free State Province", John Ntema, based on a South African case study, warns that such projects often have unintended negative consequences on the rights and well-being of affected local communities. In his case study, relocation has resulted in direct negative impact regarding SDGs 3, 4, 6 and 11 (health and well-being, quality education, sanitation and hygiene and sustainable communities respectively). He therefore recommends applying principles of participatory project planning and design to in situ informal settlement upgrading, especially because residents can mobilise one another and non-monetary assets to improve their communities. Although such interventions are slower, they tend to be more sustainable and inclusive.

As opposed to a narrative of needs and wants, in Chapter "Housing, Health and Well-Being of Slum Dwellers in Nigeria: Case Studies of Six Cities" Johnson Falade examines the contributions of communities labelled 'slum estates' to providing shelter as well as satisfying the health and well-being goals of residents. His study reveals the roles that local communities play in housing citizens and in meeting their physiological and psychological needs, in addition to protecting them from disease and injury. This important finding educates us and encourages us to re-evaluate our conceptions of slums as problem areas and to recognise the value of local community self-help efforts. This is perhaps what Falade alludes to as a 'happy ending' in his concluding remarks.

9   Coherence and Cultural Contextuality in Housing Design

The issue of culture features prominently in the urban narrative [7]. However, not much consideration is being given to culture in housing policy and programme development. An example is the quest for land/home ownership issue, wherein possession of a plot of land/house is considered a mark of 'maturity' for men in many African cultures and plays a prominent role in various activities around marriage, death and burial. When this is not taken into consideration when planning and designing homes and cities, it inevitably results in distortion of architectural designs and contravention of extant planning standards.

The notion of identity is visible in the SDGs, particularly in the mandate for localisation of the goals [6]. Cultural practices might be negatively construed and practised in a manner that excludes people economically, politically and socially, e.g., in terms of gender-biased norms concerning homeownership as well as home purchase and homeownership transfer by women. In that regard, in Chapter "Beyond a Mere Living Space: Meaning and Morality in Traditional Yoruba Architecture Before Colonialism" Olumayowa Akin-Otiko calls for the decolonisation of planning and architectural design. Specifically addressing the need to revisit the place of culture in space design at the household level in the Yoruba cities of West Africa, he argues that the morality embedded in compound design and architecture of the homestead was lost to a colonial architecture that suited the more individualistic living pattern of the colonisers. He particularly decries the immersion of colonial space design into public housing schemes, especially in a way that vests the role of cultural protection in government.

Comparing African urban forms with European ones, Saidat Olanrewaju and Olumuyiwa Adegun observe in Chapter "Urban Sprawl and Housing: A Case for Densification in Nigerian Cities" that African cities have less compact urban forms compared to many European cities with similar populations. The main thrust of their contribution is that any attempt to achieve sustainable and resilient communities in urban Africa must consider density as an essential component of the morphology of the city. For many urban sustainability thinkers, urban morphology structured around compact designs and densification should be a norm for re-directing African cities towards sustainability [5]. Here, we may ponder the relevance of densification for African cities and whether it is merely another adopted policy that does not fit the reality of African life. However, Olanrewaju and Adegun suggest that the specific details of densification projects must be such that externalities that will render land unaffordable for low-income residents are mitigated.

This is what Foluke Jegede, Eziyi Ibem and Adedapo Oluwatayo respond to in Chapter "Environmental Planning in Mass Housing Schemes: Strategies for Achieving Inclusive and Safe Urban Communities", where they challenge conventional models of layout design. They empirically test the notion of what 'safety' means for residents of public housing estates in Lagos and, based on their findings, make a case for improvements in neighborhood planning considerations. According to the authors, government presence often disappears after construction and 'safe-by-design' considerations are rarely incorporated into construction or design. As such residents are forced to secure themselves through community-led efforts that are mostly unsustainable. The researchers therefore call for enhanced uptake of sustainable design models by built environment professionals in order to create an ambience of sanity, beauty, safety and ease of mobility at the neigbourhood level. SDG 11.1 coalesces at the neighbourhood level, the forte of built environment professionals, and there is considerable capacity for them to actualise sustainable cities and communities.

10   Conclusion

In the course of answering fundamental questions on housing and its linkages to the SDGs, the authors in this volume have spoken to a wide range of stakeholders such as government officials, community groups and housing professionals. They have provided in-depth perspectives on the nature of the housing sector in various African cities, as well as on multi-level governance interconnectedness and the interrelationship between resources and material as well as human flows as they impact on urbanisation. They have tried to promote integrated solutions to address identified problems, as well as scalable actions that will provide solutions on subnational, national and sub-regional scales. They have also developed intensive policy engagement agenda in order to ensure that future urban development policies and programmes across the African continent are driven by scientific knowledge rather than conjectures.

As far as housing policy is concerned, there remains a fundamental issue to be addressed in order to fast-track a sustainable urban future for Africa: recognising the role of housing as a social and economic good that will facilitate the development of appropriate programmes to meet the housing needs and aspirations of people of all classes. In particular, there is the need to focus on programmes and projects that support affordable and inclusive housing in both the rental and ownership tenures. There is also the issue of governance and the need for more locally acceptable solutions to housing problems. This is tied to the need for more grounded policy formulation and implementation as well as to the inclusion of participatory planning and community engagement in development projects. Thus, in alignment with the New Urban Agenda, there is a call for capacity building of existing municipal governance structures that are strategically positioned to support this and actualise the localisation of the SDGs and other necessary international development mandates.

The chapters have also highlighted gaps in current practices and have recommended a redirection towards more holistic policies. Such policies should evolve into inclusive and just housing interventions that are grounded in local realities but take cognizance of international economic dynamics while leveraging on green finance and cleaner/sustainable production systems.

11 Future Research Directions

The views expressed in this book are by no means exhaustive in addressing the challenges of the African housing sector and even in actualising the goals and targets of the SDGs in Africa. To be sure, there is need for more research on how to optimise local governance for achieving the SDGs. Governments are mentioned either at sub-national or national levels, whereas localisation and mainstreaming need to be actioned and implemented especially at municipal levels. As such, the complexities of implementation, the probable constraints and potential spin-offs deserve serious research attention and practical application.

While the complexity of housing is recognised and the volume has provided discussions around several of the SDGs, there is scope for future exploration of the workings of the housing system and its linkages to the economy, especially in the promotion of decent work and economic growth (SDG 8), food security (SDG 2) and industrialisation (SDG 9). In this framing, how to deliver sustainable housing through resource-efficient and socially just methods can be pursued. Studies on urban design and infrastructure, new housing construction models, urban regeneration approaches and how to incorporate sustainability elements at the various stages of the housing development value chain can be established in future undertakings.

Given that social and cultural dynamics are key to sustainable development, future research should prioritise interrogating the housing value chain from a more nuanced gender, demographic and intersectional social lens. In this framing, issues such as homelessness—as distinct from inadequate/overcrowded housing, rural-urban interface, migration, conflict, co-production and community development, can be studied. In addition, a more direct focus on comparative research across cities will provide integrated solutions, while microscale/neighbourhood level research will remain critical to addressing more locally contextualised problems.

References

1. Bah EM, Faye I, Geh ZF (2018) Housing market dynamics in Africa. Palgrave Macmillan. https://doi.org/10.1057/978-1-137-59792-2
2. Nubi T (2015, August) Housing beyond bricks and mortar. Inaugural Lecture of the University of Lagos, Lagos Nigeria
3. Nubi TG (2009) Housing cooperatives as tools for housing affordability and availability. Housing Today J Assoc Hous Corporations Niger (AHCN), Lagos vol 1, issue No. 11, pp 81–91
4. Oyalowo BA (2018) Housing supply and co-operative societies in Lagos state. Doctoral thesis, Department of Estate Management, University of Lagos
5. The Economist (2019) The critical role of infrastructure for the SDGs. https://content.unops.org/publications/The-critical-role-of-infrastructure-for-theSDGs-EN.pdf?mtime=20190314130614
6. United Cities and Local Communities—UCLG (2019) Towards the localization of the SDGs. https://www.uclg.org/sites/default/files/towards-the-localization-of-the-sdgs-0.pdf
7. United Cities and Local Communities—UCLG (2018) Culture in the SDGs. https://www.uclg.org/sites/default/files/culture-in-the-sdgs.pdf
8. United Nations Department for Economic and Social Affairs (2018) World urbanisation prospects. 2018 revision https://population.un.org/wup/Publications/Files/WUP2018-Report.pdf
9. United Nations General Assembly (2020) Report of the special rapporteur on adequate housing as a component of the right to an adequate standard of living, and on the right to non-discrimination in this context. https://reliefweb.int/sites/reliefweb.int/files/resources/A-HRC-43-43-ADD.1-E.pdf
10. UNHABITAT (2011) Policy and regulatory framework review for slum upgrading. https://mirror.unhabitat.org/downloads/docs/11925-1-594762.pdf
11. UN Office of the Special Adviser on Africa (2017) Youth empowerment. https://www.un.org/en/africa/osaa/peace/youth.shtml
12. World Bank (2015) Nigeria slum upgrading, involuntary resettlement, land and housing lessons learned from the experience in Lagos and other mega-cities (English). http://documents1.worldbank.org/curated/en/347391472444343246/pdf/ACS13975-REVISED-P154166-PUBLIC-LMDGP-Study-20150623-Final.pdf
13. World Bank/UNHABITAT (2005) Cities alliance for action plan for moving slum upgrading to scale. Special summary edition. http://documents1.worldbankorg/curated/en/961151468181775920/pdf/809480WP0Citie0Box0379824B00PUBLIC0.pdf
14. World Health Organisation (2017) What is one health? https://www.who.int/news-room/q-a-detail/one-health

Index

Printed in the United States
by Baker & Taylor Publisher Services